Investigating defects in commercial and industrial buildings

Trevor Rushton

RICS BOOKS

Acknowledgments

I would like to thank my wife for her encouragement and support in the preparation of this book and for all of the comments, suggestions and photographs provided by my colleagues at Watts, including but not limited to Angela Dawson, Jane Dalgliesh, Steve Brewer, Allen Gilham, John Bohan and Richard Alison.

A big thank you also goes to the doctors and staff at St Helier, St Anthony's and the Royal Marsden.

Figure 119 of zinc whiskers growing on the surface of a zinc sample is reproduced courtesy of the NASA Electronic Parts and Packaging (NEPP) Program http://nepp.nasa.gov/whisker

Published by the Royal Institution of Chartered Surveyors (RICS)
Surveyor Court
Westwood Business Park
Coventry CV4 8JE
UK
www.ricsbooks.com

ISBN 978 1 84219 441 6

Typeset in Great Britain by Columns Design Ltd, Reading, Berks
Printed in Great Britain by Page Bros, Norwich

Contents

Illustrations

Introduction

The purpose of this book is to provide a practical, easy to understand guide to some of the defects commonly found in commercial and industrial property. That is not to say that it is not relevant to other forms of building; there is much common ground between all types and the scientific principles behind the defects that occur. As a practising building surveyor it is my experience that when you come across something unfamiliar, or you are required to deal with something in rather more detail than normal, a source of quick reference is invaluable. Pressures of time and fee do not lend themselves to detailed studies at a client's expense and a handy guide to the main issues is invaluable. So, I hope that this volume, a companion to *Investigating Hazardous and Deleterious Building Materials*, will be of use to all those who need to solve the problems suffered by buildings – or who want to learn how to avoid them in the first place.

Inevitably, this volume covers some of the same ground as its companion volume, but in this case the emphasis is more on the defects associated with various products and materials. The subject is immense; the further reading list at the end of this book merely scratches the surface of available texts and scientific papers. In many cases one can determine a pattern of the same facts and comments being recycled; one needs to review scientific papers to determine truly new material, yet much of this work is produced by specialists and is not easily assimilated on a day to day basis. Additionally, many texts concentrate on how to do something, rather than what happens if it is done another way.

In contemporary phraseology, defects in buildings and more particularly the study of those defects has become known as building pathology; indeed the term was adopted by RICS and the Association of Building Engineers in the late 1990s, with both institutions incorporating building pathology into their essential core areas for skill and knowledge.

I have always found the term 'building pathology' a slightly pretentious title. Its predecessor 'defects analysis' being somewhat less glamorous seemed at least to capture the subject matter of this book. However, if we pause a second to consider what we mean by building pathology it becomes clear that far from being pretentious, it sums up the process of investigation and remedy quite well. After all, pathology is simply the study of disease and like colleagues in the medical profession, we use various methods of diagnosis that would not be out of place in the surgery – samples, endoscopy, microscopy, ultrasound and sometimes x-rays, for example. We analyse the symptoms, form a prognosis and then suggest treatment. Similarly, our treatments are sometimes very good and sometimes very expensive too.

My time in the profession extends to a mere 30 years during which time a majority of my work has covered defects in commercial and industrial property. The more I am called upon to investigate problems, the more I realise the challenge in understanding the physical processes involved in failures, their sometimes complicated interaction with one another and the difficulty in keeping abreast with current construction technology. The process of diagnosis is in itself one of constant education – there is no limit to the subject unless you dedicate a lifetime's work to a particular area. I once met the production manager of a company that made nuts for the aeronautical industry – just nuts, not bolts. He told me that in the design of fixings, all of the technology was in the nut and after 50 years or so he was still learning and perfecting designs to meet new technical challenges. If there is this much technology in a nut, what hope can the average surveyor or engineer have in understanding something as complex as a modern office building?

For the most part, members of the medical profession are very highly skilled; their qualification process is long and challenging and the results of misdiagnosis possibly fatal. Whilst I by no means wish to underestimate the skills and abilities of building professionals, I do have reservations that building pathology as a competency is in danger of oversimplification – it is undermined by a lack of time in education, being just one of a whole range of competencies and soft skills that must be studied. Building pathology (defects analysis) is stock-in trade for many, but without a proper understanding of its complexities, misdiagnosis can still, occasionally, have fatal results.

The study of defects in buildings demands patience and methodology. Very occasionally there is an easy solution to a complicated problem, but for the most part there is a good deal of hard, repetitive and sometimes dull data gathering before the analysis can even start. Whilst one can often identify a problem quickly, sometimes from the barest of detail, buildings have a habit of not being constructed in the way in which one expects. Very often, one form of construction is disguised as another (for example, with the use of brick slips), giving the impression that it is constructed one way when in reality it is something quite different. A mere visual inspection could give rise to a dangerous conclusion and an even more risky contract for remedial work.

It is almost impossible to produce a book that covers every conceivable building defect, and indeed when one starts looking at a particular subject it is very apparent that the volume of material available is immense. Much information is aimed at the designer; it deals with specification and use rather than what happens if things go wrong. But very often, analysis of the actual cause of the defect is only the start of the story. The process by which the item has failed is the real question – the circumstances that lead to the failure, not the failure itself. The analysis of this aspect owes much to experience rather than academic research, but familiarity with what information the designer ought to have used or referred to is essential.

The Egan report (*Rethinking construction*, July 1998) introduced the concept of zero defects – a policy adopted enthusiastically by the government with its Achieving Excellence in Construction initiative which was launched in March 1999 to improve the performance of central government departments, their executive agencies and non-departmental public bodies (NDPBs) as clients of the construction industry. I wonder if the concept of zero defects is as likely as a true paperless office.

According to Pullar-Strecker[1]:

> '...building defects will always be with us because there will always be aspects of the building performance that are less good than the best, or even average, or less good than we feel we have a right to expect.'

The problem is that we do not always learn from our mistakes. Arguably there is such a wealth of 'how to' information available that anyone ought to be able to design something correctly first time. But construction is not like that; the vagaries of weather, shortages of skills, familiarity with the product or system and commercial pressures all combine to introduce many variables into the equation. The final result is often the product of misplaced optimism whereas what is really needed is an appreciation that the properties of materials and their assembly are never totally ideal or totally predictable.[2] Furthermore, if all designers were to stick to the tried and tested, construct in accordance with standard details and avoid innovation, progress would be stifled. Getting it wrong is part of the process of getting it right, a process of continual refinement. However, from a client's perspective that reasoning is almost bound to fail and it certainly will not cut much ice in court.

With the benefit of hindsight, one can look back at previous mistakes and see how things could be done differently. It is easy to be critical of another's performance; in the same circumstances would you have made similar decisions? Too often we can see that had common sense been applied, a different outcome would have been enjoyed. The trouble with common sense is that it is not common enough! Yet, common sense is abandoned readily when everyone else does the same thing. In some cases it is not just the hapless designer or contractor that is at fault, one has to ask what set of circumstances led to this particular decision in the first place. Was it pressure to complete by a certain date, pressure to make cost savings or mitigate overruns?

When the South Sea Bubble burst in 1720, Charles Mackay reflected on the victims' determination to find a culprit. In *Extraordinary Popular Delusions and the Madness of Crowds*, he wrote:

> 'Nobody seemed to imagine that the nation itself was as culpable as the South Sea company. Nobody blamed the credulity and avarice of the people. . . . These things were never mentioned.'

Sometimes we need to look beyond the obvious and analyse the circumstances that resulted in the eventual outcome – from a client's perspective, the news may not always be welcome.

Building pathology is a huge subject that is deserving of proper study, research and understanding; it can be both frustrating and intolerant of carelessness or ignorance. However, it can be enjoyable and rewarding, providing many challenges for the enquiring mind. Tackle the subject with enthusiasm and passion and it will provide ample reward.

Trevor Rushton

1 Definitions and processes

1.1 BUILDING DEFECTS AND BUILDING FAILURES

These two terms are not interchangeable although through clumsy drafting they are often confused. There are various definitions of defect and failure, but perhaps the most useful definition of a defect is given in BRE Digest 268 as being a shortfall in performance occurring at any time in the life of the product, element or building in which it occurs.

By contrast, a failure is the termination of a product or element's ability to perform its intended function[3].

Thus, a failure can be seen to be the physical effect of a defect. For example, the omission of wall ties from a cavity wall is a defect, if the wall then collapses this is deemed a failure. The distinction may sound semantic, but the concept is important in terms of analysis, particularly if the matter is referred to legal proceedings. It would be difficult for example to identify the damages that should flow from the omission of the wall ties – the fact that they are missing may simply create an unreliability but not necessarily a loss. On the other hand if the wall collapses (failure) it becomes much easier to define loss.

1.2 IDENTIFYING DEFECTS

Addleson[4] makes a distinction between causes and sources of defects. There are three basic causes and multiple sources (see table below).

Of course, some defects may be the product of more than one source, but the clear thread that runs through the table is moisture and indeed moisture in one form or another is often at the very heart of many problems.

From the above it can be appreciated that an understanding of the effects and mechanisms of deterioration are essential when it comes to analysing defects. There are certain basic physical principles which can be applied to try and understand what is

Causes	Sources	Examples
Dampness caused by:	Rainwater	Defective gutters
	Surface water	Penetrating dampness from high ground levels
	Services	Plumbing leaks
	Construction water	Evaporation of water from floor slabs, entrapped water in hollow cored planks
	Condensation	Dampness in cold roof voids
Movements brought about by:	Changes in temperature	Expansion or contraction of steel, brickwork, etc.
	Moisture	Expansion or contraction of timber, brickwork, etc.
	Applied loading	Furniture, equipment, people, stock, etc.
	Dead loading	Self weight of steel, concrete, etc.
	Accidents	Explosion, vehicle collision, etc.
	Ground movements	Subsidence
	Vibration	Traffic vibration
	Chemical changes	Expansion due to sulphate attack, ASR
Chemical/biological change brought about by:	Dampness	Degeneration of materials such as gypsum plaster, chipboard, mdf in damp conditions
	Corrosion	Corrosion of steel reinforcement
	Changes in temperature	Change of state – thermoplastic materials such as plastics, softening of asphalt
	Electromagnetic radiation	Fading of paint, degeneration of mastic, clouding and distortion of polycarbonate sheeting
	Fungal attack	Dry rot, wet rots
	Insect attack	Woodworm, house longhorn beetle, etc.

going on. For example, in chapter 15, I have outlined the various means by which water can be drawn into a cladding system: capillarity, kinetic energy, surface tension, air pressure differentials and gravity (see pages 128–132). Similarly we know that in most cases things flow from one extreme to another – a high temperature will move to low, high air pressure will move to low, gravity will serve to pull things down. Applying these basic physical principles will assist in understanding what is happening.

Further, knowing that certain materials are incompatible with one another either through corrosion or chemical effects can lead one to question suitability in a particular case or at least be alert to the possibility – the creative pessimism referred to in the Introduction.

1.3 GATHERING THE DATA

Clearly, the amount of data recovered from site investigation is crucial to the eventual diagnosis. Too much data is better than not enough, but the results of an ill-planned investigation will be discovered quickly. It is very important to consider the information required and to understand what others may need if help is called upon. Above all, a methodical approach is needed – clear sketches, legible notes, records of photographs as they are taken along with background data such as aspect temperature, rainfall, humidity, etc. Information from a control sample is usually helpful in making comparisons.

Collecting data can be a time consuming and often tiresome process, but if the analysis is to be done properly, the recording and evidencing of facts must be thorough and painstaking. Returning to the office with a key piece of evidence missing is not only frustrating, it is a waste of effort. Time spent planning ahead, undertaking a reconnaissance and undertaking background reading before a formal investigation will be time that is seldom wasted.

It is surprising how many surveyors fail to think three dimensionally. Perhaps it is the result of producing drawings in two dimensions, but pausing for a moment to review what is going on above, around or beyond a defect, and reviewing patterns can be invaluable. It is often worth producing plans and overlays to see if there is a particular feature at one level that contributes to a problem at another.

Non-destructive testing (NDT) methods are used increasingly and their results can be impressive. However, always ensure that the investigation is relevant and that reliable data will be obtained. Some tests are impressive and certainly fuel the argument that 'surveying needs to be seen to be done', but many are nonetheless expensive and although interesting do not in every case give the desired result. When relevant, NDT systems have their place and can yield valuable results when used in the right hands and when interpreted properly.

1.4 COMPATIBILITY OF MATERIALS

One of the mechanisms of deterioration is the incompatibility of certain building materials when they are placed in contact with one another or held in conditions where reactions can take place. Galvanic corrosion is a good example of the incompatibility of certain materials. Whilst the process of corrosion is well understood and documented, the compatibility of other materials such as paints and mastics is often overlooked. The results can be costly to rectify, but information on compatibility is not always easy to come by. Examples include the deterioration of PVC insulation to electrical wiring when in contact with expanded polystyrene insulation or the softening of asphalt or other bitumen materials when exposed to materials containing oil. In chapter 7 another example is referred to: the unexpected incompatibility between phenolic foam insulation in contact with superform aluminium in Foster's Sainsbury Centre for the Visual Arts, built in 1977 (see case study on page 55).

The problem of compatibility could be the result of chemical properties, or issues as simple as different thermal or moisture movement properties. The latter are reasonably predictable with a little work. For example, it is known that calcium silicate bricks tend to suffer an initial irreversible drying shrinkage, whilst clay brickwork displays a propensity for irreversible expansion. So long as the two materials are kept apart and the movement is accommodated there is no problem. However, combine the two in the form of clay facing bricks with calcium silicate backings and brick panels could be subjected to additional internal stresses and possibly bowing or cracking as the stresses are relieved.

Not so apparent are the effects of chemical changes – for these to take effect a series of factors need to be present:

- There must be two different materials – assume one is the material that is capable of being harmed and that the second has the propensity to cause harm.
- The material with the propensity to cause harm must be 'mobile'.
- The two materials must be in direct or indirect contact.
- The mobile material must be subject to a driving force of some kind. This will usually take the form of a concentration gradient – that is the natural tendency of different concentrations of chemical to try and reach equilibrium – much the same as two connecting bodies at different temperatures will eventually reach a stable equilibrium. The process is one of the fundamental laws of chemistry.

Some good examples of compatibility problems are illustrated by Kahles[5] who suggests a risk assessment based approach to the selection of materials. Kahles recommends that the following questions be considered before making a final selection:

- What are the relevant components?
- Will they be in direct or indirect contact?
- What are the risks to the component or product should incompatibility occur?
- Is there any test data or will accelerated tests be needed?

Whilst the obvious first port of call would be the product manufacturer, it does not always follow that the manufacturer will have appropriate test data – particularly if the product is capable of being used in a wide variety of situations. Further, minor alterations in product formulation (perhaps to overcome temporary shortages or as part of product development) could create risks that were not necessarily foreseen.

1.5 METHODOLOGIES

Your knowledge is key to resolving a particular problem therefore use what you know to find out what you do not. If you begin to understand the physics and the materials, you will begin to understand the defects.

It is also important to think in three dimensions. Architectural drawings and sections are often presented two dimensionally and thus do not yield information as to possible water paths very easily. Assembling a mental picture of how something works can be difficult but it is well worth the effort. Similarly, time spent overlaying drawings and relating defects with one another can yield an answer much more effectively than treating each defect in isolation; this sounds obvious but it is surprising how many times this simple procedure is neglected.

There are a number of texts describing particular methodologies for dealing with building defects. I do not think that there is a single correct answer; the method may need to be adapted to suit particular circumstances. However, there are some key issues which need to be considered. Above all, the process is a scientific one. Science has probably created the material and the way it is used, science should certainly be used to find out what has gone wrong. Essentially therefore we should adopt a scientific approach:

1 Define the problem
2 Collect data
3 Formulate a hypothesis
4 Test the hypothesis
5 Evaluate

Steps 1–5 provide the detail of the defect, not necessarily what to do about it. It may be the case that no action needs to be taken (although this is not usually recommended) but more commonly it may be necessary to monitor the defect or repair any damage. Sometimes more radical surgery such as full replacement is required.

The format of the actual report will depend upon the nature of the instruction – a survey report or a report for legal purposes. In the first case a simple format might be:

- What is the problem?
- What will you need to do about it?
- What are the consequences of not doing it?

Legal reports are usually structured to address specific questions and so the above format may not be suitable. However, be careful not to mix fact and opinion and ensure that the thinking behind the report and recommendations is sound. For example, a thought process that runs something like ‘a leaking gutter will cause water ingress; the ceilings are damp stained therefore the leak is due to a leaking gutter’ is unlikely to give a very reliable conclusion. Perhaps the damp stained ceiling was due to defective services or condensation. A variation of this unreliable thinking is the approach that says ‘we have found a defective widget therefore all widgets should be replaced’.

2 Corrosion

At a glance

- Metals almost universally require some form of processing to make them useable (with gold being the exception). The conversion from raw to finished product creates metastable conditions; over time and with the influence of moisture and oxygen a metal will tend to revert back to its original form; the process of corrosion.
- Corrosion is an electrochemical process relying upon a flow of electricity between two sites – an anode and a cathode. The magnitude of the corrosion current dictates the rate of corrosion.
- There are a variety of different forms of corrosion including uniform, atmospheric, galvanic, microbiological and localised corrosion (of which there are various subgroups).
- Another form of corrosion is environmental assisted cracking, again with various subgroups. Finally, flow assisted corrosion can cause significant problems for engineering installations such as petrochemical works, affecting pipework, valves and the like.
- Corrosion is not an inevitable process if appropriate precautions are taken. The correct selection of materials can avoid or reduce the effects of corrosion, whilst other measures such as surface coating, sacrificial coatings and cathodic protection can be used in appropriate conditions.
- Corrosion is usually initiated by a difference in the electrical potential of two metals in contact with an electrolyte (a galvanic couple), variations in the metallurgical state of the metal at different points on its surface or local environmental differences, for example by variations in oxygen supply.

2.1 INTRODUCTION

Although iron may not have been the first metal to be used by man, it has probably been the most used of all of the metals. By default therefore, one of the commonest experiences must be the effects of rust. The Roman philosopher, Pliny, AD 23–79, debated the problem of ferrum corrumpitur, or spoiled iron, which he concluded was a punishment of the gods.

While only gold is found in nature in its metallic state, metals used in building are almost invariably the product of the large amounts of energy used to transform raw materials (ores) into a finished product. The transformation from ore to metal creates an inherently metastable material which, if left to its own devices under the influence of water and oxygen, will gradually revert to its original form, releasing energy as it does so. Corrosion is the mechanism of that transition.

Corrosion is an electrochemical process in which a metal reacts with the environment in which it is located to form an oxide or other compound. Energy is released in the form of corrosion currents – these vary according to the metal. For materials such as magnesium, iron and aluminium, stored energy levels are fairly high, whilst copper and precious metals tend to have lower levels – a principle that is consistent with the levels of energy needed to create them in the first place. However, unlike strength, ductility and resistance, the corrosion characteristic of the metal is not an inherent property, but a product of the metal and the environment.

The process of corrosion is not inevitable, as the designer has the ability to make selections and to design protective measures in a way that can influence the life of the structure or component. General forms of corrosion (see later) are very predictable and can be accommodated in design, although some forms of localised damage can pose complicated technical questions.

There are many different forms of corrosion affecting both the structure and fabric of a building and its services installations. Corrosion of one form or another can have potentially very significant consequences, whether these are deterioration in general serviceability or the creation of actual health and safety risks. The corrosion of mild steel is perhaps one of the most significant forms of corrosion in buildings and accounts for the highest expenditure on repair.

The science of corrosion is very involved and for the most part is beyond the scope of this book.

However, a basic understanding of the process and mechanisms of corrosion is an essential prerequisite of understanding the way in which buildings and their constituent parts gradually break down and what has initiated that breakdown.

The following very broad categories of corrosion exist[6]:

Uniform (general) corrosion	Corrosion cells exist randomly across the surface of the metal such as to give the appearance of general or uniform corrosion.
Atmospheric corrosion	A consequence of alternate wetting and drying and the deposition of pollutants.
Bimetallic or galvanic corrosion	A consequence of two metals in contact with an aqueous solution.
Microbiological corrosion	The production of biofilms can serve to initiate various forms of corrosion, for example crevice corrosion.
Localised corrosion – pitting corrosion, crevice corrosion, filiform corrosion	Corrosion cells are limited by geometrical or chemical factors.
Environment assisted cracking – hydrogen embrittlement, stress corrosion, liquid metal assisted cracking and fatigue	Corrosion affected by mechanical factors which produce symptoms of cracking.
Flow enhanced corrosion	Corrosion attack is modified by the erosion caused by flow patterns.

Aside from gold, metals exposed to air develop an oxide film, which exists at molecular level. In scientific terms, it is a passive layer that is 30 to 100 Angstroms thick. Since an atom is only about 2 Angstroms in diameter, a passive layer is only about 15 to 50 atoms thick. By way of a practical example, if the thickness of the metal is represented by the height of the Empire State Building, the thickness of the passive layer would be represented by the thickness of a postcard on the roof. The passive layer, if well bonded to the base metal serves to protect the metal from corrosion; it greatly reduces the rate of passage of metal ions from the surface. In this condition the metal is described as being in a passive state.

If the metal is then exposed to an aqueous solution (electrolyte), the oxide film will tend to dissolve. Depending upon the level of acidity, the film may dissolve completely or alternatively dissolve in part. Once this occurs, and the metal is exposed, it is termed active. In its active state, positively charged ions tend to pass from the metal into the electrolyte. (An ion is an atom or group of atoms that carries a positive or negative charge as a result of having lost or gained one or more electrons. Ions with a positive charge are called cations; ions with a negative charge are called anions).

Metallic ions, because they are formed from atoms that have lost electrons, are positively charged – when an atom or ion loses electrons it is said to have been oxidised.

The location of the oxidisation process is described as the anode. Here, the oxidisation of the metal is termed an anodic reaction. This flow creates an increase in the potential difference between the metal and the solution – the potential of the metal.

But what of the electrons that are lost from the metal? The electrons are not destroyed; they are conducted through the metal to another site where they combine with a non-metallic element or another metallic ion that is in contact with the metal. In many cases, the non-metallic elements are oxygen and hydrogen. Because the charge of these ions has become smaller (the sum of more negative charges) the ion or atom which has gained the electron(s) is said to have been reduced. The site of this reduction is called the cathode hence, a cathodic reaction – it is where hydrogen evolution or oxygen reduction occurs. The accumulation of negatively charged electrons creates a potential difference between the metal and the solution – this potential difference is termed the potential of the metal.

Whilst the process of oxidisation continues, the deposition of dissolved metal ions from the solution occurs at the cathode. Theoretically, the potential differences between the metal at the anode and the electrolyte and the cathode and the electrolyte could balance each other out so that the rate of oxidisation becomes equal to the rate of reduction. In practice however this condition is not usually reached as electrons from the metal can be consumed by other chemical reactions.

The corrosion current between the anode site and the cathode site consists of electrons flowing within the metal and ions flowing within the electrolyte. The electrolyte is essential for the corrosion process to continue – it provides an escape route for the oxidised ions from the anode and a source of non-metallic or metallic ions. The electrolyte does not have to be liquid water, for example the corrosion cells that form within reinforced concrete rely upon the small amount of moisture that remains in the pores of the concrete.

Corrosion is usually initiated by the following:

- a difference in the electrical potential of two metals in contact with an electrolyte – a galvanic couple;
- variations in the metallurgical state of the metal at different points on its surface; or
- local environmental differences, for example by variations in oxygen supply.

2.2 UNIFORM CORROSION

According to the National Physical Laboratory[7], general or uniform corrosion accounts for about 30 per cent of all corrosion related failures. As its name implies, uniform corrosion occurs over the surface of the metal as a whole and is made up of a multitude of randomly positioned, microscopically sized individual corrosion cells. This form of corrosion occurs when the passive layer has been damaged or dissolved or where it ceases to offer protection.

As noted above, metals that have lost their passive film are described as being in an active state and are vulnerable to corrosion. Some metals such as stainless steels and titanium possess very stable passive films, but even these are not immune to attack in strongly acidic or hot alkaline conditions. For example, high concentrations of hydrochloric or hydrofluoric acid may cause rapid breakdown of the passive film on stainless steels and lead to high corrosion rates.

General corrosion can occur quite rapidly and is usually accompanied by the deposition of the products of corrosion on the surface – in the case of ferrous metals the deposition is rust. Rust does not protect the metal against corrosion, but other oxide forms can do, for example the tarnishing of silver or the green patina associated with the corrosion of copper. Because the mechanisms of general corrosion are well known and conform to established chemistry, it is usually possible to predict the effects and rate of corrosion and to develop strategies for dealing with it. Such measures may be as simple as selecting a suitable thickness of metal sufficient to perform for the required life of the building or component or selecting an alternative metal. However, if these measures are not practicable the following preventative measures may be considered:

- Prevent electrons from moving away from the metal by applying a surface coating such as paint or oil.
- Increase the resistance of the electrolyte, e.g. keep the metal dry or in the case of concrete, use dense well compacted concrete that has a naturally high resistance.
- Prevent oxygen reaching the surface of the metal. This can be difficult to achieve completely as many substances are gas permeable but surface coatings will help.
- Use metal in conjunction with a sacrificial coating that will corrode in preference to and so protect the base metal.
- Apply an electrical current to balance the corrosion current (known as cathodic protection).

2.3 BIMETALLIC CORROSION

Corrosion will often occur when two different metals are immersed in water containing an electrolyte such as salt or acid. Similar problems can also exist when dissimilar metals may not be in contact but are connected electrically. The problem of bimetallic (sometimes called galvanic corrosion or dissimilar metal corrosion or contact corrosion) can be profound and must be taken into account during design. Metals submersed in seawater are particularly at risk.

For corrosion to occur the following are needed:

- oxygen
- water (Air with a relative humidity (RH) greater than 70 per cent contains sufficient moisture to facilitate the corrosion of steel. Corrosion may occur at lower RH if hygroscopic salts are present.); and
- a potential difference.

If a metal is immersed in a conducting liquid (the electrolyte) it will take up a conducting potential. Different metals assume a different potential – this is recorded in the electrolytic series (see below).

If two different metals are immersed in the electrolyte a current will flow from the more positive metal (the cathode) towards the more negative metal, the anode. This current flow arises because both metals have different electrical potentials. The further two metals are separated from one another in the electrochemical series, the more powerful is the electric current produced by their contact in the presence of an electrolyte. Similarly the degree of separation influences the rate at which the metal towards the top of the series is attacked and the level at which the metal towards the bottom of the series will be protected.

When steel corrodes, the corrosion rate is usually governed by the cathodic reaction of the corrosion process, and oxygen is an important factor. As noted above, sea water is highly conductive, and generally this will permit a greater electrical flow. A high flow produces more corrosion to the anode. By contrast, fresh water is generally less conductive and may be free from dissolved oxygen. Rainwater, if polluted, can be moderately conductive, and so exposure to a wet environment in an industrial area can be more damaging than in a rural area. Exposure to marine conditions can produce highly conductive conditions.

The presence of dissolved oxygen in the water accelerates the cathodic reaction and consequently the corrosion rate increases in proportion to the amount of oxygen available for diffusion to the cathode. Where oxygen diffusion is the controlling factor, the corrosion rate tends to increase also with a rise in temperature. In acid waters (pH <4), corrosion can occur even without the presence of oxygen.

The electrochemical series records electrical potentials of pure metals in equilibrium with solutions of their ions. For the purposes of standard comparison, it is better to refer to the galvanic table, whereby metals are categorised by a list of metals and alloys placed in order of their potentials in a corrosive environment such as sea water.

The galvanic table identifies differences between metals in their active and passive states. Although many metals demonstrate active corrosion behaviour

The galvanic table

CORRODED END (Anodic)

Magnesium
Zinc
Aluminium
Cadmium
Mild steel
Cast iron
Stainless steels 18/8 (active)*
Lead
Tin
Nickel (active)*
Brass
Copper
Aluminium bronze
Cupro nickel
Silver solders
Nickel (passive)**
Stainless steels 18/8 (passive)**
Silver
Titanium

*active = oxide film disrupted, steel exposed.
**passive = oxide film intact

PROTECTED END (Cathodic)

when immersed in a suitable electrolyte, others such as nickel, titanium, and stainless steels exhibit a stable film of solid corrosion product, frequently an oxide. These films have good adhesion to the base metal coupled with good cohesive strength and a low ion conductivity which serve to reduce corrosion activity – a property known as passivity.[8]

As noted earlier, water is required to facilitate the corrosion process. However, surface contamination can absorb moisture from the atmosphere even when condensation does not form and so the fact that two metals are not *immersed* in water does not necessarily prevent corrosion from occurring. Metals buried in soil perform in much the same way as those that are immersed in water.

However, the selection of metals during the design of buildings and services depends upon a number of other factors and not just their rating on the scale. For example, even metals that are potentially highly reactive can be used successfully together given the correct precautions.

Corrosion rates can be affected by the area of the anode or cathode relative to the other. A large cathode area and a small anodic area could result in a more severe condition. For example, a copper rivet in a large steel plate is polarised rapidly and corrosion on the plate is small. On the other hand, a large cathode coupled to a small anode has the opposite effect, with rapid attack of the anode.

Detailed guidance on the subject may be found within BS PD 6484:1979[9] which contains a series of tables from which it is possible to identify the risk of additional corrosion of one metal in contact with another. However, the process of corrosion is complex and depends upon a number of variable factors and not just the presence of dissimilar metals; relative area, temperature and exposure to oxygen will all affect corrosion rates.

The data within PD 6484 may be simplified and reproduced in the table opposite. This table should be treated as a guide only; if a particular combination is critical it should be referenced to the more complete information in the Published Document (PD). The table shows the performance of each metal in contact so, to determine whether or not corrosion is a problem, the performance of both metals must be checked. For example, to check the combination of cast iron with carbon steel, first check cast iron in the vertical columns – this shows no corrosion. However, now check carbon steel in the vertical column and it can be seen that in combination, the carbon steel will corrode in preference to the cast iron.

Protection against bimetallic corrosion

The following measures can be taken to protect against bimetallic corrosion:

- Insulate the metals from one another.
- Apply a metallic coating.
- Apply a non-metallic coating.
- Use a jointing compound capable of excluding water.
- Apply a paint coating.
- Apply sacrificial protection, for example cathodic protection.
- Avoid an unfavourable anode-cathode surface area ratio by using an anode area as large as possible.
- Ensure moisture is unlikely to accumulate between dissimilar metals over long periods.

If contact between dissimilar metals is unavoidable, choose two that are close together in the galvanic table.

2.4 FILIFORM CORROSION

According to Verband der Fenster- und Fassadenhersteller e.V.[10] the problem of filiform corrosion has been known for some time to affect steel surfaces that are exposed to a chlorous atmosphere. However, in the 1970s, camouflage coatings on military aeroplanes were found to be suffering similar problems and by the early 1980s it was known that architectural aluminium was being affected, particularly where coated with organic coatings and where located in a coastal environment.

Filiform corrosion, sometimes described as filigree, snail or underfilm corrosion is a fairly common problem, particularly affecting coated surfaces. The problem

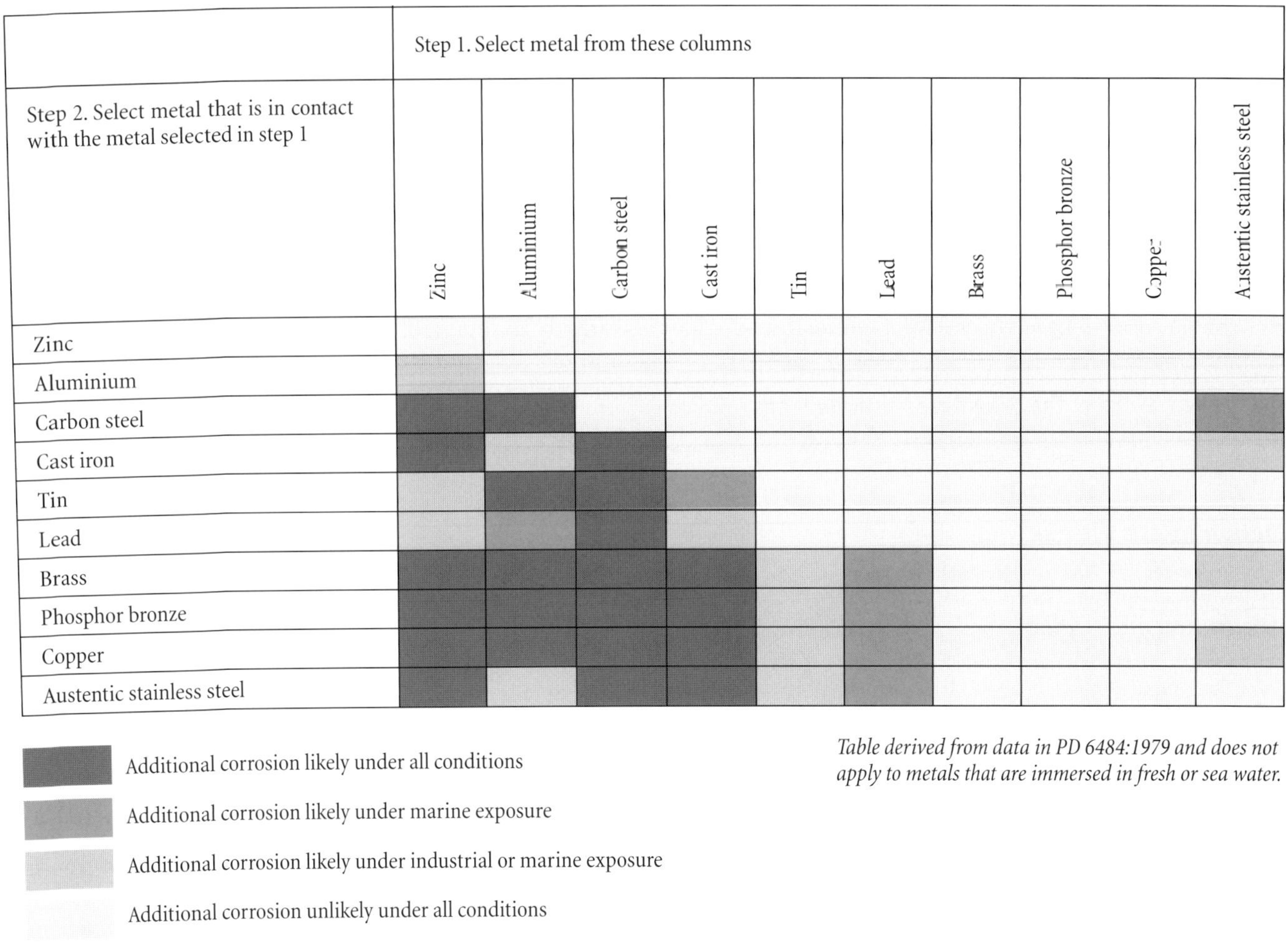

Table derived from data in PD 6484:1979 and does not apply to metals that are immersed in fresh or sea water.

is frequently associated with colour coated aluminium cladding systems, but coated steel can also be affected. The composition of the standard aluminium alloys used is of no relevance to the development of this form of corrosion.

While the form of corrosion does not affect the strength of the material, it can have a serious affect on the appearance of the metal, and so for architectural aluminium it is important to ensure proper precautions to reduce the likelihood of this form of corrosion occurring.

The corrosion commences at the 'head', a small (up to about 4mm diameter) anodic site, where there is a low concentration of dissolved oxygen. Such a location would commence at the site of a damaged coating – either a cut edge or an area that has been scratched or chipped. The cell gradually migrates across the surface of the metal, leaving a worm like filament behind. These trails can, in the correct circumstances, extend a considerable distance, although damage over a few millimetres is more often encountered. Conditions such as humidity and the water permeability of the coating can affect corrosion rates, but it is usual to find disruption and bubbling of the coating along the site of the trail. Propagation of the cell does not depend upon the metallurgical structure of the metal nor is it affected by light.

As with all forms of corrosion, it is the formation of an electrical cell that is the initiator for corrosion to occur. Potential differences between anode and cathode of up to 360mV have been measured and it is at the junction of the anode and cathode that most activity takes place. In the case of filiform corrosion, it is only the migrating head that is the active cell, the snail trail is inactive, but it does have some interesting properties.

The trails will never cross one another; if a trail meets another at a perpendicular angle, it may cease to be active, or it may split into 2 new cells each reflected at 45° to the perpendicular. Such activity means that eventually the trails become boxed in and the cells become inactive. However, the combination of multiple trails can lead to quite widespread disruption of the coated surface. The exact mechanism of filiform corrosion is still under research; it is not entirely clear why it exhibits such curious tendencies.

Factors likely to influence the development of filiform corrosion:

- the supply of chlorides – for example in a coastal environment;
- sufficient humidity – this type of attack is rare on aluminium below approximately 55% RH or above 95% RH; in natural atmospheres it occurs

most readily on aluminium between 85% and 95% RH;
- the presence of oxygen – this is fundamental because it is the primary cathodic reactant;
- the existence of machined areas – for example drainage holes, cut edges, rims or surfaces where the coating thickness is thin;
- locations such as lap joints where water may accumulate; and
- cleaning regularity – frequent cleaning can retard the development of this form of corrosion.

The risk of filiform corrosion decreases with increasing distance from the coastline, although aggressive industrial climates can influence development.

Designing out the risk

In practical terms, there is little that can be done to design out the minor faults that permit filiform corrosion to form. Designers will have little control over location whilst it would be impossible to avoid some mechanical damage during construction, or the need to machine the aluminium sections used.

In addition, experience has shown that it is very difficult to eliminate the risk of filiform corrosion in its entirety. The usual pre-treatment of aluminium using chromising is unlikely to prevent corrosion in risk areas, although pre-anodising using special techniques has been shown to help and is probably the most appropriate precaution here. Contrary to normal anodising practice, the aluminium is treated in a sulphuric acid bath at much higher current density and temperature. CWCT add that careful detailing to ensure free drainage of water and careful handling of panels can help.[11]

Repair

The extent of disruption to the coating caused by filiform corrosion means there are few options for repair. Ultimately it is necessary to remove the coating and corrosion products by grinding and then apply a zinc rich primer and a two part polyurethane paint system to match the original colour as far as is possible.

2.5 LOCALISED CORROSION

These forms of corrosion can be particularly damaging to buildings and structures, and, while to some extent predictable, are likely to cause very serious effects. One form of localised corrosion is called pitting corrosion – this occurs readily in chloride contaminated concrete and can lead to the total or at least significant loss of sections of plain and low alloy steel reinforcement without the formation of expansive rust and the tell tale signs of concrete spalling. The problem often affects those metals that were selected because of their passive nature, for example stainless steels, nickel alloys or aluminium alloys.

As its name suggests, pitting corrosion does not spread laterally over the surface of a metal, but penetrates perpendicular to the surface with a similar appearance to timber with woodworm. The problem is mainly associated with particularly aggressive ions such as chloride and fluoride both of which can initiate this form of corrosion in the absence of oxygen as long as they are present in the electrolyte.

The initiation phase of pitting corrosion is open to debate and there are a number of theories as to how it commences. As a very broad generalisation, it usually occurs at a site of localised damage to the passive layer whereby the damaged area becomes the anode and the remaining areas the cathode. Given the disparity in area between anode and cathode, corrosion can then occur very rapidly. Once established, the effects are well known and understood. Within the pits themselves the pH tends to be reduced significantly, coupled with an increased concentration of the aggressive ions to create extremely corrosive local conditions.

Pitting corrosion is often associated with stagnant conditions and may well be concealed by the products of corrosion, for example within steel pipework. Typical examples occur when pipes or components are taken out of use for a period of time.

Precautions against pitting corrosion include the provision of barrier coatings. However, these must be of high quality. Any small gaps or 'holidays' in the coating could lead to the creation of an anodic site and rapid corrosion.

2.6 ENVIRONMENT ASSISTED CRACKING

Environment assisted cracking can produce catastrophic failures in structural metals. The term is used to describe two prime methods of failure: stress corrosion cracking and hydrogen embrittlement.

Stress corrosion cracking (SCC) is a term describing the formation of brittle cracks in an otherwise sound metal located within a corrosive environment.

If an internal stress is created in a metal, for example by folding or bending, the area that is under greater stress, i.e. the worked area, will become anodic, whilst the section that is at lower or zero stress will become cathodic. The anodic regions can then become corroded in preference to the unworked areas.

For a given alloy within a specific corrosive environment and under sufficient stress, a highly branched network of fine cracks may develop. Chloride induced cracking of stainless steels, caustic cracking of plain carbon steels and ammonia damage to copper alloys are typical examples of this problem.

For stress corrosion cracking to occur, three conditions have to be fulfilled:

- The alloy must be **susceptible** to SCC; most alloys are susceptible to SCC in one or more environments.
- The environment must be **damaging** – oxygenated aqueous chloride environments are particularly damaging to austentic stainless steels and nickel based alloys.
- The stress (intensity) must be **sufficient**.

There is no particular time period for failure to occur – it could be within a few hours or even years; it is often observed in the absence of other forms of corrosion, such as general or crevice corrosion.

2.7 HYDROGEN EMBRITTLEMENT

Hydrogen often occurs as a by-product of the corrosion process as well as being introduced during fabrication, welding and brazing, finishing or machining processes using particular lubricants.

Under certain circumstances hydrogen atoms can be absorbed into the metal with a variety of possible outcomes:

- the formation of hydrogen blisters or blister like cracks (usually submicroscopic) around non-metallic inclusions in materials of low strength;
- reduction in the ductility of the metal; and
- in extreme cases of ductility loss coupled with residual tensile stresses, hydrogen embrittlement cracking (HEC) can occur.

Materials undergoing cathodic protection can be subjected to HEC, as the process can assist in the generation of hydrogen.

The concept of hydrogen entering the metal can be difficult to grasp, but aside from being the most abundant element in the universe hydrogen is the smallest known atom, therefore enabling it to absorb into the structure of the metal. For this to occur however it must be in what is called its atomic form.

HEC is of particular relevance to fasteners made of carbon and alloy steels and can cause sudden and catastrophic failures in these components.

2.8 FLOW ENHANCED CORROSION

The flow of a corrosive fluid across the surface of a metal, for example the inside surface of a pipe, can lead to rapid erosion of the passive film in localised areas. A rough internal pipe surface or a burr at the cut end of a pipe can cause turbulent flow conditions. A combination of erosion due to turbulent flow and corrosion can lead to extremely high pitting rates and pipe failure. The problem of film erosion can be exacerbated by the presence of gas bubbles or particles which impose disruptive shear stresses on the passive layer or the metal itself.

Surfaces affected by erosion-corrosion may display shallow pits or horseshoes or other local phenomena related to the flow direction. The soft alloys of copper, aluminium and lead are particularly prone to this form of corrosion.

2.9 ATMOSPHERIC CORROSION

Cycles of alternate wetting and drying created by exposure to rainfall, condensation, water spray and the like can create conditions favourable to the corrosion of metals. When the atmosphere is contaminated with pollutants such as sulphur dioxide or chlorides, corrosion rates can be greater. Exposure to this form of corrosion is not confined to external situations; if the internal environment is favourable, corrosion can still occur.

The corrosion rate can be affected by a number of factors such as:

- the duration of the wetting and drying cycles;
- the presence of contaminants in the air or on the surface of the metal;
- the relative humidity (RH) – whether condensation will occur and for how long (70–80 per cent RH is generally required); and
- the temperature and variations in temperature

The effect of moist atmospheres is to set up droplets of electrolyte on the surface of the metal. These create corrosion cells with a small anode in the centre of the droplet with a cathode at the edge. The droplets then skin over with corrosion product, either remaining as discrete scabs or spreading. If the droplet contains chlorides (as would be the case if the metal were to be exposed to salt spray in a coastal location) the corrosion cells would be more vigorous.

The effects of atmospheric corrosion on different metals are summarised in the table overleaf.

Figure 1 Atmospheric corrosion on a steel beam.

Metal	Effects of atmospheric corrosion
Copper	Produces a green film or patina of corrosion product composed of hydrated sulphate, carbonate or chloride; low rate of corrosion.
Aluminium	Usually a low rate of corrosion, but accelerated by exposure to alkaline conditions such as salts from copper roofing or wood preservatives.
Zinc	Becomes covered with a protective layer of basic carbonate or basic sulphate. Generally low corrosion rate, but this may be increased by exposure to condensation or sulphur dioxide.
Nickel	If exposed to moist air contaminated with sulphur dioxide becomes coated with a wet layer of nickel sulphate.
Stainless steels	Generally very resistant to corrosion, steels containing nickel and chrome are better and steels containing molybdenum are very highly resistant, even in industrial atmospheres.

Methods of protection against atmospheric corrosion, and their effects, are summarised in the table below.[12]

2.10 MICROBIOLOGICALLY INFLUENCED CORROSION

This generic form of corrosion covers a variety of corrosion types initiated mainly by the presence and/or activities of micro-organisms in biofilms that have formed on the surface of the corroding material. The formation of biofilms, sludge or other deposits can be particularly harmful as it can precipitate crevice corrosion, destroy corrosion inhibitors and directly influence a corrosion cell.

A number of microbiological organisms have been associated with corrosion damage in water systems. These micro-organisms can influence corrosion by effects such as the creation of differential aeration cells, the production of corrosive mineral and organic acids, ammonia, and chemical reductions.

Four environmental conditions are required for this type of corrosion damage to occur:

- metals (the host location);
- nutrients;
- water; and
- oxygen (although some types of bacteria need only very small amounts of oxygen).

Bacteria are present in virtually all piped systems. Given satisfactory combinations of all of the above factors, the bacteria will grow. If one of the conditions is not satisfied, for example the supply of nutrients is exhausted, the growth of bacteria will stop, only to commence again when new nutrients are introduced, perhaps as a result of topping up a closed system or the introduction of additional chemicals.

As noted above, crevice or sometimes pitting corrosion can occur beneath biofilms created by bacterial growth. As consumption of the metal in the pipe occurs, tubercles can be formed and flow rates can be disrupted. Particles of corrosion product can become dislodged and damage valves preventing them from operating correctly.

The problem is not necessarily confined to wet pipes – dry risers can be affected, particularly when

Method	Action required, effect and additional comments
Applying a coating	Serves to exclude water, oxygen, contamination or strong electrolytes. Could take the form of a plastic coating, paper, grease, another metal, etc.
Applying a metallic coating	Electroplating may provide a good barrier, but any local mechanical damage could result in a small anode/large cathode scenario which could create significant corrosion.
Applying a polymer coating	Industrial process which can provide a relatively thick and resistant barrier. However, as with a metallic coating local 'holidays' in the coating can give rise to a small anode/large cathode effect resulting in more severe damage.
Applying a vitreous enamel	Provides a good and robust level of protection. Impact damage to coating can give rise to a small anode/large cathode effect resulting in more severe damage.
Applying a conversion coating	Typically, chromate or phosphate. Chromate coatings are formed by the reaction of water solutions of chromic acid or chromium salts. They are usually applied to aluminium, zinc, cadmium, and magnesium products such as screws, hinges, and many hardware items. Phosphate coatings are formed by a process of chemical conversion on a metal surface to produce thin adherent phosphate compound coatings. They are used to protect carbon steel, low-alloy steel, cast iron, zinc, cadmium, aluminium, and tin.
Applying a paint coating	Paint coatings work by providing a barrier to the electrolyte rather than a barrier to oxygen. To be effective however it is essential that all corrosion products are removed prior to painting and a metal rich primer employed as this will have a sacrificial action.
Applying a sacrificial metal coating	For example the coating of steel with zinc cadmium or aluminium. These methods can be very effective.
Applying temporary protection	For example corrosion inhibitors, grease, oil or similar barrier systems to exclude oxygen or moisture.
Controlling the RH	Maintain the environment at a humidity below 60%.
Applying good design practice	Avoid water traps, permit drainage, etc.

Case study

The building comprised a new (2002) residential block for sixth formers in a large school in the south of England. The building was heated with a gas fired low pressure hot water system with high quality steel radiators. Hot water was distributed by plastic pipes run within suitable voids.

The building had been plagued with water penetration problems as a result of defective cavity trays. It was not surprising therefore when new patches of severe dampness appeared in one first floor bedroom. The external walls were opened up and tested, but this time no defect in the external walls could be identified.

Examination of detailed drawings revealed a possible water path from the inside if one of the pipe joints or radiators were to leak; the room above was reinspected to reveal a recently replaced radiator. Discussion with the school's plumber revealed that the radiator had begun to leak although lack of communication between departments meant that the replacement had not been recorded.

The radiator was dismantled to reveal severe deposits of sludge. The radiator had developed a pinhole beneath one such deposit, but it was clear that significant reductions in thickness had occurred elsewhere.

No records of water treatment at installation could be found. The system had been drained on two occasions and refilled owing to circulation problems and although treatment had been carried out since it is most probable that the bacterial sludge had formed early in the life of the system.

Further radiator failures followed leading to a need to remove all radiators from the system and to carry out thorough jetting of each to remove sludge. Uncertainties remain as to the future life of the system.

they have been charged and discharged and where small quantities of stagnant water remain as a result of the pipe not being fully drained.

The detection and prevention of conditions that are conducive to bacterial growth (and other forms of corrosion) in pipework are essential if the design life of pipework is to be achieved. Treatment measures may involve chemical treatment of the water, flushing, chemical and/or mechanical cleaning, and killing of microbes on the pipe interior, followed by a dwell time prior to reactivating the system, depending on the conditions prevalent.

If the problem has progressed to the point where the walls of pipes have become compromised or leaks occur, replacement of the pipe is the only option.

2.11 CORROSION OF METALS BY WOOD

Wood is essentially a corrosive material, with some varieties being particularly corrosive. For example, steel screws or nails inserted into oak will corrode rapidly. Some timber treatments can exacerbate the problem. Since timber contains acetic acid – a volatile organic compound, vapour release in a confined space

Figure 2 Pinholing of steel radiator due to microbial corrosion.

can sometimes corrode metals even when they are not in contact with the timber.

The rate of emission of acetic acid depends upon the species of timber, its moisture content, humidity and temperature as well as the shape and size of the timber component or member itself. Small amounts of formic, propionic and butyric acids may also be given off.

Kiln dried timber is more acid that naturally converted timber and so to begin with can be more corrosive, albeit over time it releases less combined acid in the longer term.

Since wood can absorb salt, for example when exposed to salt spray in coastal conditions or when floated to the mill in salt water, the chloride content of some timber can be very high even though it is naturally low. In some parts of the world salt is used in the seasoning process, particularly for maple. Metal contact with timber contaminated with chlorides can therefore result in increased corrosion risk in these circumstances.

Ammonium sulphate based fire retardant treatments are considerably corrosive, whilst those based upon boric acid and borax are less so. However, all salt based solutions will absorb moisture, and so the moisture content of timber at a given humidity can be higher than its equivalent, untreated condition.

Effect of timber preservatives

Copper chrome arsenic preservatives can cause corrosion problems in metal screws and fixing plates, especially when these are placed in contact with the timber shortly after treatment has been applied. Similarly, copper napthenate can form soluble copper which can coat adjoining metal to form damaging corrosion cells.

Boron compounds are fairly harmless save that prolonged exposure to damp conditions can produce alkaline conditions which may cause the timber to degrade.

Organic solvent type preservatives or water repellents have a negligible corrosion effect.

2.12 CATHODIC PROTECTION

Aside from the problems of dry rot that afflicted the Royal Navy in the eighteenth century, corrosion of the copper plating applied to ships' hulls was an expensive problem. In the 1820s Sir Humpry Davey discovered that by attaching sacrificial quantities of zinc, iron or tin he could prevent the corrosion of the copper in seawater. The principle of cathodic protection was thus discovered, although as a practical solution it was short lived as the sacrificial anodes collected marine growths and thus compromised the speed of the ships.

By the end of the Second World War, the process had been developed and refined and by the early 1950s was being used successfully to protect sections of fuel lines that had been installed during the war. Today, cathodic protection (cp) systems are common place in the building industry, being used on important early twentieth century steel framed, stone clad buildings as well as post war types suffering from the effects of chloride contamination or use.

Essentially, it will be remembered that in a corrosion cell, the anodic site oxidises, while the cathodic site does not. If an external anode is connected electrically to the 'at risk' metal, and a current applied, the metal can be made cathodic – in other words the electrical potential between the metal and the electrolyte is made more negative to a point where the corroding or anodic reactions are stifled.

There are two types of cp system:

- A galvanic cell – a sacrificial anode (often a zinc ingot but sometimes magnesium or aluminium) connected to the metal that is to be protected. This system uses the differences in potential between the anode and the cathode to create an electrical flow in the electrolyte.
- An impressed current – a system using a DC current (rectified AC) to impress a current from an external anode onto the surface of the cathode. A variety of different anode systems are possible, ranging from discrete anodes that are buried in the structure, titanium mesh anodes, ribbon tapes and conductive paint systems.

Whilst cp systems offer significant advantages over other methods of corrosion protection, they are not without risk. When used with high-strength steels for example, the evolution of hydrogen at the cathode can lead to hydrogen embrittlement and catastrophic failure.

When used in buildings, particular care must be taken to ensure that all sections of metalwork in the vicinity of the system are connected electrically. If this is not done, there is a risk that the secondary elements (those that are not connected) will be subject to enhanced corrosion as a result of stray currents. Such an example could occur where there is a desire to protect the steel frame of a stone clad building but where iron cramps are used to restrain the individual ashlar blocks. If the cramps are not connected to the system, they would be at risk of additional corrosion.

3 Deterioration of concrete

At a glance

- In its natural state, with a pH level of around 12–13 concrete provides a good measure of protection to embedded reinforcement. The steel forms a passive layer which serves to protect it against corrosion.
- Once the pH level of concrete falls to 9–10, the passive layer is disrupted leading to the risk of corrosion. The actual risk depends upon a number of factors such as the presence of sufficient moisture and oxygen.
- Good quality, well compacted concrete with a low water to cement ratio provides better protection than poorer, more porous mixes.
- When exposed to the atmosphere, concrete is subject to attack from atmospheric carbon dioxide and sulphur dioxide in moist conditions. These acidic gases serve to lower the pH value of the concrete although sulphur dioxide usually only affects the surface. Carbon dioxide on the other hand diffuses into the body of the concrete, reducing its alkalinity in a zone gradually extending back from the surface. The process usually takes many years to reach the reinforcement, but when it does, disruption of the passive layer places the reinforcement at risk.
- Carbonation will advance quicker when the RH is in the region of 50–75%. Above or below this range the rate usually decreases. For this reason the rate of carbonation will usually be greater within internal conditions.
- A significant cause of deterioration in concrete is chloride attack. Chloride ions create conditions of high electrical conductivity in the concrete and thus increase the risk of corrosion. Chlorides attack the passive layer and are difficult to deal with using conventional hand-placed concrete repair methods.
- Chlorides were used as an accelerator in concrete up to about 1978 when their use was discontinued in the UK. However, concrete of any age is vulnerable to attack from chlorides arising from salt spray, de-icing salts and the like. Contamination from these sources can be potentially more damaging than with chloride that was used as an accelerator.
- If the chloride concentration (usually expressed as a percentage by weight of cement) is high enough, pitting corrosion can occur. In this form of corrosion, the reinforcement is effectively eaten away without the formation of expansive rust. Pitting corrosion could be particularly dangerous in prestressed or post-tensioned structures because of the thinner sections involved.
- The effective treatment of chloride affected structures is difficult and it is probably best to use electrochemical methods such as cathodic protection to minimise the risk of further corrosion activity.

3.1 MECHANISM OF CORROSION

Reinforced concrete was used to replace steel during the shortages of the war years, with products as diverse as boat hulls, military camps and fortifications. The 1940s saw the development and introduction of prestressed concrete. The post-war rush to replace housing and buildings that had been destroyed in the war lead to the widespread use of concrete as a relatively cheap and plentiful material. However, the corrosion of reinforcement in concrete is now a significant problem, particularly for concrete buildings constructed in the early 1960s, now nearly 50 years old.

The preceding chapter explains the various generic types of corrosion and their causes. The illustrations of corrosion given in chapter 2 have been based upon the notion of two different metals immersed in an electrolyte. General corrosion can also be initiated by changes in the metallurgical condition of a single section of metal such as would occur if the concentrations of oxygen, temperature or other physical factors are different on its surface.

In concrete there is usually sufficient moisture present within the pores to create the electrolyte in a corrosion cell. This moisture is termed the pore fluid and permits a flow of ions between the anodic and cathodic sites. The speed of the ions and the size of the charge that they carry determine the conductivity of the electrolyte and hence the rate at which corrosion will occur.

Normally, embedded steel is protected from corrosion by virtue of the highly alkaline environment that exists within the body of the concrete. A very thin but highly compact passive oxide layer develops around the steel to reduce its corrosion rate to an insignificant level. The passive layer (see section 2.1) is only about 15–50 atoms thick and can be damaged by chloride ions or acidic conditions such as occur when the concrete becomes carbonated. Once damaged, and assuming that other conditions are suitable, corrosion can occur.

Rust, one of the products of corrosion, occupies a greater volume than the original steel. Rust will therefore exert tensile forces on the concrete and because concrete is not strong in tension even a relatively small amount of corrosion product can be sufficient to overcome the tensile strength and cause bursting in the form of a spall or crack. A poor quality or porous concrete, whilst offering little protection against corrosion, may also absorb some of the products of corrosion into the pores making it less likely to crack.

Carbonation

Atmospheric carbon dioxide reacts with the hydration products of hardened cement to form carbonate materials. The reaction is associated with a loss of alkalinity of the concrete, gradually reducing from about pH 12.6 down to pH 9–10 in concrete that is fully carbonated. Below pH 9–10, the protective passive layer is destroyed. Since the source of carbon dioxide is the atmosphere (about 0.03 per cent CO_2 by volume) the areas of concrete directly exposed will react, or carbonate first. The reaction will gradually spread inwards over time, its rate of progress governed by factors such as temperature and humidity, the standard of the original mix and physical factors such as cracks, honeycombing and the like.

If the zone of carbonation reaches the steel reinforcement in reinforced concrete, it does not necessarily follow that corrosion is inevitable, but it is a possibility. Whilst design codes are intended to keep reinforcement within uncarbonated concrete by specifying a minimum depth of cover it is frequently the case that reinforcement becomes displaced or through inaccuracy the shutter fails to create sufficient cover so that modest carbonation results in the steel being contained within an unprotected zone.

Steel located within carbonated concrete does not automatically corrode, it is merely at risk of corrosion. If the combination of moisture and oxygen is correct, corrosion will take place but usually only when the RH is greater than about 70 per cent. This is why external concrete is more likely to corrode than internal areas despite the fact that the rate of carbonation will usually be greater internally.

Like many building materials, concrete is essentially porous, containing pores that vary in size from nanometres to micrometres or larger. This property permits the absorption of liquid and gas into (and out of) the concrete. External faces will reach RH levels that are at equilibrium with the surrounding air, but further into the body of the concrete levels may reach 100 per cent. The RH around the steel reinforcement is likely to be different to the surface RH as well as that of the inner core.

Thus, with different levels of humidity, vapour transport through the body of the concrete will occur, with evaporation taking place at the surface during dry weather and water absorption taking place during wet weather. Water absorption can be rapid, and if that water is contaminated with other chemicals such as calcium chloride from de-icing salts or from salt sea spray, chlorides can be transported into the concrete. According to BRE Digest 491[13], studies have shown that the first 1–2cm of concrete were rapidly penetrated by chloride-bearing water and that absorption and capillary suction were the predominant mechanisms, with ionic diffusion taking place within the body of the concrete over a much longer period of time.

The rate of absorption and diffusion will depend upon pore size, with dense, well-compacted concrete of good quality generally performing better.

The rate of carbonation will depend upon a number of interrelated factors, with temperature and humidity being significant in addition to of course a source of CO_2.

Physical factors affecting carbonation result from the structure and permeability of the concrete, which in turn is governed by:

- water/cement ratio – a low water, high cement content mix will reduce the rate of carbonation in hardened concrete;
- cement content and type;
- aggregate type;
- degree of compaction; and
- method and effectiveness of curing.

Environmental factors affecting carbonation include:

- Source of CO_2 – this is usually atmospheric (Permanently wet conditions hinder penetration so carbonation will be low.).
- Relative humidity – since water is needed for carbonation to take place, a low RH (permanent hot and dry conditions, <25% RH) reduces the rate of carbonation or halts it altogether. Similarly, a very high RH reduces the rate of carbonation as the excess moisture slows the rate of CO_2 entry into the concrete. Optimum conditions for carbonation are RH 50–75%, although this is interrelated with porosity.

- Temperature – warmer temperatures increase the rate of carbonation subject to RH.

From the list of factors above it will be appreciated that wet conditions do not necessarily lead to increased carbonation. Indeed the rate of carbonation internally or in internal environments is likely to be greater than in external situations.

> The rate of carbonation can be estimated using the formula:
>
> $d=k\sqrt{t}$
>
> Where:
>
> d = depth of carbonation (measured by site testing)
>
> k = constant
>
> t = age of the concrete
>
> So, for example, if testing on a 20 year old structure reveals a depth of carbonation of 5mm, the constant for that mix would be:
>
> $k = \frac{5mm}{\sqrt{20}} = 1.12.$
>
> Therefore to estimate the depth of carbonation in 30 years:
>
> $d = 1.12\sqrt{30} = 6.13mm$ approx.

Since carbonation extends from the outside in, cracks in the surface will permit the entry of gas further into the body of the concrete. Most if not all concrete will exhibit surface cracking (usually up to about 0.3mm) but if the depth of the crack reaches or exceeds the reinforcement it is possible that localised corrosion could take place. If the cracks run perpendicular to the line of the reinforcement the corrosion sites will be very localised and may never cause a problem. Conversely, if the cracks run parallel with the reinforcement, the potential for corrosion could be greater.

Figure 3 Lack of cover and carbonation extending inwards on three faces of this reinforced concrete decorative fin have resulted in corrosion of the reinforcement.

Testing for carbonation

Testing for carbonation in ordinary Portland cement (OPC) concrete is relatively straightforward and can be undertaken on site or if greater accuracy is needed in the laboratory using petrographic analysis.

Site tests usually involve exposing a fresh sample of broken concrete and applying an indicator solution of 1g phenolphthalein in 50g alchohol diluted in 100ml of deionised water. The indicator solution will turn pink on contact with suitably alkaline (uncarbonated) concrete and remain clear on carbonated concrete. The depth of penetration can then be measured. Care is needed however, that the sample is not distorted or contaminated. It is usual to drill two holes 10–15mm apart and to break the concrete between them to expose a fresh surface. As an alternative a core can be split laterally to provide a suitable surface. Measurement of the core itself (i.e. an unsplit core), a sawn edge or a corner knocked off a column is likely to produce a misleading result.

Calcium chloride

In the UK, the introduction of chloride ions into reinforced concrete can, if sufficient in quantity, result in a high risk of corrosion of embedded steel. There are three possible sources of chloride:

- it was introduced as an accelerator during the mixing process;
- it was introduced naturally, for example from the use of unwashed marine aggregates; or
- it was introduced as a result of external contamination, e.g. de-icing salt, exposure to salt spray.

The introduction of calcium chloride as an accelerator was common up until around 1978 and was particularly prevalent in the 1950s and 1960s. Similarly, buildings of this age often exhibited exposed concrete frames or elements of structure, which are susceptible to external contamination.

Calcium chloride significantly reduces both the initial and final setting time of concrete, enabling for example precast work to be removed from the moulds earlier than would otherwise be the case. The maximum rate of increase in strength gain could be attained within one to three days of curing, its development depending on curing conditions, water/cement ratio, type of cement, amount of admixture, etc. In comparison with a reference concrete (without calcium chloride), strength gain may vary between 30 and 100 per cent in the first three days, the value falling with further curing. In most cases calcium chloride was known to increase drying shrinkage, the magnitude depending on the amount of calcium chloride added, the type of cement, the period of curing, and environmental conditions. For concreting in cold weather, the addition of calcium chloride meant that the concrete or mortar would harden quickly, thus developing early resistance to damage by freezing and thawing.

Chloride induced corrosion can be particularly difficult to deal with effectively. In certain circumstances 'pitting' corrosion can occur, with the steel effectively being eaten away without the formation of expansive rust. By cutting out affected areas and undertaking traditional hand placed concrete repairs, a new corrosion cell is formed, with the consequent risk of corrosion occurring at the interface of the repaired area and the existing concrete.

The introduction of sufficient quantities of chloride ion into the concrete can stimulate reinforcement corrosion, as the passivity of the concrete around the steel is reduced. Chlorides form solutions with potentially high conductivity and the higher the conductivity, the greater the rate of corrosion.

Sufficient levels of chloride ion can also disrupt the passive layer and reduce corrosion protection. The risk of chlorides breaking down the passive layer increases as the chloride concentration increases or as the alkalinity of the concrete diminishes. Carbonation, the gradual lowering of pH levels as a result of exposure to atmospheric carbon dioxide, is thus instrumental in determining corrosion rates. Once the pH level of concrete falls to a level lower than ten, the oxide film can breakdown and corrosion of the reinforcement can occur given sufficient moisture and oxygen.

Chlorides that are cast in during the mixing process are to a point chemically bound and may produce slightly less of a threat. However, a certain amount of chloride ion will remain mobile. The gradual reduction of alkalinity resulting from carbonation can serve to liberate more of the bound ions and so increase the risk of corrosion activity.

Chlorides introduced as a process of external contamination can be potentially more damaging because:

- the concentration can be erratic; and
- the ions are not chemically bound.

The ways in which chlorides may be present in concrete are as follows:

- free chloride ions in the aqueous phase (pore water) of the concrete;
- chloride bonded more or less strongly with the calcium silicate of hydrated cement (physically absorbed); and
- chemically bound as definite compounds, e.g. chloroaluminate or chloroferrite (also known as fixed chloride).

Whilst concrete in damp conditions can be more susceptible to chloride attack, it does not follow that simply because an element is contained within a dry internal environment it will not suffer. High levels of chloride contamination can permit corrosion regardless of environmental conditions.

Under BS EN 206-1:2000 (which also prohibits the use of chloride admixtures) the maximum chloride ion content by mass of cement varies from 0.2%–0.4% according to the class of concrete and in the case of prestressed concrete from 0.1%–0.2%. These standards apply to new structures, where chlorides could be introduced from the aggregate.

A number of factors control the propensity of the concrete to absorb chlorides from external sources:

- depth of cover;
- mix design, including the cement type;
- use of admixtures, e.g. water reducing or water proofing;
- degree of curing;
- moisture content (relative humidity); and
- temperature.

If chlorides are from external sources, it follows that the concentration of chloride will be highest towards the surface and gradually diminishing into the depth of the concrete. For this reason it is necessary when sampling for chlorides to take several readings at different depths so as to establish the concentration at different levels and so help to determine whether or not the chloride content is as a result of airborne contamination or other sources.

3.2 METHODS OF ASSESSING CHLORIDE CONTENT

For the most part, sampling of concrete components on site will be necessary. The first method involves the removal of cores for analysis in the laboratory. As this process is fairly costly, it tends not to be used for bulk sampling, but may be conducted where the core might yield other information – perhaps for petrographic analysis, strength testing, etc.

The most popular method is to use dust drillings, in a manner first identified by the BRE[14]. In this process, about 25–30g of powder or dust is extracted using a rotary percussion drill rotating at about 750rpm and fitted with a 10–25mm diameter bit. Dust arising from the first 5mm depth is discarded, the remainder being collected in a purpose made chute held under the drill bit. It is usually better to take the sample from several drillings in a cluster to achieve more reliable results. The samples are then tested in a laboratory or with a site test kit to yield the chloride content – usually measured by weight of cement (this involves making an assessment of the cement content). An alternative is to provide information per weight of sample (a threshold level of 0.4 per cent usually corresponds to a chloride content of 0.06 per cent by weight of concrete).

The dust method is not foolproof: readings can sometimes be distorted by contamination or poor representation of the actual concrete. Taking a single reading will tell no more than the chloride content at one particular depth; where ingressed chlorides are suspected a series of incremental readings or profiles will be necessary.

A more accurate method of taking profiles is to use a specialised grinding machine called a profile grinder. Under this process a grinding bit grinds out concrete powder at selected, exact depth increments, selected between 0.5mm to 2mm. The bit is attached to a grinding machine held against the surface by a grinding plate. The powder produced is collected with a portable vacuum cleaner containing a reusable filter, except on vertical faces where the powder collection is made in a plastic bag attached to the grinding plate. For every depth increment of 0.5 mm approximately 5 grams of powder is available for analysis.[15] The dust samples can then be analysed in a laboratory or with the use of a rapid test kit for on-site use.

Sometimes it can be convenient to obtain a sample simply by knocking off the edge of a column or beam, rather than arranging for the test methods described above. Such a procedure might be relevant where chloride activity is suspected, perhaps during a pre-purchase survey and where suitable equipment is not to hand. The practice is not to be commended, but if it is used, a sample of about 50g is needed. Since the sample is usually taken from a corner, it may have been subjected to chloride ingress on two adjacent faces and may therefore yield inaccurate results. Concrete from a depth of at least 25mm is usually preferred. Ideally, the sample should be a single lump and not a collection of fragments.

Simple site tests can be undertaken using an indicator solution on freshly exposed concrete. The site is first sprayed with silver nitrate followed by potassium chromate. If the liquid turns brown significant chlorides are present; if it turns yellow/white, chlorides may be present and further investigation is warranted. Other related methods involve the use of flurosceine and silver nitrate.

Physical measurement of chloride content can be achieved using a number of techniques including scanning electron microscopy, spectrophotometry and x-ray fluorescence.

There are also various electrochemical methods of assessing corrosion activity in hardened concrete in-situ, including half-cell mapping and linear polarisation resistance. Of these, half-cell potential mapping is frequently used. This method involves making an electrical connection to the steel reinforcement in the concrete element under assessment. This is connected to a high impedance digital millivoltmeter, often backed up with a data-logging device. The other connection to the millivoltmeter is taken to a copper/copper sulphate or silver/silver chloride half cell, which has a porous connection at one end that can be touched to the concrete surface. This will then register the corrosion potential of the steel reinforcement nearest to the point of contact. By measuring results on a regular grid and plotting results as an equipotential contour map, areas of corroding steel may readily be seen. Using 3D mapping techniques, a more graphical representation of the corrosion can be shown.[16]

The use of the technique is described in an American Standard, ASTM C876–80, *Standard Test Method for Half Cell Potentials of Reinforcing Steel in Concrete*, although some doubts have been expressed concerning the accuracy and reliability of this method for use on buildings as opposed to bridge decks, for which the standard was designed.

A useful source of guidance on non-destructive testing can be found in BRE Digest 434.[17]

3.3 RISK ASSESSMENT

Guidance on assessing the risk of reinforcement corrosion is provided by the BRE in Digest 444 Part 2[18]. Here the risk of corrosion for structures of various ages is presented in the range 'negligible' to 'extremely high risk'. Factors effecting the risk assessment are either a dry or damp environment, the depth of carbonation and of course the level of chlorides present.

Until fairly recently it used to be the case that chloride ion content below 0.4 per cent was considered to be a low risk, 0.4–0.6 per cent presented medium risk and 0.6–1 per cent a high risk. However, the risk of corrosion increases with age and it is better to factor this, together with the nature of exposure, into the equation. The following examples, as detailed in the digest, highlight how the evaluation of risk posed by chloride ion content is affected:

- In a 25-year old building, the risk of corrosion in dry, uncarbonated concrete with a chloride content less than 0.4 per cent is negligible. However, if the depth of carbonation has reached the reinforcement, the risk becomes moderate.
- In a similar building, with a chloride content of 0.6 per cent, damp conditions and where carbonation has reached the steel, the risk of corrosion becomes high. Above 1 per cent, the risk becomes extremely high.

3.4 REPAIR METHODS

There is a wealth of information available on methods of repair of concrete structures affected by chloride ingress. BRE Digest 444-3[19] sets out valuable information to assist in the selection of appropriate methods. The European Draft for development DD ENV 1504-9 also provides information on the various methods of repair, setting out a series of principles and methods for dealing with damage to concrete and the effects of reinforcement corrosion. See also chapter 12 of *Investigating Hazardous and Deleterious Materials.*

Patch repairs

Carbonation induced reinforcement corrosion is typically repaired using isolated patch repairs which are mainly long lasting and durable. The successful repair of chloride induced corrosion is however notoriously difficult because of the tendency for new corrosion cells to form at the boundary of the repair. This mechanism is called 'incipient anode effect' and should be minimised by removing, wherever possible, all concrete with significant chloride contamination. In recent years the introduction of proprietary sacrificial zinc anodes embedded within the patch repair and attached to the reinforcement can help to reduce this effect. For high levels of chlorides and long term protection this may not be sufficient.

Electrochemical methods – chloride removal

For heavily chloride contaminated structures, particularly car parks, the only tried and tested long-term solution is cathodic protection. A variation of cathodic protection is desalination (chloride extraction) – a short-term process using higher current densities than cathodic protection. Migrating corrosion inhibitors (penetrating surface coatings applied under strict conditions) have found some success.

As stated earlier, chloride ions create conditions of high conductivity and thus corrosion rates are higher. It follows that as the chloride ion is negatively charged, electrochemical processes can be used to repel the chloride ion from the steel surface and move it towards an external anode.

Trials of chloride removal systems were undertaken by Norwegian Concrete Technologies in the early 1970s and the systems were developed later in the decade. Between 1987 and 1992 research was undertaken as part of the Strategic Highway Research Programme (SHRP); this programme was completed in 1994.

The system works by fitting a temporary anode mesh close to the surface of the concrete. An electrical connection is made between the reinforcement in the concrete, a power supply and the anode mesh. An electrolyte is applied around the anode mesh comprising either a sprayed cellulose fibre or a warm water bath.

A current of about one amp per square metre is applied for a period of time – this may be as much as 40 days. The chloride ions are transported from the concrete to its surface where they are then carried out by the water or removed in conjunction with the temporary electrolyte.

The success of the operation can be monitored by:

- establishing a point of diminishing returns – as the resistance of the concrete increases the amount of chloride removal reduces;
- taking physical samples and testing them for chloride content using conventional methods;
- taking regular measurements of the anode system or electrolyte to determine when the chloride level has stabilised; and
- measuring the charge duration (amp hours per square metre) and using experience gained on similar work to establish a limit.

Tests have indicated that removal of high amounts of chloride from around the first 20mm is relatively easy, whilst chloride that has penetrated deeper into the structure is harder to remove.

Given that total extraction of all chlorides would be an impossibility, there remains the risk of future migration of chloride ions and the risk of reoccurrence of corrosion activity.

The provision of an anti-chloride coating on to the surface of the concrete will deal with fresh contamination. Field tests show that protection can be achieved for at least four years, although there is some uncertainty as to how much longer. Some test results suggest ten years but the actual period could eventually be very much longer than this.

Generally speaking, the removal of chlorides from external or environmental sources (salt spray, de-icing salt, etc.) appears to be more successful than chlorides that were bound in the mix from the date of construction. Where high amounts of chloride have been added to the concrete mix, stopping the corrosion may be difficult to achieve.

Worries have been expressed over the likelihood of generating alkali silica reaction (ASR) and research is currently in place to see if this is a problem. If the corroding structure is thought to be liable to ASR problems (stimulated by an increase in the alkalinity close to the steel surface) a detailed investigation of the nature of the subject concrete should be undertaken first.

Worries have also been expressed over the bond strength although it appears that significant reduction in bond strength can be occasioned if high current densities are used during the treatment and if smooth, rather than deformed reinforcement bars, are used. It is therefore important to limit the current density.

Investigations undertaken by Taywood Engineering into the problem of microcracking of concrete structures suggests that the treatment produces no significant increase in the level of microcracking in concrete, and that structures repaired using conventional methods are likely to suffer more from it as a result of the effects of percussion tools.

As with most systems, very careful design of the remedial treatment is advisable and if low cover occurs there is a risk that rapid chloride removal will take place around the first bar, but penetration further in to the concrete may not be so effective.

The treatment cannot however be applied to prestressed concrete because of the risk of hydrogen embrittlement.

Desalination appeared initially to be a popular system, but it is doubtful whether it will supplant cathodic protection as a method of dealing with chloride affected concrete.

Cathodic protection

Cathodic protection is essentially similar to the method described above, but current densities are generally lower and the system is designed for continuous use, not a short period of time.

Anode systems can take the form of a conductive paint coating, embedded (discrete) anodes or external meshes. The anodes are usually connected to a data-logging system so that current densities and corrosion rates can be monitored and corrected where necessary.

Corrosion inhibitors

Corrosion inhibitors are designed to penetrate the concrete and to create a very thin layer (usually a few molecules thick) around the steel.

Various types of inhibitor exist. Some will deal with corrosion activity at the anode, some at the cathode and some at both anode and cathode. Calcium nitrite has been used as a component of fresh concrete or repair mortar to act as a corrosion inhibitor.

Anodic inhibitors can, in certain circumstances prove troublesome. Insufficient materials on the steel can lead to the formation of localised corrosion cells and subsequent pitting attack. Cathodic inhibitors reduce the amount of oxygen around the cathode site and thus inhibit the electrochemical reaction. These materials are more tolerant of low doses, but are generally less efficient than anodic types.

Commercial systems are likely to be of the combined type – vapour phase inhibitors (VPIs). These are volatile compounds that can be transported via a variety of waxes, oils or gels. When applied to the concrete by roller or spray, they will diffuse through the pores in the body of the concrete to the steel surface.

VPIs will not work under all conditions, and an upper limit of chloride content will need to be established before assessing the likely benefits. Inhibitor molecules are generally fairly large (by comparison with chloride ions) and therefore transportation into the concrete may be slow, particularly where the concrete is of good quality. In poor quality concrete, the rate of diffusion will be improved. Drilling holes into the concrete, scabbling the surface or forming grooves will speed transportation, but at the risk of mutilation – something to be avoided if possible. Application in damp conditions may also reduce the speed and effectiveness of the treatment.

Initial evidence from independent bodies suggests that there have been some improvements in corrosion rates, but quality control measures and careful application are essential. The only way of determining success will be to carry out corrosion rate monitoring before and after treatment, preferably by reference to a control area. Building in a corrosion monitoring system would seem to be an appropriate precaution. The system is probably not suitable on its own, but when used as a component of a treatment strategy, there is evidence to suggest that a marked and lasting reduction in corrosion can be achieved.

When incorporated into repair mortars, inhibitors can diffuse into the concrete and reduce the risk of the formation of new corrosion cells at the interface of the repaired area with the adjoining original; this is a particular problem when dealing with traditional repairs to chloride contaminated concrete.

Corrosion inhibitor treatments may in some conditions offer a cost-effective alternative to conventional repairs. Success will depend upon a variety of factors, not the least of which is ensuring correct application under correct conditions. Corrosion monitoring is essential to determine the effectiveness of the treatment. Natural scepticism has been fuelled by mixed results of field trials, but further developments can be expected over the next few years.

3.5 SULPHATE ATTACK

Sulphate attack can cause serious damage to brickwork and concrete, creating expansive forces which are or can be sufficient enough to disrupt mortar and create significant expansion, bowing and cracking of concrete floor slabs together with related collateral damage.

The salts of gypsum, magnesium sulphate and sodium sulphate are all water soluble and are typical of the range of chemicals that are known to cause problems. Being water soluble they are readily transported into cement mortar or concrete whereupon they react with the tricalcium aluminate to form a calcium sulfoaluminate hydrate (ettringite). Ettringite is characterised by the formation of acicular crystals which generate high expansive forces in the mortar or concrete. For sulphate attack to occur there must be sufficient sulphate to begin with and sufficient and long-term water.

Concrete ground bearing slabs and foundations are at risk when in soils that contain high sulphate levels and where they are not separated from contact with the ground with suitable polythene sheeting or membranes. Similarly, contact with contaminated hardcore (particularly that containing high levels of black ash, burnt colliery shale, blast furnace slag and similar materials can, if combined with sufficient moisture, create the conditions for sulphate attack.

A second form of sulphate attack is known as thaumasite (TSA); in this case the product of the reaction is calcium silicate carbonate sulphate hydrate. Under this form of attack, concrete turns eventually to an incohesive mass, but for this to occur several factors must be present[20]:

- persistent wetness;
- a sufficient quantity of water soluble sulphate;
- calcium carbonate, e.g. from limestone aggregate;
- a pH of 10 or more; and
- a temperature below 15°C.

Simultaneous occurrence of all the known risk factors will be unusual, and according to the British Cement Association (BCA) the majority of identified cases have been restricted to concrete buried in the Lower Lias or Kimmeridge clays of the West Country. Some cases have also been identified in the North East, where concrete is in contact with pyrite-containing colliery waste ('red shale').

Since sulphate attack can only happen if suitable salts are transferred into the concrete or mortar it follows that if measures are taken to keep the concrete dry, attack cannot proceed. For example, if a floor slab is laid over a damp-proof membrane (DPM) (now required by Building Regulations but only fairly recently) there will be a barrier to moisture sufficient to prevent salt migration.

The same would be true for brickwork, but in this case it is much more difficult to keep brickwork dry. Measures such as rendering could potentially make matters worse since dense cement renders often crack and let water in, but restrict it from drying out. Soluble sulphates in the bricks themselves can then dissolve out to react with the cement mortar, causing it to expand. (See also figure 50 on page 90.)

Visual indicators

Signs of expansion of brickwork along brick joints both horizontally and vertically particularly in rendered brick are usually a good indicator that sulphate attack may be taking place. In floors, because they are restrained, upwards bowing coupled with map pattern cracking and possibly displacement of brickwork at slab level are common indicators. Structural failure as a result of TSA is unlikely and would be preceded by early warning signs, such as cracking and deflection.[21]

BS 1881-124:1988 provides methods of testing concrete samples. Assuming that concrete contains 15 per cent by mass of cement, sulphate content can be expressed as a percentage of the mass of cement with anything over five per cent being likely to be harmful. In severe cases, 10–15 per cent has been identified. Visual indicators of affected concrete may be the existence of a flaky or crumbly concrete, and in the case of thaumasite, a white mush.

3.6 AGGREGATE REACTIONS AND DEFECTS

Not all aggregates are stable when in contact with cement paste and moisture. Some types of stone can react unfavourably by expansion or contraction, whilst others react chemically to produce expansive by-products capable of disrupting the concrete. For example, particles of coal or lignite can cause localised surface defects such as pop-outs, whilst aggregates containing clay can be prone to expansion and contraction depending upon the amount of moisture present.

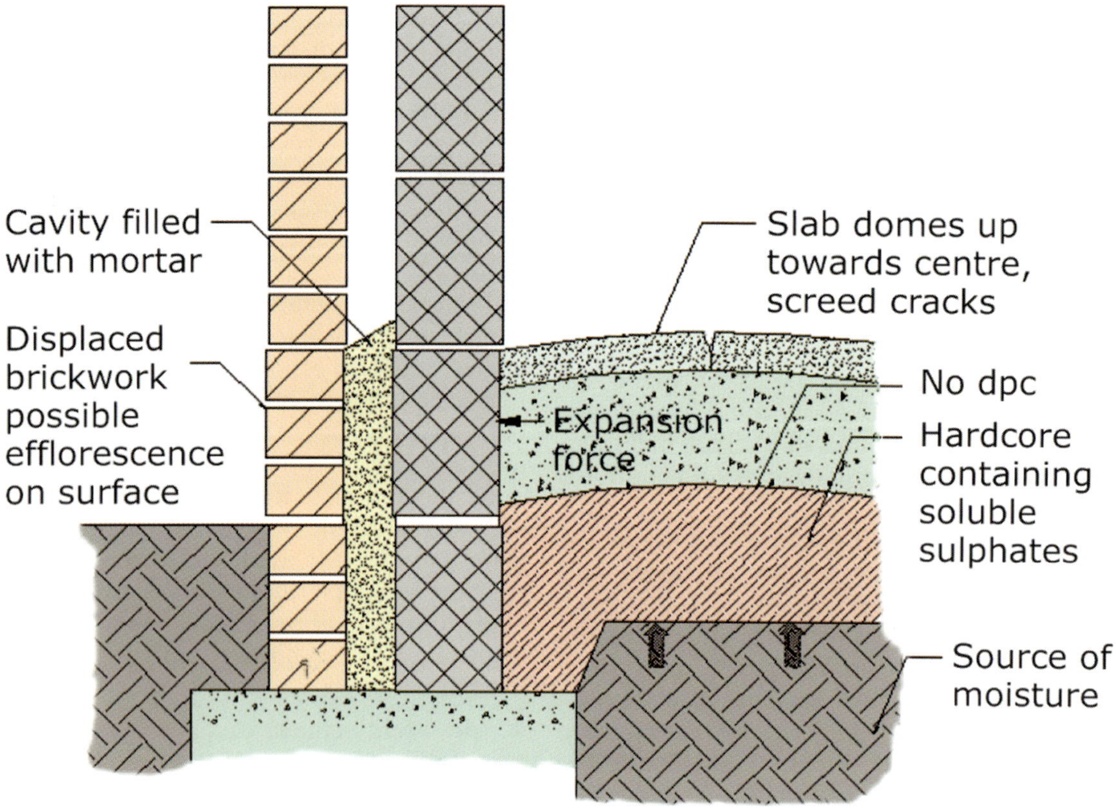

Figure 4 Effects of sulphate attack on a floor slab and enclosing walls.

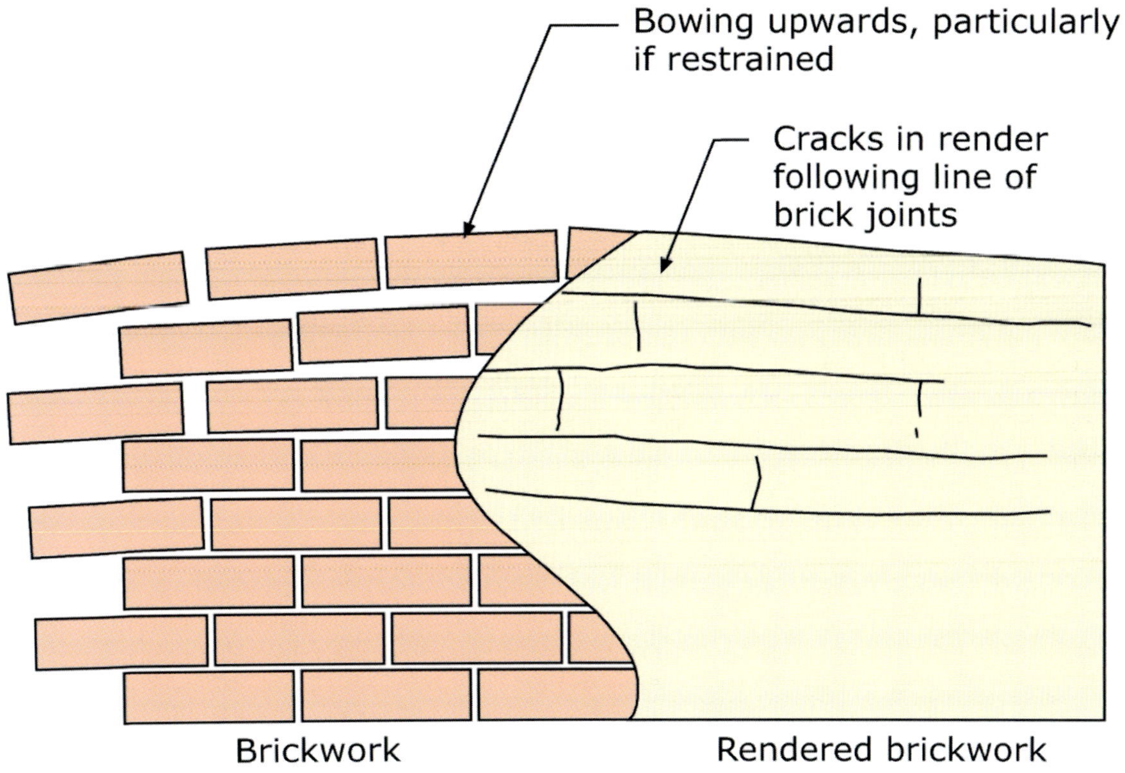

Figure 5 Effect of sulphate attack in walls.

Defective particles of aggregate close to the surface will often create a characteristic three-legged crack pattern sometimes referred to as Isle of Man cracking (after the island's symbol). A large number of Isle of Man cracks are usually described as map cracking.

Alkali silica reactions

Alkali aggregate reactions, of which alkali silica reactions (ASR) are only one variant, are relatively uncommon in construction in the UK and tend to occur in civil engineering structures rather than conventional buildings. For most purposes it is sufficient to consider the risks of ASR in preference to the other forms of reaction as these are not known to occur with most types of aggregate used in the UK.

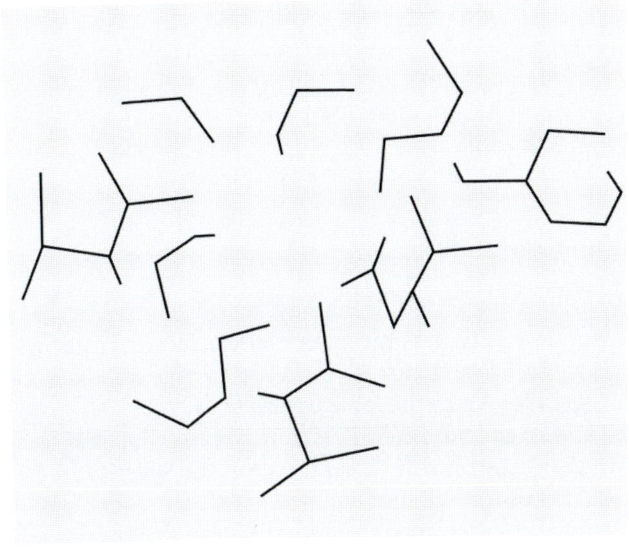

Figure 6 Isle of Man cracking.

ASR is a reaction produced when highly alkaline pore water in the concrete mixes with the silica molecules of any of the following rocks or minerals:

- imperfectly crystallised siliceous rocks and minerals (e.g. opal, chert and cristobalite);
- highly siliceous volcanic glasses;
- highly strained or granulated siliceous rocks (e.g. the metaquartzites);
- other stressed silicates; and
- exceptionally fine-grained siliceous rocks (e.g. siltstones and phyllites).

The pore water contains dissolved salts of sodium and potassium. With the exception of the pure limestones and dolomites, almost all rocks contain some silica. Certain types of silica are more reactive than others and generally, the more soluble the form of the silica, the faster and more intense is the reaction.

The chemical reaction produces a gel, which imbibes water, expands and can cause the concrete to crack or disrupt. Cracks then allow more moisture to enter to fuel the reaction, thus producing greater amounts of gel; this accumulation will eventually result in disintegration. Sometimes a pattern of 'map' cracking occurs but in other occurences small 'pop-outs' or detached lenses of concrete can occur. Examination of

Figure 7 Surface pop-out caused by particle of coal within the aggregate.

a spalled lens will often reveal a reservoir of gel behind it. The durability of the concrete can thus be compromised and in extreme cases the tensile strength of the concrete component can be reduced.

For ASR to occur, three interrelated factors must be present:

- high alkalinity;
- sufficient moisture; and
- critical silica in the aggregate.

When designing concrete mixes to avoid the risk of ASR, the underlying principle is that precautions become more demanding as the alkali levels of the cement increase. Most of the aggregates used in the UK are considered to be 'normal' rather than highly reactive.

Organisations such as the BRE and the Concrete Society have prepared guidelines and flow charts to assist the designer in the selection of appropriate measures to reduce the risk of ASR. By adopting a 'sodium oxide equivalent' scale, cement can be classified according to low, moderate or high alkalinity. Under the scale, the equivalent sodium oxide in the concrete is expressed as kilograms per cubic metre (kg/m^3). High alkalinity cement means an equivalent of 0.75 per cent sodium oxide or greater. Good construction practice dictates that for normally reactive aggregates the total alkali content of the concrete should not exceed 3kg/m^3.

According to the guidance issued by The Institution of Structural Engineers[22], the amount of alkali in concrete depends not only on the concentration of alkali in the cement but also on the total amount of cementitious material in the mix. For particular aggregate types and cementitious materials a broad relation exists between the total amount of alkali in the concrete and the risk of damaging ASR occurring. The most serious cases of ASR damage in the UK generally relate to mixes which had more than 4.5kg of alkali per cubic metre of concrete. These high alkali contents are associated with mixes with high cement content between 450–550kg/m^3 and/or high alkali cements.

Visual indicators

If ASR is suspected, the analysis will usually be confirmed by petrographic examination. However, certain visual indicators can give a clue to the existence of ASR, as shown in the table opposite.

The existence of ASR is not necessarily fatal to a structure, but much depends upon the severity of the attack and the element that is affected. A risk based analysis is often recommended before deciding upon replacement strategies. The structural significance of an element is a function of the consequences of its failure. These are judged to be slight or significant as defined below:

- Slight – the consequences of structural failure are either not serious or are localised to the extent that a serious situation is not anticipated.
- Significant – if there is risk to life and limb or a considerable risk of serious damage to property.

Alkali carbonate reactions

Alkali carbonate reactions are not generally as well understood as ASR and are thought to be very rare as they only occur with certain impure forms of dolomitic limestone. The aggregate reacts with the dissolved potassium and sodium alkalis within the pore fluid, altering the crystal structure of the aggregate and causing it to expand.

Visual indicators of ASR

Indicator	Description
Damp conditions or a source of moisture	For example an external structure in contact with the ground, foundations, floor slab, etc. An RH of around 80% is needed and this will be found in almost all ground bearing elements.
Cracking parallel with the reinforcement	Applies to reinforced concrete only as the reinforcement restrains the expansion of the concrete.
Map cracking	Applies to unrestrained, i.e. unreinforced concrete. The cracking will resemble the type of cracking that occurs when mud shrinks on a dried out river bed. This suggests that the surface has shrunk relative to the body of the concrete although with ASR the reverse is true.
Gel	Sometimes a sticky substance exuding from cracks or broken aggregate particles. When carbonated, the gel appears as a whiteish opaque coating rather like bad efflourescence.
Pop-outs	Sometimes small lenses (30–50mm diameter) spalling away from the surface, with sticky gel behind.
Expansion of the element	Evidenced for example by closure of movement joints, displacement of enclosing walls, etc.

4 High Alumina Cement (HAC) concrete

At a glance

- High Alumina Cement (HAC) (now calcium aluminate cement) was originally produced in the UK from about 1926.
- The material developed high early strength, high temperature resistance and resistance to sulphate attack; its cost meant that it was more suited to precast operations than in-situ work.
- The greatest use of HAC occurred in the 1950s to 1960s, predominantly in the manufacture of prestressed 'X' and 'I' beams for floor or roof decks. However, in the early 1970s several structural failures occurred and it was determined that HAC undergoes a process known as conversion which is associated with a loss of strength and increased porosity. By 1976, the material had effectively been banned for structural applications.
- Further research has shown that HAC can be seriously affected by contact with other materials such as gypsum and in damp conditions a condition known as alkaline hydrolysis can have a profound impact on strength.
- Strength and durability assessments of HAC structures are necessary before deciding upon an action plan. In dry internal environments, even with loss of strength, HAC may continue to give satisfactory service. However, damp conditions could be very serious, and in some circumstances, the strength of a fully converted component may well be below that which is necessary for it to provide its structural role. In such circumstances, removal or strengthening may be needed.
- To assist with the analysis of X or I beams, a set of rules has been published by the Building Regulations Advisory Committee (BRAC). The BRAC rules are the standard text for analysis in the UK, but they only apply to specific forms. Care is needed when analysing non-standard structures.
- Like ordinary concrete products HAC is affected by carbonation, and because thin prestressing wires often had little cover, it is possible that the corrosion protection afforded by the concrete could, in 40-year old structures, now be reaching a point where the steel is at risk.

4.1 INTRODUCTION

Concrete made with High Alumina Cement was first produced commercially in the UK in the mid-1920s and although used in various structures from that date its main use was in the form of precast units for buildings constructed during the 1950s and 1960s. Various other countries employed HAC, most notably Spain, with uses in France and Germany where it is known as Tonerdezement. It is estimated that around 50,000 structures contain HAC in the UK.[23]

Whilst HAC was, comparatively speaking, an expensive product, its benefits were derived in the form of its high early strength development; beam castings could be removed from their moulds within 24 hours of placing – a clear benefit for the manufacturer who could produce castings over a 24-hour cycle.

HAC is however an unusual product, it does not perform in the same way as ordinary Portland cement. During hydration, chemically unstable products are produced which undergo a mineralogical change known as conversion. Conversion causes a 53 per cent decrease in the volume of hydrated material although this occurs without a dramatic alteration of the overall dimensions of a concrete element. Inevitably, for this to occur, the concrete must become more porous and this results in a reduction in compressive strength[24]. The process of conversion is governed by temperature and can occur over periods of time from hours to years.

The loss of strength associated with HAC is significant – possibly 50 per cent or more – and this can lead to structural inadequacy and possibly collapse. Three roof failures in the 1970s led to detailed investigations (albeit that weaknesses had been predicted some time before) and in 1976 HAC concrete was banned for structural use both in the UK and in a number of countries, or in some cases permitted only subject to very stringent conditions. As a result of this, HAC was consequently only used for special applications.

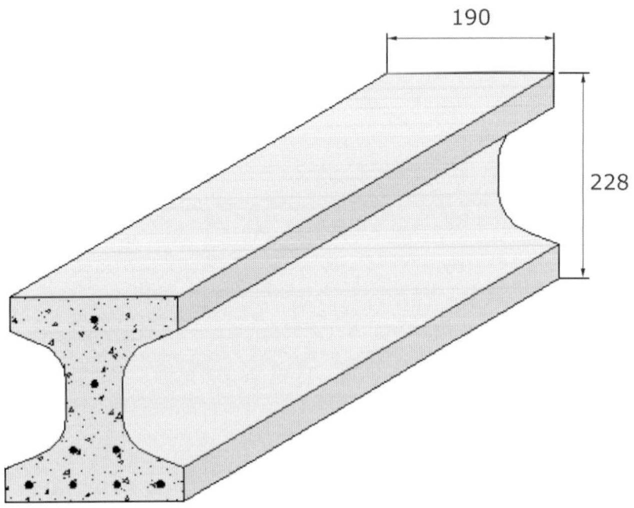

Figure 8 Typical profile of an HAC X beam, in this case manufactured by Pierhead.

In recent years, concerns have been raised over durability as well as strength, with the BRE publishing guidance in 2000 on problems of carbonation.[25]

Aggressive marketing campaigns by the cement manufacturers have lead to a gradual interest in the re-adoption of HAC into buildings with various attempts to introduce HAC back into structural codes.

Calcium aluminate cement (CAC) is the new term for HAC – perhaps to avoid some of the stigma associated with the material. The Cement Association of Canada cites the properties of CAC as early strength gain and resistance to high temperatures, sulphates, weak acids, and seawater. Typical applications for CAC concrete include:

- chemically resistant, heat resistant and corrosion resistant industrial floors;
- refractory castables; and
- repair applications.

CAC has also found use in the mining industry for filling masses and anchoring mortars, where its properties of rapid setting and high strength are utilised.

Current guidance for the use of CAC suggests that concrete should be proportioned with a water (including water that has been absorbed by the aggregate) to cement ratio of not more than 0.40 and a minimum cement ratio of 400kg/m^3. However, many engineers question how easy it would be to restrict the water/cement ratio to 0.40 given site conditions. CAC is an expensive product when compared with ordinary Portland cement (OPC): not only does it cost three to four times more, but the recommended minimum content of 400kg/m^3 is considerably higher than the cement content of OPC concrete. Given the cost and the risks involved it seems unlikely that CAC will find its way back into mainstream structural use.

4.2 FAILURES

Whilst the defining period for HAC use in the UK was 1973 to 1974 when three roofs collapsed, a structural failure of a roof at Sheephouses Reservoir in 1962 was put down to failure caused by loss of strength associated with prolonged storage under humid conditions.[26] No fatalities resulted from the UK collapses, but in Spain a collapse in November 1990 resulted in one fatality. Because of the insight that is provided by the examination of past failures, it is worth looking at what went wrong with the three principal failures at Stepney, Camden and Leicester.

The Stepney failure occurred at Sir John Cass's Secondary School when two beams in a swimming pool roof collapsed. The subsequent investigations identified two primary causes:

- conversion and associated loss of strength; and
- reduction in resistance to chemical attack.

Initially, it was thought that the high temperatures and high humidity within the pool building had contributed to the problem, but later studies showed that significant loss of strength had also occurred in the adjoining gymnasium building, which was not subject to the same conditions. Whilst chemical attack may have been contributory in the pool building, the margin of strength remaining in the gym beams was such that they were very vulnerable to failure as a result of chemical attack or the effects of fire.

Camden High School for Girls was completed in 1956 and its composite precast column and beam construction proved a model for many subsequent buildings. The assembly hall at the school was constructed of precast perimeter columns and edge beams supporting prestressed roof beams at 600mm centres spanning some 12m. The beams were notched at their bearing end onto a continuous ledge about 32mm wide[27]. Failure of the roof structure was attributed to:

- insufficient bearing of the roof beams;
- conversion of the HAC concrete roof beams and associated loss of strength;
- termination of the prestressing wires within the span of the roof beams; and
- insufficient structural cross tying.

The Roof of the Bennett Building at Leicester University was the third of the failures to be investigated. Here it was argued that severe loss of strength of some concrete edge beams (originally specified at 55.2MPa and found to be 6.9MPa) had led to an inability to deal with the tensile stresses required.

In his discussion on the reintroduction of CAC into design codes, Neville observes that over the years there have been various attempts to rewrite history, deflecting the causes of failure away from HAC to other causes such as poor manufacturing conditions or chemical attack – observations often made by materials scientists as opposed to engineers.[28] The essential point

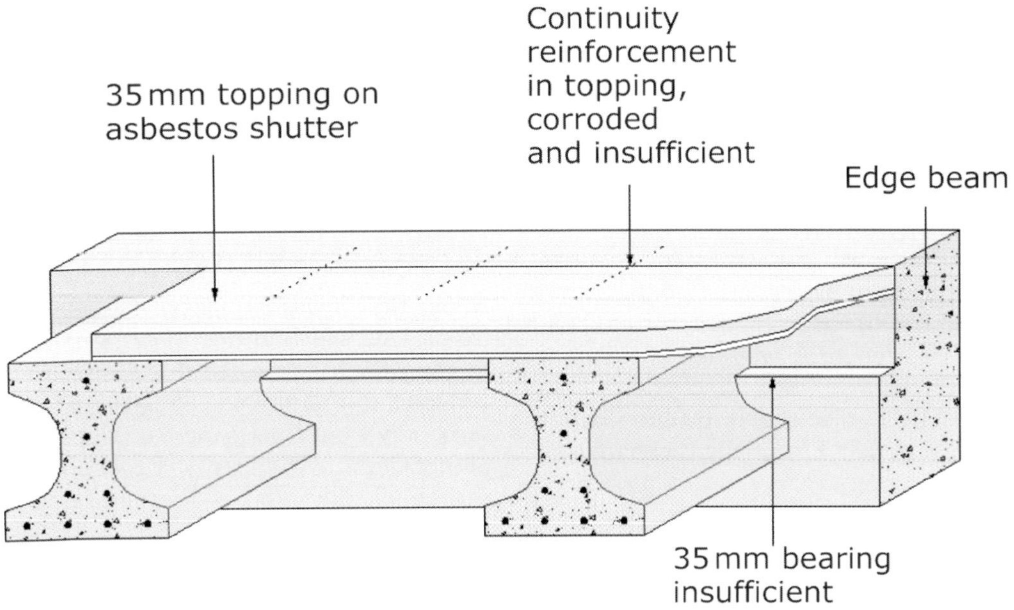

Figure 9 Some of the characteristics of roof beam failure.

must remain that HAC does suffer a loss of strength and that proper engineering judgment must be exercised when assessing structures constructed from this material.

4.3 COMPLIANCE WITH THE REGULATIONS

In 1976, amendments to BS 8110 removed the use of HAC for structural purposes. This situation has remained the case for many years – the prohibition on the use or specification of HAC has been a component of collateral warranties and appointment documents at least since the 1980s. However, current practice is to move away from long lists of prohibited materials in favour of a practice that permits use if in accordance with British Standards, Building Regulations and the like and on the basis that the material is used within its capabilities. In other words, if a material is used and specified correctly and in those circumstances it will perform perfectly satisfactorily, then there is no reason to caution against its use. As noted earlier, a sustained campaign to reintroduce HAC (as CAC) has resulted in several changes which could pave the way for a greater use of this material in the future.

In the UK, Regulation 7 of the Building Regulations 2000 deals with materials and workmanship. Regulation 7 is quite succinct:

> 'Building work shall be carried out:
>
> a) with adequate and proper materials which:
>
> (i) are appropriate for the circumstances in which they are used;
> (ii) are adequately mixed or prepared; and
> (iii) are applied, used or fixed so as adequately to perform the functions for which they are designed'.

Crown copyright material is reproduced with the permission of the Controller of HMSO and the Queen's Printer for Scotland.

The Approved Document that supports Regulation 7 specifically mentions HAC as one of a range of materials that may change their properties when exposed to certain environmental conditions, which may affect their performance over time. The document continues:

> 'Such materials can be used in works where these changes do not adversely affect their performance. They will meet the requirements of the Regulations provided that their final residual properties, including their structural properties, can be estimated at the time of their incorporation in the work. It should also be shown that these residual properties will be adequate for the building to perform the function for which it is intended for the expected life of the building.'[29]

Crown copyright material is reproduced with the permission of the Controller of HMSO and the Queen's Printer for Scotland.

BS EN 14647:2005[30] contains an appendix listing the principles for the correct use of calcium aluminate cements. In addition, the document contains several key pointers towards the reintroduction of HAC for structural purposes. Specifically, it calls for design to be based upon the strength of concrete after conversion (in effect in compliance with the Approved Documents). The standard goes on to consider durability, but, according to Neville[31], in a way which could create a sense of false security.

Notwithstanding the above, it seems unlikely that an engineer would be prepared to commit to what could be a significant risk of specifying CAC for structural purposes, particularly if guidelines are not totally clear. However, surveyors and engineers are frequently called upon to examine buildings that may contain HAC components, in which case a pragmatic stance is needed – one can form a more positive prognosis with the knowledge that all HAC is not necessarily bad.

4.4 DETERIORATION MODES

The preceding paragraphs have referred to the process of conversion and the loss of strength associated with it. However, loss of strength is only one of the issues that affects HAC concrete, and to conduct a full appraisal of an HAC structure it is necessary to consider durability as well. Essentially there are three areas to examine:

- loss of strength associated with the process of conversion;
- carbonation; and
- damage by chemical attack – mainly sulphates or alkalis.

A huge amount of research material is available on the subject of strength and conversion and for detailed information the reader is directed to the texts given in the further information section of this book. However, a general outline of the issues is given below.

Conversion

It is now accepted that HAC concrete undergoes a mineralogical change known as conversion – a feature that is influenced by factors such as temperature, water/cement ratio and humidity. Essentially, the addition of water to cement during the mixing process creates metastable hydrates. Metastable means that they gradually return to their stable state over a period of time. Stable hydrates are denser than metastable hydrates, so the conversion process results in an increase in porosity and a consequent loss of strength. A water/cement ratio in excess of 0.40 is known to cause a greater loss of strength, whilst a lower ratio will result in a lower loss of strength. At one time it was thought that an analysis of the degree of conversion would be beneficial, and although this may remain the case for unusual circumstances, it is common to adopt a residual strength of 21Mpa.[32]

However, the strength of HAC concrete can be seriously affected by chemical attack and so great care must be taken to eliminate this possibility before undertaking strength calculations.

Over the years there have been various attempts at reducing or eliminating the problem of conversion. For example, BRE developed a material, BRECEM, which mixed HAC with ground granular blast furnace slag, but this product took a year to reach the same strength that HAC could achieve in one day. In Canada, the approach of the National Research Council was to add siliceous pozzolanic powder and small amounts of inorganic salts. At the time of writing, Canada's National Research Council's Institute for Research in Construction (NRC-IRC) is actively seeking to market the technology to cement firms in the USA, Britain and France.[33]

Carbonation

At one time it was thought that concrete manufactured from HAC was immune to carbonation, but research by the BRE has revealed this not to be the case. HAC concrete is vulnerable to carbonation in much the same way as normal concrete (albeit that the method of detection is different). Exposure to atmospheric carbon dioxide reduces the natural alkalinity of the concrete and can place the steel reinforcement at risk of corrosion in the right circumstances.

X and I beams were often reinforced with small diameter stressing wires, typically with 16mm or so of concrete cover. The effects of carbonation after say 40 years exposure could therefore be significant in any durability assessment. Research by BRE suggests that many HAC structures will now have reached the point whereby carbonation has reached the reinforcement strands.[34] Given the small diameter of the strands and the fact that they are prestressed, the effects of corrosion could be serious, leading to loss of strength and consequent deflection of the component exhibited by transverse cracks as opposed to longitudinal cracks following the line of the reinforcement.

However, mere carbonation is not in itself sufficient to allow corrosion to occur. Water and oxygen must also be present to allow a corrosion cell to form. It follows therefore, that provided the High Alumina Cement concrete (HACC) component is kept within protected, dry conditions, there is a reduced risk of deterioration through carbonation alone. Unfortunately though the matter is not clear-cut; studies in Spain have yielded data on 1,400 joists cast between 1940 and 1950 and examined in 1991: around 20 per cent of the joists exhibited an active corrosion rate whilst around 9 per cent revealed cracking. Given that in many cases corrosion of steel due to carbonation-produced expansive formations of rust and consequent spalling of the concrete cover, how is it that corrosion in HAC did not necessarily result in cracking – the visual evidence that something is amiss? Bloomfield[35] considers that the increased porosity of converted concrete permits some of the corrosion products to diffuse into the concrete, thus reducing internal pressures. If deterioration is marked by transverse cracking, serious consideration should be given to urgent support measures as this form of damage could be a precursor to sudden failure.

On the basis of the above, it is suggested that any investigations into the degree of carbonation of the concrete are accompanied by careful visual inspection for cracking, and identification of corrosion either non-destructively or by exposure.

Given the sensitivity to moisture the greatest risk therefore lies in the use of roof members or locations such as cleaner's cupboards where prolonged wetting could occur. It is therefore important to appraise the condition of the concrete and any waterproof coverings before making any formal judgment as to the remedial work required. Do not be tempted to speculate that simply because an HACC component is located in an environment that should normally be dry it will never be exposed to water or moisture.

Chloride attack

The mechanism of chloride attack in concrete has been dealt with in chapter 3. Chlorides create conditions of high electrical conductivity within concrete and where this occurs corrosion cells can be more vigorous. It would be most unusual to find HAC-containing concrete being manufactured with calcium chloride concrete additives, but contamination from de-icing salts or sea spray could occur. As with OPC materials, contamination in this way can be potentially damaging because of the greater proportion of free ions, which are highly conductive, and the variability of levels of contamination. Should contamination occur, the reinforcing strands in an HAC beam would be vulnerable to corrosion in much the same way as they would in conventional mixes. By default, salt contamination is most likely to occur in a moist or wet environment, and this should flag up questions of durability.

Resistance to chemical attack

In its unconverted form, concrete made with HAC is generally more resistant to chemical attack from weak acids and pure water down to about pH 4. However, HAC is just as susceptible as ordinary Portland cement to damage from stronger acids.

Aside from deterioration arising from conversion and carbonation there are two further modes of deterioration that must be considered:

- sulphate attack; and
- alkaline hydrolysis with carbonation.

Resistance to sulphates is generally good in unconverted HAC concrete. However, if, or rather, when conversion takes place, the increased porosity of the concrete means that sulphates can penetrate the body of the concrete through tiny fissures. Having penetrated the body of the concrete, they may react with the hydrogarnet that is present in converted HAC concrete to form crystals of ettringite. Consequently, the expanding crystalline growth within the voids of the concrete would create internal stresses that disrupt the concrete.[36]

For sulphate attack to take place, water, as well as the chemicals, must have been present persistently over a long period of time at normal temperatures. Usually, if HAC concrete has been affected it will be due to contamination by sulphates from external sources, rather than sulphates in the body of the concrete itself.

Alkaline hydrolysis is a relatively rare form of attack, but one which can have a profound effect upon HACC, sometimes resulting in its degeneration into powder.

Essentially, sodium or potassium salts from Portland cement products (such as floor screeds) interact with atmospheric carbon dioxide to attack hydrated HAC to form calcium carbonate and hydrated alumina, reducing the cement paste to powder. For alkaline hydrolysis to occur, persistently wet conditions are needed.

If chemical attack has taken place (especially attack by sulphates and alkalis), the strength assessments recommended within the BRAC guidance (see below) could well be unreliable.

4.5 INVESTIGATION

BRAC Guidance

The authorative text for the assessment of X and I beams is a document published by the Building Regulations Advisory Committee shortly after the Stepney and Camden failures. The guidance was prepared with reference to CP3 (the old loading code) and CP110 – *Structural Use of Concrete* – which was current at the time, but has now been replaced by BS 8110. The BRAC rules as they have come to be known have been re-evaluated over the years but have still been found to be valid and not overly cautious.

However, it is important to note that the rules only apply to certain, quite specific circumstances and should not be used for the analysis of other types of HAC component. The components must be:

- X or I beams cast using the long-line casting process; and
- free from chemical attack.

There are three generic stages in an investigation, namely:

- Stage 1 – identification;
- Stage 2 – strength assessment; and
- Stage 3 – durability assessment.

A rapid chemical test has been developed – this is usually sufficient to distinguish High Alumina Cement from other cements, although it must be used with care and can give misleading results. Samples taken from carbonated concrete can give a false indication that HAC was used. Similarly, contamination of a sample with gypsum plaster could give a negative impression even though HAC was present. For conclusive tests, detailed laboratory analysis is required.

Examination of constructional drawings can yield clues as to the nature of the concrete. For example,

Bison did not use HAC whilst Pierhead was a major supplier. A full list of HAC component suppliers may be found in the BRAC rules, together with a chart showing the characteristics of each component.

A visual assessment of the suspect component may not, in itself be reliable. A dark colour or brownish tinge may indicate HAC (especially if the shape of the component resembles an X or I beam), but some Portland cements can look similar.

Stage 1 – identification

Visual – Not a particularly reliable diagnosis, but it may give an indication of the possibility of HACC. Taking into account the age of the building, e.g. pre-1976 and the nature of the construction (for example X or I beams) should be enough to establish if a risk is likely. The colour of the concrete can be an indicator – dark grey, perhaps with a brownish tinge. The formation of powdery deposits on materials subjected to long term water ingress should invite questions, as the possibility of alkaline hydrolysis and severe loss of strength would need to be investigated.

If available, an examination of the original construction drawings may be sufficient to prompt further investigations. The drawings might not specifically state HACC but reference to a manufacturer known to have utilised HAC may be helpful. To this end, interrogation of local authority archives might be a useful exercise.

Rapid chemical test – This form of testing is useful for gaining a 'snapshot' of the nature of a building, and will often be carried out as part of a suite of tests during a pre-acquisition survey. Ideally, samples should be taken properly, but it is sometimes possible to knock an edge of a suspect component and to send this to a testing lab for analysis.

The method of testing is described in BRE Special Digest SD3[37]. Essentially the test involves mixing a 1g sample of the material with 10cm^3 of sodium hydroxide solution. After filtration, the solution is then treated with oxine. If the mixture forms abundant yellow precipitates this is a good indication that HAC is present. Little or no precipitate firmly indicates that HAC is not present. In other words, the test is a pretty good indicator that HAC is not present, but it is prone to false positive results – suggesting that the sample contains HAC when in fact it does not.

It is very important to avoid contamination of the sample, particularly with gypsum plaster as this can give false readings. If the sample is taken from carbonated concrete containing Portland cement, this can also form yellow precipitate, thus giving a misleading impression.

If a definite diagnosis is required, alternative laboratory tests will need to be undertaken.

There are several alternative and more reliable methods of diagnosis. Perhaps the most common is the technique of differential thermal analysis or differential scanning calorimetry. Other methods are petrography or x-ray diffraction, both of which will give a reliable answer, although these are very specialised methods needing expensive equipment and skilled personnel.

Stage 2 – Strength assessment

A stage 2 strength assessment is required to determine if the precast concrete members have sufficient structural capacity, even at the reduced fully converted strength, to safely withstand the applied loading. Subcommittee P sets out the guidelines for concrete strength based on 21N/mm^2. The strength assessment requires the section properties of the beam to be established. Thereafter the structural strength of the element can be calculated. In cases where the section properties are unknown, or cannot be determined by investigation, then assessments are limited to determining the concrete strength using near-to-surface tests.

In order to complete the calculations, it is important to establish the following data:

- section profile;
- number and diameter of reinforcing strands and depth of cover;
- condition of the strands (to validate that corrosion has not taken place);
- degree of fixity at the ends of the beams; and
- whether there is a bonded screed, or a screed laid on quilt or a simple unbonded screed.

This information can then be compared with the section properties in the BRAC data sheets.

If it is possible to do so, the bearing end of a typical beam should be exposed, possibly by removing part of the outer leaf of an enclosing cavity wall. Such an examination will show the exposed ends of the reinforcement as these will be visible. However, such an examination is not always possible and so in this case, a series of intrusive investigations over several beams should be sufficient to reveal the strand patterns. If breaking open is needed, it is of course better to make a series of small, local investigations, each exposing a separate strand, rather than trying to expose all strands at one site.

Few manufacturers incorporated shear reinforcement in X or I beams. It is therefore important to establish the bearing conditions and to identify any signs of horizontal cracking in the web close to bearing positions. Additionally, the condition of the web should be noted to determine if there is any cracking likely to be caused by corrosion of reinforcement.

In many cases, the strength assessment will demonstrate that the beams will continue to perform their function satisfactorily – this is especially true of beams with a span of five metres or less. However, if the assessment shows an underperformance of less than ten per cent it may be still be possible to justify the structure by applying a more rigorous calculation using BS 8110. Similarly, petrographic analysis might provide valuable information as to the water/cement ratio. If less than 0.45 a higher strength may be appropriate for use in the calculations.

If the underperformance is between 10–25 per cent, static load testing could be undertaken, coupled with strength assessment by taking core samples. However, if this approach is to be adopted, the testing laboratory must be involved at an early date as it may not be possible to extract the size of core normally needed for strength assessments. The core will need to be taken from the top of the beam. (Normally a 100mm diameter core is needed.)

If the underperformance is greater than 25 per cent then it is unlikely that refinements in the strength calculations or load tests will provide the improvements needed. In these circumstances, strengthening of the structure (or its removal) will need to be considered.

Remedial work, where necessary, may involve the removal of the HAC elements altogether. Alternatively removing or replacing part of its original load-bearing function by adding a secondary support may be satisfactory, but increasingly common is the method of carbon plate bonding which is likely to be far less disruptive than other, more conventional systems. Carbon plate bonding involves preparing the bottom surface of a beam by needle gunning or vacuum grit blasting and then bonding sheets of carbon fibre to the underside of the member.

Stage 3 – Durability assessment

A stage 3 durability assessment is required to determine the long-term durability where affected by chemical attack and reinforcement corrosion. Testing can be undertaken to determine the presence of alkalis and sulphates. Laboratory testing may be supplemented by a detailed visual inspection and the removal of lump samples for petrographic examination. A durability assessment should also include a visual examination of the reinforcing steel where lump samples are removed.

5 Glulam and box beams

At a glance

- Current laminated timber is constructed from stress graded longitudinal strips of timber, planed to a close fit and glued together using phenolresoreinol formaldehyde (PRF) adhesives. Individual strips are usually finger jointed.
- Its relative low weight by comparison with steel and concrete means that self-weight is less of a constraint in design than other, heavier materials. A steel beam of equivalent bearing capacity would have a self-weight around 20 per cent greater than a laminated timber beam.
- Being timber, glulam is likely to be affected by the same agencies that would affect ordinary timber products: poor detailing, frequent wetting, exposure to aggressive chemicals etc. can have serious effects. Overloading of timber beams is a potential issue, either by excessive snow loads or unintentional means such as providing alternative, heavier roof coverings or roof mounted plant. If this is coupled with decay, the effect on the structure could be serious.
- Heavy snow in Germany and adjoining European countries in the winter of 2006 brought about the collapse of a large number of timber buildings, some constructed with laminated timber and some being several decades old. Preliminary research suggests that regular planned maintenance inspections of glulam structures (and for that matter related structures such as plywood box beams) should be undertaken. Investigations suggest a common problem with restrained shrinkage, curved beams and connections of cross girders.
- Strengthening of failed beams (or improvement of their load bearing characteristics) can be undertaken by external plate bonding or post-tensioning with steel wires or rods. Carbon fibre or steel reinforced polymer are being used to good effect for plate bonding although these materials, because they are bonded to the timber, alter its appearance.

5.1 INTRODUCTION

Timber has long been well regarded as a structural member for long spans. Although early laminated timber beams were manufactured over 100 years ago, developments in adhesive technology during the Second World War resulted in improved bonding techniques and laminated timber or glulam is now a mainstream building product. According to the Glued Laminated Timber Association, the consumption of standard glulam beams in the UK doubled between 1985 and 1995.[38]

Current laminated timber is constructed from stress graded longitudinal strips of timber, planed to a close fit and glued together using phenolresoreinol formaldehyde (PRF) adhesives. Individual strips are usually finger jointed.

Earlier systems utilised casein glues and undercut jointing methods for the individual strips. In some cases stress graded timber may not have been used, so older systems may be somewhat less predictable under engineering analysis than contemporary sections.

Its relative low weight by comparison with steel and concrete means that self-weight is less of a constraint in design than other heavier materials. A steel beam of equivalent bearing capacity would have a self-weight around 20 per cent greater than a laminated timber beam, whilst concrete would be circa 600 per cent heavier. Laminated timber beams and portal frames are common forms of construction throughout the world, particularly in Europe and North America. The technology exists to create very large clear spans and so the structures often find favour for use in sports halls, skating rinks, industrial and warehouse buildings.

Glulam beams can be manufactured in a wide variety of shapes and with spans commonly of up to 50m. In North America, some sports stadia have been constructed to spans very much greater than this.

Figure 10 Glulam beams in a temporary pavilion.

Figure 11 Deep glulam beam concealed within a void above a refrigerated warehouse.

5.2 PRACTICAL ISSUES

One of the limiting factors in design terms is simply the ability to transport materials to site. Whilst a 50m span is perfectly feasible, road transport will create difficulties as will physical problems of lifting it into place. For this reason jointing of one form or another is necessary. Commonly, joints are formed from dowelled connections (a connection that could be bolted, screwed, nailed, etc.). Because there is a need to transfer forces effectively, bolted connections are often designed.

Bolted connections in glulam require special consideration as compressive failure of the timber must be avoided. Few, large bolts may have a level of rigidity that is greater than the embedment capacity of the timber; this creates a risk of tension or shear failure in the beam.[39] To avoid compression failure numerous slender bolts are usually employed, the principles of design being set out in Eurocode 5.

Being constructed of organic materials, glulam is not necessarily an install and forget product. In much the same way as conventional timber is affected by dampness, decay and beetle attack, glulam is also vulnerable to these agencies. Since large parts of the structure are often concealed from view, an out of sight out of mind policy can sometimes apply, leading in some cases to significant structural failure. There is nothing inherently defective about the technology and use of laminated timber, but building owners must be alive to the need for maintenance inspections and early indicators of a developing problem.

Aside from potential defects in the original construction, typical causes of deterioration and decay of glulam beams result from exposure to environments such as:

- extreme heat, cold or moisture;
- exposure to aggressive chemicals;
- damage from equipment or machinery;
- decay caused by poor ventilation and trapped moisture;
- material defects caused by low-grade timber;
- overloading caused by additional roof mounted plant or suspended ceilings;
- exposure to splash zones;
- condensation;
- poor detailing or protection of timber in contact with the ground or below damp-proof course level;
- saturation of walls in contact with timber arising from environmental conditions, defective rainwater goods, etc.; and
- valley gutter beams.

Chemical treatment of structural timber is usually a given, particularly in the UK if not Europe, but the physical size of laminated timber products makes timber treatment using conventional means impracticable. Eurocode 5 sets out some guidance although this is mainly aimed at the selection of appropriate strength values for timber rather than environmental risks. The following table is derived from Eurocode 5 and the UK national Annex:

Service Class	Environmental factors	Examples of construction
1	Timber to have no more than 12% moisture content when exposed to air at 20°C and 65% RH for no more than a few weeks per year.	Warm roof Intermediate floor
2	Timber to have no more than 20% moisture content when exposed to air at 20°C and 85% RH for no more than a few weeks per year.	Cold roof Ground floor Timber framed external wall External timber protected from direct wetting
3	Timber exposed to conditions more onerous than classes 1 or 2.	Fully exposed external timber

Timber is not at risk of decay if its moisture content is kept below 20 per cent, so the ideal first step is to design out the risk of this being exceeded. If such measures are impracticable, the next alternative is to select timber that is naturally resistant to decay. Finally, if conditions are still unsatisfactory, preservative treatment will be needed. Generally, in environments where the moisture content of the glulam is less than 20 per cent (Service Classes 1 and 2), it will not be necessary to treat the laminates, or the finished component, against fungal attack. The risk of insect attack for internal timbers will require separate assessment.[40]

In the UK, glulam is usually made from European whitewood, which is 'resistant' to treatment. European redwood, which is 'moderately resistant', is used where slightly greater risks are envisaged, but other sources are usually much more expensive and so less regularly utilised.[41] If treatment is required this is usually by flood coating to the point of refusal using organic solvent based preservatives.

Pre-treatment of the individual laminations is not normally undertaken (except for bridge beams) because the process of planning the timber during manufacture removes the outer faces where a majority of the preservative will be contained.

5.3 STRUCTURAL ISSUES

Heavy snow in Germany and adjoining European countries in the winter of 2006 brought about the collapse of a large number of timber buildings, some constructed with laminated timber and some being several decades old. Whilst snow loads may have been a contributory factor, preliminary research by the University of Karlsrhue[42] indicated that of 140 analysed failures, sports and ice rink structures were most frequently affected, with defects occurring in simple or continuously supported beams of 6–48m spans. Around 70 per cent of the failures indicated cracking in the direction of the grain, with 6 per cent exhibiting shear failure, 6 per cent decay problems and 6 per cent tension failures. Alarmingly around one third of the analysed buildings were in urgent need of structural intervention or in danger of collapse.

At the time of going to press, detailed conclusions from the Karlshrue study are not available, save that it is clear that regular planned maintenance inspections of glulam structures (and for that matter related structures such as plywood box beams) should be undertaken. The study suggests a common problem with restrained shrinkage, curved beams and connections of cross girders.

Unintentional overloading of a roof structure can have serious consequences for laminated timber beams. For example, in 2004 a 27m roof beam failed above the sports hall in a school in Ohio. The beam, which was around 1.2m × 220mm in section, developed a 25mm wide crack at mid span, the crack originating at a knot position in the bottom lamination. Some 25 years before, the roof had been re-covered with a loose membrane held in place with ballast, a covering that replaced the earlier, lighter covering. Subsequent analysis revealed that the addition of ballast had exceeded the design loading of the roof by some 62 per cent.[43]

5.4 REPAIR OF GLULAM

Possible options for the restoration or upgrading of strength for failed glulam include the addition of steel plates, external post-tensioning, sectional replacement, and composite systems, such as fibre reinforced polymer (FRP) and steel reinforced polymer (SRP).

External post-tensioning is an attractive technique in that the outward appearance of the beam is not changed significantly. The post-tensioning is achieved by means of standard prestressing tendons or high-strength steel rods, usually located outside the original section. The tendons are connected to the structure at anchor points, typically located at the ends of the member. End-anchors can be made of steel fixtures bolted to the structural member. The desired uplift force is provided by deviation blocks, fastened at the high or low points of the structural element. Prior to external post-tensioning, all existing cracks are epoxy-injected and other needed repairs are performed to ensure post-tensioning forces are distributed uniformly across the member.[44]

Carbon FRP comprises thin fabric sheets bonded to structural members with an epoxy adhesive; the material has tensile strength approximately ten times that of steel and when applied as an external plate bonding system can improve bending strength considerably. Typically, installation is achieved by applying an epoxy adhesive to the prepared surface, installing the FRP fabric into the epoxy, and then applying a second layer of the epoxy adhesive.

Steel reinforced polymer is the generic term for a family of reinforcements made from ultra-high strength twisted steel wire some 11 times stronger than a typical steel plate. The product can be moulded into or on virtually any structure or part.

An alternative to FRP and SRP systems is the bonding of steel plates or channels to the timber. In addition to epoxy adhesive, mechanical anchors are typically used to ensure the steel element will share external loads in case of adhesive failure.

6 Asbestos cement and cement fibre roofing

At a glance

- Asbestos cement roof sheeting was first used in the UK in 1913 and for 60 years or so was the covering of choice for industrial buildings.
- The sheeting was manufactured in a variety of sinusoidal profiles, 'big six' being most common for larger buildings.
- Normally the sheeting contained chrysotile (white) asbestos at a concentration of around 10–15 per cent. Since 1999 new roof sheeting should be asbestos free.
- The roofing is durable and if installed properly relatively maintenance free. Cracking of the sheeting associated with corrosion of fixings or shrinkage on ageing is commonly found in older installations.
- Surfaces are relatively porous and attract moss and algae growth which is not particularly desirable. However, cleaning of the sheet will probably release asbestos fibres which could enter the drainage system or contaminate internally.
- Older roofs are sometimes repaired with glass fibre sheets and although asbestos cement coverings are fragile and should never be walked upon, concealed glass fibre sheets, particularly if painted over, could be particularly hazardous to maintenance operatives.
- Older roofs can be coated with liquid applied membranes to seal their surface and restore watertight performance. However, applications of liquid membrane are not in every case successful particularly where there is a risk of the sheets chattering in the wind. Options such as overcladding with lightweight steel roofing or replacement altogether are likely to give better assurance of weather integrity.

6.1 INTRODUCTION

Until 1913, the design of industrial buildings was constrained by the restricted spans that could be constructed at reasonable cost. Roofing materials were often slate, with heavy trussed rafter design needed to achieve even modest spans. Since the lowest practical roof pitch for large slates was around 30°, large spans also produced high roofs and a resultant increase in dead load, gable wall area and the like. But in 1913 this changed with the opening of the first asbestos cement factory, Turner Brothers in Manchester.

Because the new roofing sheets were large and required only short end and side laps, they provided a single skin covering (as opposed to slates). In turn, this produced a covering that weighed about 50 per cent less than slate, with corresponding economies in the support structure. The obvious economies led to a form of roof construction that dominated the design of industrial buildings for 50 years or more.

By the 1960s, some thought was being given to the provision of improved thermal insulation to roofs – this became especially important in the 1970s with the oil crisis playing a major role in the design of buildings. The demand for insulation meant that it was often necessary to provide some form of insulated spacer between the roof sheeting and the purlins, commonly producing a cold roof (as opposed to a warm roof where the insulation fits into the profile of the weather sheet). Timber battens were not uncommon, but often a rigid foam batten or sheets of rigid foam were used. Whilst improving thermal performance, the provision of insulation also brought about problems of condensation, which in turn dictated much better attention to the provision of vapour barriers and vapour control measures.

The problem with insulated spacer fittings was that they could become compressed, leading to a loosening of the fixings and water ingress.

Chrysotile (white) asbestos was the primary reinforcement in asbestos-cement products; often used in concentrations of 10–15 per cent it produced durable, watertight sheeting that could last successfully for 40–60 years. Concerns over the use of asbestos led to a search for alternatives, with a variety of man-made fibres being tried – often suffering from poor durability or flexural strength. This coupled with the development of organic coatings on steel and aluminium meant that asbestos rapidly lost ground as a roofing material, with profiled metal gaining the market share.

The *Asbestos (Prohibitions)(Amendment) Regulations* (S.I. 1999 No.2373) 1999 imposed a total ban on the importation, supply and use of chrysotile (white asbestos) effective from November 24, 1999 and so cement fibre products post-dating 1999 should be asbestos free. Whilst non-asbestos products had been marketed for some time before 1999, various roof fittings and mouldings were still imported and may have found their way onto buildings, even if the main roofing was asbestos free.

Visual inspections of cement fibre roofing may give a good indication of whether or not asbestos has been used (particularly for older buildings), but the only way to determine this with certainty is to take bulk samples.

6.2 THINGS TO LOOK OUT FOR

The following headings provide details on what defects to look for when assessing cement fibre roofing.

Cracking parallel with the long edge

This form of cracking is often found in sheets that have been decorated on one face (possibly a factory finish or site applied treatment). Sheets painted in dark colours are particularly vulnerable to this type of failure, usually caused by bowing of the sheet initiated by expansion on one face and contraction (by carbonation) on the reverse or inner face. Like all cementitious materials, cement fibre sheeting is affected by carbonation – a chemical reaction occurring in the surface of the exposed sheet as a result of exposure to atmospheric carbon dioxide. The process results in shrinkage accompanied by an increase in porosity. The increase in porosity also results in a loss of impact strength within the sheet – by as much as 50 per cent over time.

Figure 12 Longitudinal and transverse cracking caused by shrinkage and impact loads.

Cracking across the width of the sheet

Cracking across the width of the sheet could have been caused by structural movements or overloading. Cement fibre roofs should never be walked upon as they can collapse without warning (the author speaks from first hand experience), but foot traffic does sometimes occur and could cause the sheets to deflect and crack. According to the National Federation of Roofing Contractors, falls through asbestos cement roofs are a regular source of deaths in the construction industry.[45] Alternative causes could be wind loads, possibly coupled with defective or insufficient fixings or unexpected live or impact loads.

Corrosion of fixings

Corrosion of fixings may have three effects:

- cracking of the sheeting around the fixing;
- water penetration around the fixing; and
- structural failure.

Traditionally, hook bolt fixings were used to secure fibre cement sheets to supporting purlins or sheeting rails. The earlier fixings would have a flat steel washer and rubber seal to distribute loads over the sheet and would often be capped with a dab of bitumen as weather protection. Later fixings would have been fitted with protective caps. It is quite common to find the protective caps missing or decayed with the result that atmospheric corrosion occurs.

Figure 13 Corrosion of hook bolt fixing resulting in cracking of cement fibre sheet.

A second source of corrosion could be the production of condensation within the roof void – especially if a cold roof has been constructed. Expansion of the fixing as a result of the corrosion process will cause cracking in the sheet if the pilot hole was too small to accommodate it. This could in turn cause more water ingress.

Cracking around fixings

Cracks radiating from fixings are usually an indication of corrosion (see above) but may also be due to structural movement and the sheets being held too firmly. Moisture movement can be quite significant, with around three per cent of their initial dry length being possible. Thus, for a sheet measuring 1.8m length, elongation due to moisture absorption could be around 5.4mm. Again, if the pilot hole through the sheeting is too small, and the fixing is overtightened, small movements in the structure (from temperature, moisture, etc.) may not be accommodated by the sheeting. According to BS 8219, fixing holes should be at least 2mm larger in diameter than the diameter of the fixing.

From above it will be remembered that carbonation of the asbestos causes shrinkage; this apparent conflict between expansion on the one hand and shrinkage on the other depends upon the age of the sheet and the number of wetting and drying cycles that it has undergone. The older the sheet the more likely it is that shrinkage will be the predominant factor.

Contemporary practice is to use self-drilling self-tapping screws for fixing cement fibre sheeting. These fasteners are designed to create holes that are 2mm larger. These fixings (and indeed conventional fixings) should be installed through the crowns of the profile.

If fixings are under or overtightened, the weather performance of the seal may be affected. In a correctly installed fastener, it should be possible to squeeze the sides of the seal together slightly between thumb and forefinger.

Each sheet should have two fasteners per purlin.

Rooflights that have been overpainted

A common treatment for aged asbestos cement or cement fibre roofs is the application of a liquid waterproofing membrane designed to improve both the appearance and the weathertightness of the sheeting. Sealing the surface of the sheet also has the benefit of preventing fibre loss. Such coatings should never be applied over glass reinforced plastics (GRP) or PVC rooflights, but regrettably, instances are found where through ignorance or laziness the rooflights are concealed in this way. As noted above, fibre cement roofs are not strong enough (or should be regarded as not strong enough) to withstand foot traffic and, whilst they may appear robust, can fail suddenly. Plastic is even weaker and if it is disguised as an apparently solid covering the risks are much greater, even if crawling boards are used.

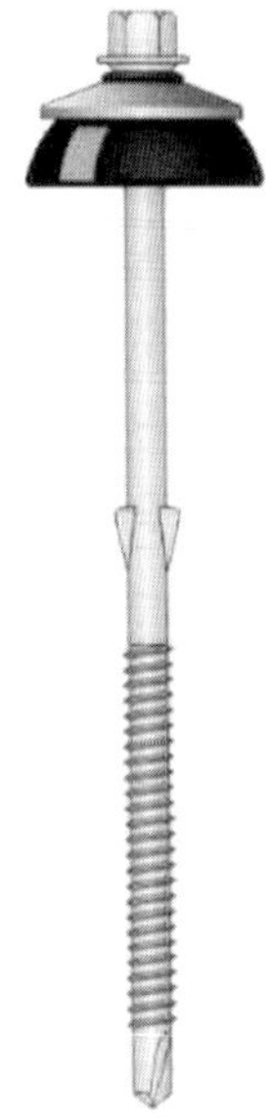

Figure 14 A self-drilling fastener for cement fibre to metal purlin. The wings on the shaft enlarge the fixing hole in the sheeting.

Sheets replaced with glass fibre

Because of its versatility as a material, GRP can be readily manufactured to replicate any cement fibre profile, and so replacement of damaged cement fibre sheeting using this material could be an attractive option if the cement profiles are unobtainable. Whilst GRP is a durable material in its own right, mixing GRP and cement fibre products in situations that are not immediately apparent could give rise to additional health and safety risks. Again, patching a cracked and defective cement fibre sheet with a GRP sheet may restore its weathertightness, but it does not deal with the loss of strength.

Plastic rooflights, whether PVC or glass fibre, suffer from similar rates of deterioration when used with cement fibre roofs, especially if those roofs are finished in dark colours. To prevent softening and distortion, BS 8219 recommends that where dark colours are used, they be painted with light colours at lap positions.

Compression of spacer battens

As noted earlier, the introduction of insulation into roofs brought about a raft of condensation and water penetration problems. A common form of finish in buildings of the 1960s through to the early 1980s was the use of plastic faced plasterboard held within galvanised steel tee bars spanning between purlins.

Figure 15 Roof sheeting replaced with GRP. The painted finish is beginning to wear away exposing the GRP base.

The most convenient location for roof insulation was above the lining system, and it was here that a layer of mineral wool or glass fibre was often applied. In order to make the insulation continuous, the roof sheets had to be raised above the purlins using either timber or quite often spacers made from rigid foam, this reducing the effects of cold bridging.

One consequence of foam spacers is that although they offer good insulation properties, they also compress. Some compression may be reversible, but if the material is crushed it will not recover and so roof fixings will cease to hold the sheets rigidly and some 'chattering' of the sheeting can occur in windy conditions, leading to water ingress at laps or fixing positions.

Missing or defective fillers and fittings

Moulded eaves fillers, ridges, verges and other fittings are vulnerable to mechanical damage arising, for example, during routine maintenance of gutters or from wind damage or deterioration of secondary fastenings through corrosion. Verge and ridge members can be prone to distortion, with resultant damage if insecurely fixed.

Moss growth and surface deterioration

As noted earlier, the process of carbonation increases the porosity of the sheet. In damp conditions, the surface provides good nutrients for the development of moss and algae; these can produce acidic rainwater run off that could have a harmful effect upon lead flashings or other metal components. However, removal of algae by washing down is an unsatisfactory measure since asbestos fibres can be washed away or disturbed in the process. Washing down with high pressure washers should never be performed, although specialist contractors have developed machines for cleaning that retain dislodged fibres, which can then be disposed of as asbestos waste.[46]

Leakage at laps and end laps

Unlike profiled steel, which tends to be laid to lower pitches, asbestos cement and cement fibre roofs do not always require sealants at laps and end laps. Generally, sheeting laid at pitches of 22.5° or more and with end laps of 150mm do not need additional sealants even in conditions of moderate to high exposure. However, for lower pitches, sealants are required in most cases, although increasing the end lap to 300mm can avoid this depending upon exposure. For pitches down to about 5° double sealing of end laps and a line of sealant at side laps is essential. Detailed guidance can be found in BS 8219.

Sealants need to maintain their flexibility and remain in contact with the sheeting for the life of the roof. Earlier sealants tended to be polyisobutylene or oleo resinous types although butyl tapes are now common.

Provision for movement

As noted above, provision for movement is essential given the potential for shrinkages and moisture movement. Aside from the need for enlarged fixing holes, buildings longer than 45m should have one joint of circa 30mm through the centre of a row of sheets and covered with a suitable capping. For every additional 30m in length a new joint is needed.

6.3 CEMENT FIBRE SHEETING

While technically similar in profile and method of laying, cement fibre sheeting is generally used to describe roof sheeting manufactured from manmade fibres – asbestos substitutes. Roofing using this material is described under BS 8219:2001[47].

The search for a reliable alternative to asbestos involved a number of less than satisfactory manmade fibres. In Europe, the most successful products often used a mixture of PVA fibres and cellulose fibres with about 80 per cent cement. These sheets were produced in a similar way to asbestos cement using a process first developed by Ludwig Hatscheck in the late 19th century. (For more detail on the history and manufacturing processes involved see *Investigating Hazardous and Deleterious Building Materials*, RICS Books, 2006).

The durability of fibre cement roofing was examined by the BRE in 1991[48]. They concluded that:

- bending strengths of fibre cement sheeting was generally less than for its asbestos cement counterparts;
- there was a variation in strength between wet and dry conditions and in transverse and longitudonal directions; and
- poor site practice and poor design would lead to reduced life expectancy in much the same way as it would for asbestos cement.

In addition, it should be noted that during the mid-1990s a series of fibre cement roofs in Denmark failed, owing it was thought to an excess of fly ash in the mix.[49]

Case study

The building comprised a single storey portal framed warehouse of about 3,000m^2 constructed in the 1970s. The roof was of asbestos cement sheeting incorporating GRP rooflights. Internally the roof was lined with plastic faced plasterboard with a 75mm thick layer of glass wool insulation laid above the purlins. There was no additional vapour control in the roof void. The roof sheets were raised above the purlins and supported by 'Corofil' rigid foam insulation battens.

The building had been refurbished prior to letting to an IT company. Part of the refurbishment work had involved the cleaning down of the roof and the provision of a silver liquid applied bituminous based coating applied to stabilise the surface and, it was hoped, make the roof watertight (although no previous leakage had been reported).

Shortly after taking occupation, the tenant complained of water leaks during rainfall. Detailed visual inspections of the roof surface revealed no obvious defects, but the leaks persisted. A condensation analysis was performed but did not reveal conditions at the time that were likely to be causing condensation. Water tests on the roof failed to replicate the leakage.

Following a careful review of symptoms it became apparent that the leakage was related to wind direction and wind speed. The liquid membrane treatment had been applied by operatives using crawling boards and (it was suspected) by walking over the roof sheeting – a highly dangerous practice. The loads so applied had crushed the Corofil battens in areas with the result that the hook-bolt fixings no longer provided a rigid restraint and the sheets were free to 'chatter' in windy conditions. Water penetration was probably occurring at fixing positions; the liquid coating was also restricting the freedom of water to drain out at sheet laps allowing it to build up in certain conditions.

The roof was condemned, but to enable continued occupation a new profiled steel roof was laid over the top as an over-roofing option. The new roof also enabled the level of insulation to be increased, with the liquid membrane acting as a vapour barrier.

7 Profiled metal sheet roofing

At a glance

- Almost universal for industrial, out-of-town retail and warehouse buildings in the UK.
- Coated steel roofing is the most popular followed by aluminium. Sheets can be manufactured in long lengths to reduce end laps and can be curved to suit contemporary roof profiles. Composite panels as opposed to site assembled systems are gaining in popularity.
- One of the most common finishes is plastisol, a PVC based coating of around 200μm (microns) thickness. Care is needed in the choice of colour as dark colours are affected by high temperature as a result of solar radiation; they may not function as well as lighter colours and can fail within a few years.
- Cut edge corrosion of plastisol coatings is still a common problem; this is where the coating fails at lap and eaves positions, often peeling back and creating a small reservoir of water. The base metal then corrodes if the coating is unrepaired. Several proprietary silicone paint systems are now available for treatment of the problem.
- Low pitched roofs are vulnerable to water ingress at side and end laps so careful positioning of sealants is needed.
- Earlier site-assembled roofs could suffer from condensation problems, particularly as a result of heat radiation to a cold night sky. This problem could cause condensation to freeze on the underside of a metal roof sheet, only to melt and drip into the premises as temperatures rose the next day.
- Composite panels are generally non-combustible. However, panels dating from around 2000 and before may not be non-combustible and could contain cores of PUR or in some cases EPS (particularly in cold stores) which can burn fiercely in the event of a fire and be difficult to control. For this reason, buildings constructed with a high proportion of non-fire rated composite panels are likely to attract high insurance premiums.
- Panel fixings must be designed to suit particular site conditions. Leakage around fixings can occur if they are of an inappropriate type or if they are overtightened so as to overcompress the sealing washer.
- Rooflights are commonly of GRP, polycarbonate or PVC, although in most contemporary installations, GRP is the most likely in view of its durability and ability to satisfy health and safety requirements. PVC rooflights do not perform particularly well and will often be a source of leakage as a result of thermal movements and distortion.

7.1 PRODUCTS

The outer sheet of a built-up or composite roof sheeting system is usually termed the weather sheet. Its purpose is however rather more than simply the exclusion of the elements; it provides a means of transferring wind and snow loads to the structure, as well as assisting stability.

Although aluminium sheet is common, the greater market share is enjoyed by galvanised sheet steel, often used because of its strength and relative cost. One of the most common roof finishes for profiled, galvanised steel is plastisol, which is the generic name for a wide number of organic PVC based coatings that are applied in liquid form to hot dipped galvanised steel coil. (Organic coatings are derived from animal or vegetable matter or a compound of carbon).

Other materials may include:

- polyester;
- polyurethane; or
- polyvinylidene fluoride (PVDF).

7.2 LIFE EXPECTANCY

Cladding manufacturers will often quote a minimum service life – what they term a 'period to repaint decision'. Effectively, this is the period of time that elapses before major redecoration is required. The decision is reached when the protective system has broken down and no longer protects the base metal or alternatively when the aesthetic appearance of the cladding becomes unsatisfactory. The latter is of course a very subjective decision whilst if the coating has failed, the position is purely factual.

The minimum service life will depend upon atmospheric conditions, location, exposure to sunlight and humidity as well as a host of service conditions. However, there is general acceptance that light coloured roofing sheets will perform better in terms of service life than their dark coloured counterparts. The correct choice of colour can have a very significant impact upon the durability of the coating, in some cases increasing (or decreasing) its service life by as much as 100 per cent.

Examples of premature failures of organic coatings can be seen around poorly designed flues which deposit exhaust gases or fumes onto the surface of a coating, or in airport buildings where exposure to jet fuel and combustion products can damage plastisol coatings leading to staining and breakdown of the coating.

Whilst coating formulations have improved enormously over the last ten years, many thousands of buildings are still clad with older first and second generation organic coatings. These coatings were generally very responsive to ultraviolet light – this is one reason why the coating on roofs tends to last a shorter period than wall cladding. However, orientation also pays a significant part in the weathering process, as exposure to prevailing winds (generally to the south-west or north elevations where water may be slower to evaporate) can lead to a shorter life on those elevations.

The early versions of plastisol sheeting tended to be used for wall cladding, whilst the use of asbestos cement, and later cement fibre products, for roofing continued well into the 1980s. The early versions of plastisol suffered durability problems, commonly encountered on walls. By the mid 1990s, the situation had reversed, with most failures occurring at roof level. According to research at that time, as much as 70 per cent of early plastisol failures resulted in delamination, dropping to around 24 per cent with later generations of the product.[50]

7.3 EXPOSURE CONDITIONS

The data in the table below has been compiled from BS 5427-1:1996 and allows for a comparison of exposure categories in the UK.

Notwithstanding the likely period to first painting, the performance of the coating will, like most materials, be reflected in the degree of care and attention that it receives during its lifetime. Exposure to heavy pollution for example, or a failure to remove silting, could lead to the creation of corrosion cells if the roof is not cleaned regularly. Heavy foot traffic or mechanical damage can similarly damage coatings, particularly some of the thinner ones like PVDF with the result that localised deterioration can occur. Local damage can of course be treated locally before more serious deterioration takes place (see section 7.4 on cut edge corrosion), but in general terms the deterioration of a coating will follow a recognised pattern, leading to the need to recoat.

For plastisol based coatings the initial appearance will degrade gradually, with general soiling that is capable of removal with suitable detergents or surfactants. Following this, chalking of the surface will occur – often improvements can be achieved at this stage by washing down as above. Chalking does not affect the functional performance of the roofing or cladding but can be unsightly. Repainting may sometimes be necessary when the appearance of the cladding is important.

The next stage of deterioration is the formation of microcracking (very fine crazing of the surface) which

Comparison of exposure categories for the UK

	Coastal	Industrial and urban	Rural and suburban	Wet (internal)	Dry (internal)
Features	Can extend inland for several kilometres unless steep rise in land and offshore winds, in which case a few hundred metres more is appropriate. Western seaboard of UK due to reduced UV and higher rainfall which removes salt deposits.	Worst conditions arise in areas of heavy industry, gradual reduction to urban areas. However, it is not unknown for significant localised areas of pollution in rural areas due to the location of manufacturing plants.	Areas that are remote from coastal or urban locations.	Swimming pools, chemical process plants, livestock buildings – all with high humidity.	RH generally below 50%; temperature 15–25°C.
Risks	Chloride contamination leading to enhanced corrosion damage. Erosion of surfaces and coatings from windborne particles.	Corrosion due mainly to sulphur dioxide or specific chemical deposition from manufacturing process.	Generally less aggressive conditions, but corrosion most likely to be affected by humidity and rainfall rather than by other mechanisms.	Concentration of fumes may lead to breakdown of organic coatings.	Benign conditions.

Type of coating	Typical coating thickness	Minimum service life (years)	
	μm(microns)	Inland location	Coastal location
HDX coated steel (Trade name for organic coatings manufactured by Arcelor)	50–60	15	10
HPS 200 coated steel (Trade name for plastisol organic coatings manufactured by Corus)	200	20–30	15–25
PVDF coated galvanised steel	15–22	15	10
Architectural polyester and silicone polyester (dependent upon colour)	25	5 (dark) – 10 (light colours)	5 (dark) – 10 (light colours)
Celestia (trade name for organic coating manufactured by Corus) – zinc/aluminium-alloy coated steel	120	12–17	10–15
Mill and stucco finished aluminium	0	40	20

occurs over progressively larger areas. The process is closely followed by the breakdown of the coating itself to expose the galvanised steel beneath. At this point, recoating is required in order to preserve the zinc. However, since the zinc is sacrificial and serves to corrode preferentially in favour of the steel, the roof can still be expected to perform satisfactorily.

Upon exposure to the atmosphere, the zinc coating will start to corrode – this is often evidenced by the formation of white corrosion deposits together with peel back of the coating.

Once the zinc has degraded and the steel begins to corrode repair is still possible so long as there remains sufficient base metal. Once perforation has occurred no amount of decoration can rectify the situation.

As noted earlier, the life expectancy of dark colours can be much lower than lighter, more reflective surfaces. Essentially, darker colours are able to absorb more

Figure 16 The deterioration of protective coatings is not necessarily worse on the weather side. Here, a build up of contamination on the unwashed soffit has lead to deterioration of the organic coating.

radiated heat from the sun giving rise to potentially higher surface temperatures. Similarly, exposure to ultraviolet radiation from the sun can lead to the disruption of chemical bonds in coating materials, particularly plastics. At night, darker colours radiate heat effectively leading at times to super-cooling effects (see section 7.8 on cold night-sky condensation). The range of temperatures so experienced means that expansion and contraction cycles are more pronounced than with light materials and so details such as fixings and rooflights may be placed under greater stress – elongation of fixing holes being a common symptom.

According to CIRIA Technical Note 107 (1981)[51], lightweight cladding laid above insulation can be expected to perform over a temperature range of −25°C to +60°C, a range of 85°C for light coloured materials. For dark coloured materials in similar conditions the maximum temperature is probably more like 80°C, giving a range of 105°C. Thus, for a roof spanning 10m, total movement in a steel roof of some 12.6mm would theoretically need to be accommodated. Of course, it is highly unlikely that the sheeting would have been laid at the minimum temperature, and so a range of contraction and expansion movements would be expected.

Thus, for the dark roofing described above, one could expect installation occurring somewhere between 5°C and 25°C. Therefore a range of 5°C to 80°C should be assumed for expansion and from +25°C to −25°C for contraction. The maximum movement would then be around 9mm (contraction of 6mm). The coefficient of thermal expansion of aluminium is about double that of steel sheet and so temperature related movements are potentially significant. Interestingly, surface temperature data published by British Steel in 1993 recorded maximum temperatures, for dark materials, of 72.1°C (olive green) but with a declaration that this was not necessarily the maximum temperature that could be reached in the UK as the measurements took place on only one day.

Across the width of the roof movement is more likely to be accommodated because of the ability of the roof corrugations to distort slightly. Lengthwise however the problem could be more serious although the length of the shank to the primary fixings may permit some measure of movement. Such movement must be contemplated during the design stage.

A further result of roof sheet expansion is that the lap seals could be broken allowing water to penetrate (especially on a shallow roof) as a result of wind assisted capillarity or air pressure differentials.

Another factor is the propensity of a combination of high temperatures (above 70°C) and entrapped moisture such as may occur at a lapped joint. Whilst in normal circumstances the steel weather sheet is protected against corrosion by the sacrificial cor-

Figure 17 Total delamination of plastisol coating on the south facing crowns of a profile sheet roof. The coating was originally Van Dyke Brown; this generally performed badly in roofing applications. (Photo courtesy of John Bohan).

rosion of the zinc in preference to the steel, above 70°C the situation is reversed, with the steel corroding preferentially.

But high surface temperatures can also have effects on secondary materials such as rooflights (see section 7.12 on PVC rooflights), which can be subjected to temperatures that are close to the limit of their normal working range or softening point leading to reduced durability.

In the 1970s and early 1980s the early generations of plastisol coated sheets were used without huge respect for the effects of colour and durability. One common colour was a dark brown – Van Dyke Brown. It was quite common to find this material used for roofing applications in the UK – sometimes with quite disastrous results. The then manufacturer, British Steel, was caught by a number of significant claims whereby the colour coating failed to live up to its expected period to repaint decision.

Case study

The site comprised two substantial parallel warehouses occupied by a major supplier and manufacturer of domestic goods and automotive products. The roofs to each building were of composite panel construction finished with a leather grain plastisol finish of Van Dyke Brown and pitched at around 6°. Curved eaves were formed, with rainwater draining into concealed eaves gutters. PVC rooflights in long strips provided a measure of day lighting to the interior of the building.

Within ten years of construction the building was plagued with water penetration problems, whilst large areas of plastisol coating flaked off and corrosion occurred to the roofing sheets. When viewed from the north, the roof coatings appeared intact, but when viewed from the south, virtually all of the coating to the exposed sides of the trapezoidal roof profile had flaked off. PVC rooflights were discoloured and distorted and corrosion at lap joints was extensive.

The roof sheet manufacturer accepted that the plastisol had failed within its expected period to major decoration, possibly due to manufacturing faults. However, the use of that material was contrary to design guidance at the time of construction and so the resolution of the problem became embroiled in argument.

The high temperature build up permitted by the use of dark cladding had resulted in excessive expansion of the roofing along its length, (the curved eaves construction provided further restraint) thus stressing and distorting fixing positions and subjecting rooflights to extreme temperature conditions that exceeded or were very close to their operating and performance limits.

Eventually, the building was re-roofed utilising new materials supplied by the manufacturer and a new over-roof on spacers and sheeting rails. The construction enabled improved insulation levels and the resolution of vapour barrier problems that had plagued the original construction.

It is easy to think that protected areas, such as the soffit of an overhanging eaves detail would have a longer coating life. In reality the areas are unwashed by rain-water and so a build up of airborne contaminants (particularly salt spray in a coastal environment) can lead to aggressive conditions and early coating failure. Washing down with clean water at least once a year would be desirable.

7.4 CUT EDGE CORROSION

One of the most common defects found in industrial roofing is that of cut edge corrosion whereby the protective coatings peel back at lap joints in the roof sheeting, exposing the base metal to risk of corrosion. Commonly, the problem can be seen to commence on roofs 10–15 years old, but sometimes, deterioration can occur on roofs that are much younger than this.

In 1994, in their publication *Durability of Cladding a State of the Art Report*,[52] the authors observed that in a sample of some 1,700 plastisol failures, nearly 50 per cent exhibited corrosion at the cut edge. The problem was not confined to plastisol as a high proportion of PVDF sheets displayed corrosion at cut edges too.

Initially there was much speculation as to the cause of the problem, with the general consensus being that the method of cutting the roof sheets to length damaged the coating, thus causing microcracking, the admission of moisture and then corrosion. However, it is more likely the case that the problem is due to water becoming entrapped at a lap joint, particularly within a roof that is pitched at 6° or less, and being held there as a result of capillarity, pressure differentials and silting. Dirt traps occurring at the bottom edge of a sheet can be a common factor (hence the practice of turning the bottom edge down slightly to shed water that would otherwise be held by surface tension). Quite often the

Figure 18 Corrosion at the cut edges of a PVDF coated sheet. Note also missing protective caps to primary fasteners and resultant corrosion. (Photo courtesy of Christopher Gibbons.)

very edge of the plastic coating begins to lift, trapping more water and exacerbating the problem. It is not uncommon to find the coating peeling back 100mm or more, creating small reservoirs of water on a roof. Often, the location of such a problem can also be identified by silting or tidemarks.

The presence of moisture permits corrosion to occur and once the protective coatings have deteriorated, section loss and eventually perforation of the base metal can be expected. If left untreated, peel back and the resulting corrosion can be serious and so it is important to treat symptoms as soon as practicable.

Repairs are usually effected by cleaning the affected area, removing the defective coatings (for example by chemical stripping, abrasive blast cleaning, high pressure water equipment or a combination of those methods) and treating the sheet lap with a silicone based paint or a pigmented synthetic polymer topcoat applied over specific primers selected for the particular substrate in question.[53] The usual width of the repair is no less than 75mm, with a dry film thickness of 350μm. Similar treatments can be provided at eaves level, with both the top and under surfaces of the sheet being fully encapsulated.

If high pressure water jetting is used there is a risk that water could be driven up between the sheets, to trickle out again later – if this is allowed to happen, the new paint coating or seal could be affected, so it would be very important to ensure that the laps are fully dry before applying the coating.

To prevent the problem occurring in the first place, current guidance is that the cut edges or laps should always be painted. However, painting a cut edge is difficult to achieve in practice (without overcoating the adjoining sheet) and is often neglected.

7.5 STRUCTURAL CONSIDERATIONS

When walking on a trapezoidal profile roof it is most often the case that the person will step on the crown of each profile rather than the trough. A person weighing 90kg will exert a force on the crown of somewhere in the region of 1.3kN over a length of about 100mm. This force is somewhat greater than the nominal 0.9kN allowed in BS 5427, but because the sheet should resist this at any point on its surface (a difficult calculation to perform) it would be more usual to design to a value of 1.5kN/m – and this will probably result in a steel thickness of 0.7mm. On some older roofs (or indeed roofs where the contractor has attempted to cut corners), thinner sheets may be found – these being more vulnerable to distortion and buckling as a result of foot traffic.

If the design loads are exceeded significantly, profiled steel could theoretically fail as a result of:

- web buckling (could occur over intermediate supports);
- compression buckling (likely to occur);
- shear (unlikely to occur); or
- tensile failure (unlikely to occur).[54]

Thankfully, catastrophic failures of roof sheeting are rare, but deflection damage is quite common, leading potentially to:

- reduction in drainage capability at low pitches;
- strain on fixings;
- damage to components that are attached to the roof; and
- leakage at joints due to de-bonding of sealants.

For the above reasons, good roofing practice would seek to limit the amount of deflection to around 1:150 for steel roofs.

Steel has a Young's modulus approximately three times that of aluminium; by comparison therefore it is a stiffer material and better able to resist deflection. Aluminium sheets are prone to distortion and even with a limiting deflection of 1:90 they can still be noticeably flexible.

The roof sheeting needs to transfer down-slope loads to the purlins and to restrain the purlins against wind uplift. In an uninsulated roof, the sheeting will be in direct contact with the purlins, but for the most part, in order to introduce insulation into the assembly, some form of spacer will be required. Over the years, and particularly recently, changes to the Building Regulations in England and Wales have resulted in a need for greater depths of insulation, and hence greater depth of spacer. In the 1980s for example a typical thickness of insulation would have been 80mm; in 2006 it was approaching 200mm.

The traditional assumptions as to the performance of the roof sheeting is now the focus of some concern as is the tolerance necessary in secondary steelwork that is necessary to achieve airtightness standards.

The most popular spacer system currently in use within the UK comprises a bar supported by a series of brackets above the purlin. The greater the thickness of insulation, the longer the brackets and hence in structural terms the greater the lever arm. To prevent the spacer assembly from racking it is not always sufficient to rely upon the weather sheet itself, but to provide additional 'sway' brackets to bring rigidity to the system.

7.6 NOISE

Whilst contemporary practice is likely to use composite sheeting or site assembled construction with spacers and insulation linings, some older roof constructions utilised a layer of foil faced rigid polyurethane foam laid directly under the roof sheet.

Such a practice could lead to severe cracking or popping noises from the roof according to sudden changes in temperature (for example on a bright sunny day). The effect is the result of the expansion or contraction of the roof sheet sliding over the foil surface of the insulation, amplified by the diaphragm nature of the roof. Lifting the roof clear of the insulation with the use of spacers is the only way to avoid this problem.

7.7 CONDENSATION

The mechanisms of condensation are dealt with in section 7.8 but in brief the problem occurs when warm moist air comes into contact with a colder surface. The air is unable to retain as much moisture as before and this is yielded up in the form of condensation. Condensation occurring in profile sheet applications can occur in several areas although predominantly on or within roofing systems. Site assembled roofing (as opposed to composite panels) are particularly vulnerable to interstitial condensation if insufficient attention is given to vapour control.

Small amounts of condensate will commonly occur on most nights, even in the summer, producing in the region of 0–10g/m^2 as a fine mist; however this amount of water will usually evaporate rapidly during the day, without leading to any long term accumulation. Problems will occur when the rate of condensation is high enough to allow accumulations sufficient to cause running or dripping of the condensed water. According to the MCRMA guide[55], at least 70g/m^2 must accumulate before it will run down a sloping surface and 150g/m^2 before dripping will occur from a horizontal surface.

Referring back to the table on page 46 it can be seen that aggressive conditions arise internally in buildings such as swimming pools, livestock buildings or buildings where manufacturing processes generate large amounts of vapour. However, other quite normal operations can produce high risks, such as:

- certain types of gas fired warm air heaters;
- warm air blowers directed at colder exterior surfaces;
- air-conditioning systems with operational humidity control; and
- exposure to cold night skies (see section 7.8).

Codensation can cause the following defects on roofing:

- water condensing on the underside of a roof sheet can run down the slope to become entrapped at lap joints, causing the unprotected edges to corrode;
- corrosion of the fixings owing to the retention of condensate together with corrosion of the steel around the fixing hole;
- pattern staining on damp surfaces;
- mould growth;
- staining of building materials; and
- risk of damage to stored goods, etc.

7.8 COLD NIGHT SKY CONDENSATION

The effect of cold night sky condensation can cause significant water problems within a roof constructed of site assembled sheeting and insulation. Changes in the Building Regulations that took place in the 1970s were the first of a long line of thermal insulation improvements to industrial buildings that led to many unexpected problems of water ingress into buildings – a problem that was compounded by the early failure to apply sufficient attention to basic measures such as vapour control and ventilation.

Essentially, during a cold night, the surface of a roof acts as a heat emitter, radiating heat to the atmosphere. Radiation cooling to the night sky is based on the principle of heat loss by long-wave radiation from one surface to another body at a lower temperature. Because heat always flows from hot to cold, a roof will radiate heat towards a cold sky, i.e. it radiates to the sky but receives no corresponding short-wave solar energy. This radiation will cause the surface temperature of the roof to fall below that of the surrounding air, whereupon heat will then start to flow, by convection, from the internal air into the roof. If the air is still, then this convective heat transfer is less effective than the heat loss through radiation and so its temperature falls further until an equilibrium point is reached where the inward convection heat flow balances the outward radiation heat flow.

Measurements by researchers investigating the possible benefits of night sky cooling have identified that on a clear desert night, a typical sky-facing surface will cool at a rate of about 75W/m^2. This means that a metal roof facing the sky can be significantly colder than the surrounding air temperature. Trapped moisture within the roof void will condense on the underside of the sheet and may freeze. The process is similar to the formation of condensation and frost on a car windscreen during a cold clear night. The provision of insulation within the roof compounds the problem as it reduces the rate of heat transfer. According to the BRE, a temperature differential of as much as 8°C can occur between ambient air temperature and roof surface temperature.[56]

In the morning, when solar radiation heats the surface of the roof, and the building's heating system is switched on, the ice melts, water drips from the underside of the sheet onto the insulation or under roof lining whereby it can run down and penetrate the interior of the building via a suitable break in the lining. Whilst most buildings can tolerate small levels of interstitial condensation, significant amounts can be damaging to the roof sheeting, leading to advanced corrosion on the under surface of the sheet particularly if it does not have good corrosion protection.

Cold roof construction (which is essentially what site assembled roofing systems often are) is prone to condensation and so it is essential that the roof void be properly ventilated (see also BS 6229:2003[57]).

Case study

The development comprised a small estate of business starter units, portal framed and clad with a site assembled cladding system comprising an underlining of plastic faced plasterboard in a tee grid frame, a layer of glass wool insulation and an outer roofing sheet of plastisol coated profile steel sheet. The buildings were, when inspected, a few months old. Tenants complained of serious water penetration problems, occurring mainly for an hour or so in the morning from about 9am onwards and stopping after an hour or so. The problem was first identified in the October of the year of construction.

Investigation of the problem revealed that the fitting of the underlining system was poor, leading to vapour loss into the roof void. Air leaking into the void was condensing on the roof sheeting and freezing during the cold clear nights that were occurring at the time. The tiles were refixed and the joints taped to reduce vapour loss, while the eaves and ridge were amended to allow water vapour to be ventilated. Given the age of the building, a significant amount of water vapour was being extracted from the concrete ground floor slabs; this gradually stabilised, so by the following year the problem was significantly improved.

However, uninsulated roofs are also at risk as water vapour in the air can condense on the under surface causing particularly acute problems on roof canopies, domestic garages, car ports and the like.

The problem can be eased by ensuring that vapour control measures at ceiling or lining level are adequate; that lining sheets are properly sealed or, in the case of plasterboard type plastic faced lining systems, the joints are properly sealed. It used to be the case that the provision of a good breather membrane on the top surface of the insulation was essential although this practice has now changed. The provision of ventilated eaves and ridges enables moist air to be drawn out of the roof void, but it does not necessarily prevent this form of condensation, as water can condense out of the circulated air and freeze in the same way as described above. The purpose of the breather membrane was to prevent any water that so formed from saturating the insulation and allowing it to either drain away or be held until it evaporated normally.

The *Profile Metal Roofing Design Guide*[58], published by The Metal Cladding & Roofing Manufacturers Association (MCRMA), observes that with proper vapour control measures it is seldom necessary to install breather membranes in site assembled built up roofing systems. This is largely due to the changes in construction that have occurred over the last 20 years or so and the now more or less universal use of profiled liner sheets which make proper vapour control very much easier to achieve.

It is probably true to say that the problems of cold night sky condensation were worst at the time when effective vapour control measures were either not understood or simply ignored. The case study above was based upon an investigation in 1983. In contemporary construction, one would not necessarily expect to see the effects with the same level of severity. However, that is not to say that the problem can be treated lightly.

BS 5250:2002 *Code of Practice for the control of condensation in buildings* recommends control measures and observes that the problem can be acute in buildings housing high humidity processes or in poorly heated structures such as garages or industrial buildings even when there is good ventilation.[59]

7.9 ROOF FIXINGS

Primary fasteners are those that are used to attach the sheet material to the support structure and where necessary to fix spacer bars to the structure. By default these fasteners are designed to transfer loads and to resist wind uplift on the roof. Since they have a structural function, the frequency of primary fasteners will need to be specified.

Secondary fasteners are those that are used to make sheet to sheet or flashing to sheet connections.

There are numerous different patterns of primary and secondary fastener, although generically, most are now self-drilling, self-tapping types designed to eliminate the need to drill a pilot hole first. Most types incorporate a neoprene washer that is designed to seal the top surface and prevent water ingress around the shank of the fixing. The exception to this is the use of blind rivets at side laps, which must be inserted through a pre-drilled hole.

Fasteners are usually made from carbon steel or stainless steel. Carbon steels are generally plated to reduce corrosion risks, although some damage to the plating is inevitable during installation.

Generally, plated carbon steel is adequate in most conditions as long as the heads of the fixings are protected from exposure to the weather. This can be achieved with the use of plastic caps or integral caps – the latter being preferable.

In order to prevent bimetallic corrosion stainless steel fixings must be used for aluminium sheets.

7.10 POSSIBLE DEFECTS

Typical problems with fasteners are detailed in the table opposite.

7.11 WEATHERTIGHTNESS

It is probably fair to say that in terms of weathertightness, the risk of water ingress will increase as the roof pitch decreases. In the 1980s six degrees was a common pitch; this has now fallen to about four degrees, but roofs down to one degree are sometimes encountered – the latter requiring very special treatment.

Typical problems with fasteners

Symptom	Possible cause
Water leakage around fastener. Primary fasteners may be installed either through the crown or through the trough of a trapezoidal sheet	Fastener incorrectly tightened. If the neoprene washer can be seen squeezed out around the metal head, the fastener may have been overtorqued. Similarly, an under-torqued fastener will leak.
	Fastener inserted at an angle, insufficient compression of neoprene washer.
	Fastener inserted into a timber spacer or purlin. The moisture content of the timber will have reduced leading to shrinkage and possible distortion, causing the fixing to loosen.
	Possible extreme expansion of the roof sheeting leading to tearing or enlargement of the hole around the fixing.
	Fastener located downstream of the sealant in a lap joint, water able to bypass the neoprene washer by travelling between the sheets at the lap.
	Loosening of the fastener as a result of localised compression of a rigid foam spacer batten.
	Loosening of the fastener as a result of deflection in the roof sheet, possibly as a result of foot traffic on the roof.
	Incorrectly formed blind rivet or secondary fastener at side lap – sheets not properly joined together.
Corrosion of the fastener	Possibly resulting from leakage (any of the above causes).
	Interstitial condensation on the underside of the roof sheet.
	Incorrect selection of fastener metal – galvanic corrosion.
	Colour matched plastic cover caps missing.
Inversion of the washer	Incorrect diameter of head coupled with wind uplift on the roof sheet.
Compression and distortion of the roof sheeting	Missing or compressible spacer material, for example rigid foam.
	Incorrect choice of fixing. Fixings for composite panels usually have a second thread around the head designed to support the outer sheet and to prevent distortion from occurring.
Staining of roof sheeting around fixing	Drilling swarf not removed.
	Roof at shallow pitch preventing proper drainage and dirt retention.

The use of standing seam roofing systems enables a lower pitch to be achieved, primarily since these systems can be manufactured in long lengths, thus avoiding laps, while the traditional side lap details are by their nature far less vulnerable to water ingress.

For lower pitches therefore, particular attention needs to be paid to both side lap and end lap joints to ensure that they are properly sealed. The MCRMA guide provides useful detail as to the design and installation of lap seals, but essentially, end laps should have a double row of sealant (usually 6 × 5 or 6mm diameter butyl sealant strip), with the first row set as close as possible to the edge of the sheet to prevent water entrapment through capillarity. The second row should be set back by about 150mm – its purpose is to prevent condensation running down on the inside into the lap. Fixing positions must be upstream of at least one row of sealant, and the fixings themselves must provide suitable compression to the seal. During site inspections, a small length of steel packing tape makes a useful addition to the surveyor's tool kit as it can be inserted as a probe beneath the lap to test the location (or existence) of the first row of sealant.

On side laps, proper design and installation of seals is again critical, with the run of tape being placed centrally along the crown or slightly to one side of it. Proper compression is achieved by inserting secondary fasteners at 500mm centres, but it is essential that the sheets are compressed together during laying to make sure that the fastener pulls the two sheets together properly.

Historical guidance recommended that the side laps should be arranged to point away from the prevailing wind. However, this is not always practical and provided the seals are properly installed it makes little difference to the performance of the roof.

7.12 ROOFLIGHTS

Rooflight areas covering about 10–15 per cent of the total roof area are a common feature of many industrial building types. Usually provided to give a measure of natural light to the interior of a building, many installations using low-melt PVC were designed to act as vents in the event of a fire.

Broadly speaking there are two types of rooflight: in-plane and out of plane. Out of plane rooflights tend to be framed or moulded systems, often supplied as single units. In-plane rooflights, as their name suggests, are fitted flush with the roof surface. Problems of water ingress, distortion and clouding are common with these systems. According to the National Federation of Roofing Contractors (NFRC) problems with rooflights are normally the result of:

- inappropriate fixings; and/or
- inappropriate selection of material.[60]

By comparison with metal (or fibre cement) roofing systems, rooflight materials are usually:

- of much lower strength;
- more prone to deflection;
- more likely to suffer pull through of fixings; and
- have a higher coefficient of thermal expansion.

In-plane rooflights are of several different types, being essentially site assembled or factory made. Single and double skin rooflight types are self-explanatory, with the profile being fabricated to match the profile of the roof sheeting and, in the case of the double skin types, the profile of the inner lining sheet.

Factory assembled dual or triple skin rooflights are now much more common and are often used with composite sheet roofing systems or standing seam systems and are designed to offer improved thermal insulation – some have a translucent insulating layer sandwiched between the inner and outer leaf.

Rooflights are commonly of GRP, polycarbonate or PVC, although in most contemporary installations, GRP is the most likely in view of its durability and ability to satisfy health and safety requirements. The adjacent table provides a comparison of these materials.

As noted above, all plastics materials have a high coefficient of thermal expansion and unless this is taken into account during design and installation problems can and often will result. PVC and poly-

Comparison of materials

PVC	Common rooflight material in the 1970s and 1980s. Now virtually discontinued other than for the DIY market. High coefficient of thermal expansion, poor strength and durability. Prone to discolouration (particularly cheaper varieties) and embrittlement due to exposure to UV light. May be softened on exposure to high temperatures (for example when used in conjunction with dark coloured roofing). Unlikely to satisfy HSE requirements in terms of accidental foot traffic unless additional safety measures are incorporated. Usually clear when installed.
Polycarbonate	Often used in out of plane rooflight systems. Clear thermoplastic with a higher softening temperature than PVC. Later variations have additional UV protection, but during the 1970s, early 1980s, deterioration due to UV light was common leading to discolouration and distortion. Polycarbonate can be extruded or moulded to match a large variety of profiles. Multiwall polycarbonate, for example five wall, can offer very good thermal insulation properties.
GRP	Glass reinforced plastic is by far the most popular material in contemporary construction and has been used successfully for many years. In order to provide weathering protection, sheeting has either a polyester surface film or a gel coat. Both are sacrificial and in older buildings it is common to find erosion of the gel coat and exposure of the fibres without necessarily the failure of the rooflight itself.

Figure 19 Discolouration and distortion to polycarbonate barrel vaulted rooflight.

carbonate roofing materials are particularly prone to failures as a result of expansion and so it is usual to specify enlarged fixing holes and larger diameter washers (usually 29–32mm in diameter) to enable sheets to move. Distortions frequently occur if the plastic is clamped in position, whilst tearing or ovalling of fixing holes can lead to leakage problems. It is also important that the rooflights are fixed through the trough and not through the crown as the plastic is rarely strong enough to take crown fixings without distortion.

In order to maintain weathertightness, rooflight fixings have to maintain an adequate seal when the sheet is subjected to its maximum inward imposed load. When this occurs, the substrate will compress down, tending to loosen the seal between the sheet and fastener, thereby increasing the risk of water penetration. In site assembled rooflights it is usual to provide intermediate spacer bars to reduce the span between purlins; if these are omitted, flexure of the rooflight will occur. Factory assembled units usually incorporate suitable hard fillers to achieve the same purpose.

Sidlelap fixings for rooflights are generally located at around 400 centres; their primary purpose is to compress the side lap seals, but they also serve to transfer loads to the adjoining sheeting. It is better to lap the rooflight over the adjoining metal on both sides, but if a rooflight has been laid such that one side is under the adjoining sheet it is important to ensure that self-tapping fixings have not been used – these will have possibly damaged the sheet during fixing and will not have created an enlarged fixing hole to allow for thermal movement.

To prevent leakage around fixings, washers should have EPDM seals and must be tightened correctly. Because of the large diameter, washer inversion can be a problem unless the washers are sufficiently robust.

7.13 COMPOSITE PANELS

Over the last 10–15 years, composite panels have tended to predominate over built-up systems owing to the cost effectiveness of installation and the removal of concerns over interstitial condensation relating to poor site practice. However, the panels have not been without controversy as following 9/11 the insurance industry saw composite panels, predominantly those containing polyurethane insulation, as being a high fire risk and began to load premiums or impose high deductibles on the insured. For more details on this aspect of performance, see *Investigating Hazardous and Deleterious Building Materials*.

Today, it is usual for composite panels to be manufactured from polyisocyanurate foam or mineral wool, thus avoiding the adverse attention of insurers.

Composite panels derive their strength and rigidity from the three layers used in their construction acting compositely together. Taken individually, the components (weather sheet, insulation core and inner liner) have far less stiffness and resistance to bending than when they are bonded together. To cope with increasing demands on thermal insulation values, core thicknesses have been increased, possibly to a point where future improvements will have to involve changes to engineering services rather than greater insulation.

Whilst a variety of core insulants are available, it is reasonable to consider the following as being representative:

- beaded polystyrene foam (XPS) – used mainly for insulation in thick, self-supporting composite panels in cold stores;
- polyurethane foam (PUR) – virtually standard in panels up to the early 1990s;
- polyisocyanurate foam (PIR) – similar to PUR in appearance but containing a higher ratio of isocyanurate to polyol making it able to sustain higher temperatures than PUR; and
- mineral wool.

PUR and PIR are both thermosetting materials which means that they can only be formed once due to cross linking between molecules. These products are self-bonding to the metal outer leaves of the sandwich panel.

XPS is by contrast thermoplastic, meaning that it can be reformed. However, thermoplastic materials must be adhesively bonded to the metal faces of the panel.

Case study

Another type of thermosetting core is phenolic foam. Whilst this type of foam performs well in fires, it is rarely used in mainstream products. Phenolic foam is manufactured from formaldehyde resin mixed with a light volatile solvent as a blowing agent. Early variants of this type of foam were problematic as it was difficult to process, friable and produced water, which needed to be dried out before use. Any residual water was particularly acidic and therefore corrosive.

The material was used as an insulating core in bespoke aluminium composite panels used to clad Foster's Sainsbury Centre for the Visual Arts in 1977, but ten years later the cladding had to be replaced because of a destructive reaction between the phenolic foam cores and the superplastic alloy that forms the outer shell. Superplastic aluminium alloy is designed to flow at certain temperatures and to be capable of extension up to ten times its original length to enable complex shapes to be formed. The cladding panels at the Sainsbury Centre were not strictly composite panels as we know them, having been constructed in layers and pop-riveted together. They were replaced with standard composite panels.

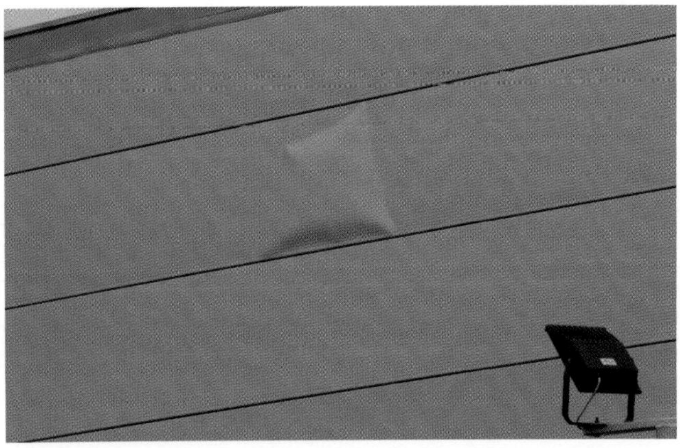

Figure 20 Debonding of a microribbed composite panel.

Mineral wool, because it contains no adhesive properties, must be bonded to the two sheets with additional adhesive layers.

Durability

Composite panels are generally considered to be durable materials and as long as attention is paid to the method of corrosion protection and its subsequent protection, modern composite panels should perform well in service. However, a number of factors can affect the life and ultimate durability of panels as shown in the table below.[61]

Causes and effects of the deterioration of composite panels

Cause	Effect
Long term loading such as snow loads, self-weight and equipment loading.	Can result in increased deflections and a reduction of strength.
Repeated loading, for example foot traffic, wind loading, snow loads or elongation due to temperature variations.	Gradual degradation of strength. Panels are only really suitable for foot traffic during construction and very occasional maintenance access.
Internal moisture variations.	In certain conditions moisture that becomes entrapped in a panel, for example as a result of mechanical damage, poor detailing, poor joint details, etc., can permit interstitial condensation on the inside face of the outer sheet. In this location, corrosion, delamination or blistering could occur.
Exposure to high temperature.	Can cause the core material to degenerate and peel back from the face causing blistering and loss of strength. Blistering, or local delamination results from rapid increases in internal cell gas pressure, causing gases to migrate into areas where the cell structure has been compromised or damaged.
Exposure to moisture.	Reduction in thermal insulation properties.
Exposure to solar radiation.	Deterioration and colour change of external finishes.
Exposure to high humidity	Corrosion of surface layers or fasteners.

8 Roof drainage

At a glance

- Broadly, there are two types of roof drainage system in common use: gravity systems and siphonic systems.
- Gravity systems rely upon larger and more frequent downpipes than siphonic systems but are the traditional 'low technology' approach. Siphonic systems are able to discharge water at a faster rate and rely upon smaller diameter outlets connected via short tail pipes to horizontal connector pipes, which lead to a downpipe. Water flows into the pipework via a special outlet designed to exclude air, the pipe fills and the force of gravity pulls water downwards thus creating a vacuum and forming a siphon.
- Both systems rely upon proper gutter design and a sensible assessment of the probability (expressed as a return period) of a two minute storm intensity, the value of which varies according to geographical location. Flat roof and eaves gutters can usually overflow without causing harm (albeit temporary inconvenience) but parapet and valley gutters are likely to cause leakage into a building if they overflow and must therefore be given a higher risk rating. This means selecting a longer return period.
- Older codes of practice and indeed at one time the Building Regulations Approved Document gave a design value of 75mm/hr but this is now proven to be inadequate for buildings where the consequence of overflow cannot be tolerated.
- Early siphonic systems ran into difficulties owing to the underestimation of rainfall intensity and the consequent inundation of the system. In the 40 or so years that the systems have been in production, design and performance standards have become better understood and documented, although specialist computer aided design is generally required. Whilst there are several specialist contractors and suppliers, all systems work in fundamentally the same way.
- Overflows are essential for valley and parapet gutters; however, the capacity of an overflow must be designed and not left to chance.
- Gutter maintenance (and access for maintenance) is also essential for both gravity and siphonic systems. With their smaller diameter pipes and outlets, siphonic outlets are prone to silting and blockage with organic materials and so can lose efficiency quickly.

8.1 INTRODUCTION

Rainwater drainage from pitched and flat roof areas is in some cases an underspecified and misunderstood function, too often taken for granted or based upon rule of thumb. Even contractors used to installing roofs and gutters on a regular basis sometimes adopt a cavalier approach to design, perhaps ignoring it altogether in the belief that the capacity is being calculated by someone else.

With a conventional eaves gutter, or a flat roof that is able to discharge over its perimeter, the consequences of under capacity are less significant. However, if there is any risk that the water could be contained in some way, design becomes critical. A failure to design for particular rainfall intensity can bring disastrous consequences, not only to the fabric of a building but to the entire business operation as well.

Within the UK, Part H of the Building Regulations deals with the design of rainwater disposal systems in buildings. Helpful guidance is given for the sizing of eaves gutters, but for buildings containing parapet or valley gutters or where siphonic systems are used, reference must be made to BS EN 12056-3, which contains design methods. The key to proper sizing is the selection of appropriate rainfall intensity for a two minute peak intensity (two minutes being the average time for the entire roof area to contribute to flow into the gutter). It is a general rule that the longer the event, the lower the intensity of rainfall. Gutters must be sufficient to deal with periods of high summer rainfall in which large amounts of water fall in a short period of time. The likely frequency of these events can be used to estimate the probability of an event taking place during the lifetime of the building.

Of course, one could design on the basis of the worst possible scenario, but this would for most purposes result in a considerable overdesign; there is a need to balance risk and cost and so standard design practice is to consider the return period of a storm event and the chance that this may be exceeded in any given year. For normal eaves gutters and flat roofs, the return period is usually taken as one year – these roofs create a low risk of water penetration in the event of a sudden storm.

In the UK, for valley and parapet gutters the return period is usually based upon 1.5 times the anticipated life of a building (for example 1.5 × 30 years) since the consequences of a sudden storm would be severe if the gutters overflowed. If a higher degree of protection is needed (for example for a data centre or archive) the return could be based on 4.5 times the anticipated life of the building or in the alternative the maximum possible intensity.

Rainfall intensities vary according to geographical location, with lowland areas most likely to suffer sudden intensive storms. BS EN 12056-3 provides a series of rainfall intensity maps corresponding to different return periods to enable maximum flows to be calculated; other European countries publish their own data to accord with specific circumstances.

Example rainfall intensity for a notional site in Nottingham

From BS EN 12056-3 the intensity of rainfall could be taken as:

Return period 1 year = 0.020 l/s per m^2

Return period 5 years = 0.032 l/s per m^2

Return period 50 years = 0.048 l/s per m^2

Return period 500 years = 0.072 l/s per m^2

Maximum = 0.160 l/s per m^2

Let us assume that the design life of the building is 30 years, but that it has paparpet gutters. Given the increased risk of water penetration arising from parapet gutters, allow an additional factor of safety, i.e. 1.5 × anticipated life = 45 years. There are no charts for 45 years, the nearest data is for 50 years at 0.048l/s per m^2, so use this as a basis for design.

So for a drained roof area of $100m^2$ the maximum flow into the gutter could be 4.8 l/s, which equates to a rainflow intensity of 17.28 mm per hour. (To convert l/s to mm/hr multiply by 3,600).

In figure 21, outlet B drains a total area of $138m^2$ and so will take the greatest amount of rainfall. Therefore, for the example given above, the outlet would be required to discharge at 6.6l/s. This example ignores the pitch of the roof which strictly speaking should be factored in to give an effective catchment area. To calculate the effective catchment area when measuring on plan adjust the calculation by taking the eaves to ridge measurement (A_h) when measured on plan and adding this to half of the height of the pitch in metres.

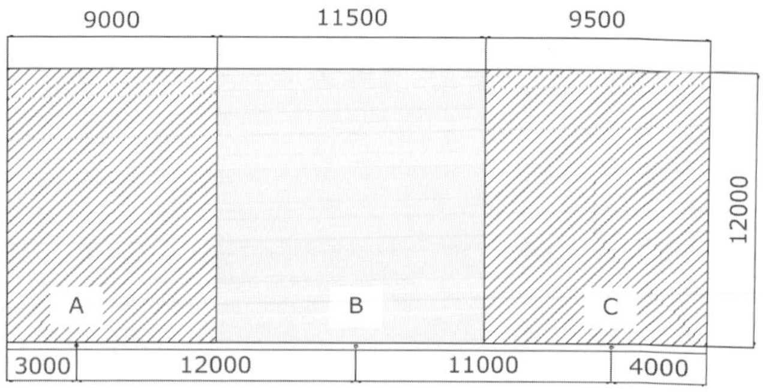

Outlet B positioned centrally takes the largest catchment area and represents the worst case for design purposes.

Figure 21 Drawing showing three outlets in a gutter draining a monopitched roof.

8.2 GUTTER AND OUTLET SIZING

Having established the suitable rainfall intensity for a given area, the gutters and outlets must be sized. The NFRC publish simplified tables[62] but otherwise detailed calculations need to be prepared using the methodology set out in BS EN 12056-3. The calculations are not particularly straightforward, although a number of software programmes are now available to simplify the exercise. It is important to remember that the size and configuration of the gutter outlets are often the governing factor in gutter design. The speed at which a gravity drainage system can discharge water depends upon the head of water at the outlet and the profile of the outlet, a tapered outlet generally being more effective than one with straight sides. Improved flow can be provided by setting the outlet in a sump or box receiver as this both increases the head and the length of the weir over which water can flow. Outlets into the side of a gutter are less efficient.

For valley and parapet gutters it is usual to provide at least two outlets. However, roof drainage is not a simple matter of 'if one outlet becomes blocked the other will take over' since the depth of water in the gutter will be much greater and there is the risk that the gutter could be overtopped. Gutter design involves the provision of a freeboard to provide security against water splashing and wind induced waves, but whilst it offers some notional emergency capacity it may not be sufficient to deal with what is in effect an increase in the area of roof drained by one outlet.

To safeguard against accidental overtopping therefore, a weir overflow or overflows should be provided. These are often installed without much thought as to capacity, so for example a 150 × 75mm rectangular outlet is often found at the end of a valley gutter so arranged as to discharge water clear of the inside of the building in the event of outlet blockage or sudden peak rainfall in excess of that designed. The problem with this approach is that the overflow may have a capacity that is much less than that of the normal outlet meaning that the gutters could still be insufficient. To provide a reliable alternative, the capacity of the overflow should be calculated; this may reveal a need

Figure 22 Poor outlet detail, strainer reduces flow, no box receiver.

to increase the dimensions of the overflow to 450mm or more.

In some cases a weir overflow may not be practicable, perhaps due to structural obstructions. As an alternative, vertical pipes can be used, fitted to the sole of the gutter and projecting above the normal water design level in the same way as the overflow to a conventional WC cistern.

Gutters to industrial buildings are usually of metal, either galvanised steel or aluminium. Some roofing systems still employ cement fibre gutter sections, whilst older buildings may have asbestos cement gutters.

Galvanised steel gutters may be galvanised after manufacture and range from 3–6mm in thickness. Alternatively, galvanising before manufacture is possible, but these gutters are usually 1.6–3mm in thickness. Galvanised steel provides good protection,

Figure 24 At just over 1m centres the support brackets for this parapet wall gutter are too far apart for comfort. Design conditions will vary, but in this case, 600mm centres were preferable.

but given the aggressive conditions that exist in gutters, not to mention foot traffic, it is usual to coat the inside surface with bitumen paint after the initial weathering has taken place. (If applied too early or in wet conditions the paint may lose adhesion.)

Aluminium gutters are usually fabricated from 1.5–3mm thick sheet; joints in these gutters can also be welded to reduce the potential for leakage at joints.

With all types of gutter, proper and rigid support is necessary. A large box or boundary wall gutter will

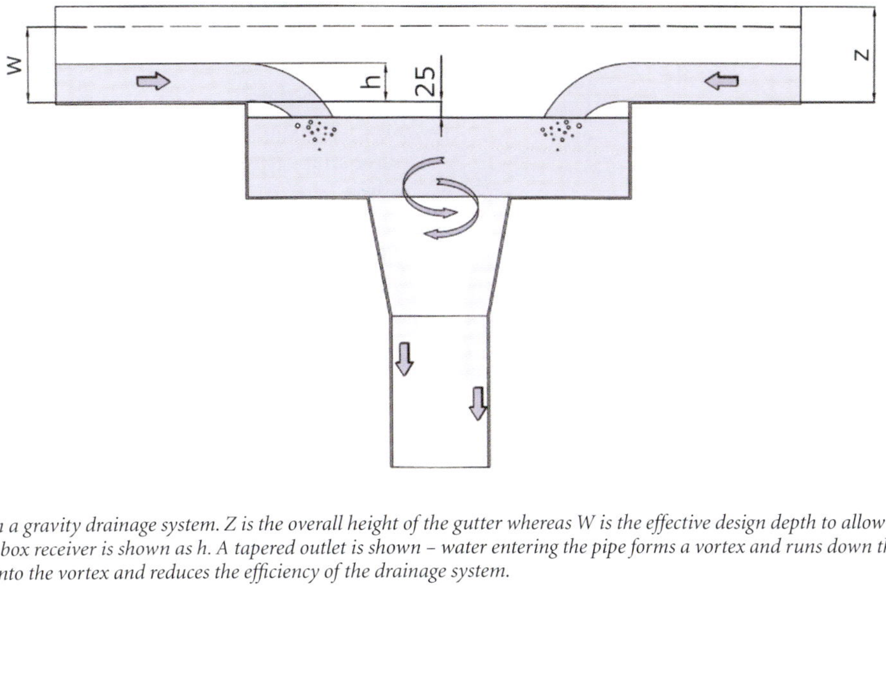

Figure 23 Box receiver in a gravity drainage system. Z is the overall height of the gutter whereas W is the effective design depth to allow a freeboard margin. The head of water at the box receiver is shown as h. A tapered outlet is shown – water entering the pipe forms a vortex and runs down the outside walls of the downpipe. Air is drawn into the vortex and reduces the efficiency of the drainage system.

Figure 25 Ineffective valley gutter owing to lack of maintenance.

hold a considerable volume of water and therefore weigh an appreciable amount. Support bracketry must be robust enough to support this and also the weight of a person or persons cleaning the system.

Generally, gutters in industrial buildings are laid level and are not therefore self-cleaning. As mentioned elsewhere, regular inspection and cleaning is essential.

8.3 SIPHONIC DRAINAGE SYSTEMS

Siphonic drainage systems were first designed in the late 1960s in Sweden and introduced to the UK in the 1980s. The system is now in use throughout the world, although still treated with caution by some owing to a legacy of poor design decisions in the 1990s. When the systems were first introduced, rainwater drainage design was based on conventional gravity systems which were covered by BS 6367:1983. The code contained advice on rainfall intensity (based on mm/hr) but was open to interpretation. The incorrect selection of design rainfall intensity could lead to serious under capacity and according to the Siphonic Roof Drainage Association it was the selection of an inadequate intensity rather than a fundamental flaw in siphonic drainage that gave the system a bad name.[63]

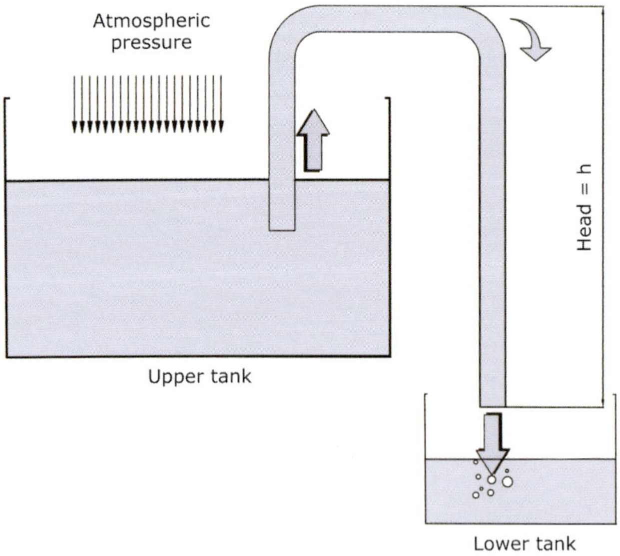

Figure 26 Siphonic action. The siphon will only work if the outlet is at a lower level than the inlet. The higher the head, h, the more efficient the system.

Early design code (now superceded)

BS 6367:1983 *Code of practice for drainage of roofs and paved areas* provided advice in respect of rainfall intensity.

The code advised that it was not possible to ensure complete safety from flooding or overflow. A design rate of rainfall of 75 mm/h was considered generally satisfactory for roof gutters where overflow was unlikely to occur inside a building and for other gutters where some risk to the contents of the building might have been acceptable.

By current standards and by painful experience we know that a design rate of 75mm/h is inadequate for valley and parapet gutters; who was to decide whether or not risk to contents might be acceptable, particularly if a speculative building is being designed and the end user is unknown?

The standard went on to set out five categories of design rate:

- **Category 1:** A design rate of rainfall of 50 mm/h for paved areas on which ponding can be tolerated during a heavy storm and for a few minutes after the storm has ceased.
- **Category 2:** A design rate of rainfall of 75 mm/h for sloping surfaces where ponding cannot occur and for flat surfaces where ponding cannot normally be tolerated.

  Note: 'In some cases (i.e. some valley gutters) the risks associated with categories 1 and 2 will be considered too great and higher design rates of rainfall should be used (categories 3 to 5).'
- **Category 3:** Provided a method of computation 'in cases where the building or its contents require an additional measure of protection'.
- **Category 4:** Provided a method 'that should be used if a higher degree of security than that provided by category 3 is desirable'.
- **Category 5:** To be used for cases where 'the highest possible degree of security is required'.

All siphonic roof drainage systems function in the same way. They comprise a special baffled roof outlet which is connected to a vertical tail pipe. The baffle is designed to prevent air from being drawn into the system so that during rainfall the outlet drain fills above the baffle, cutting off air flow into the pipe. This lack of air, coupled with the downward pull of the water under gravity, creates a vacuum in the tail pipe.

The tail pipe or pipes are then connected to a horizontal collector pipe and thence to a downpipe. Again the collector pipes and downpipe are intended to run at full bore to maximise the effectiveness of the system. For

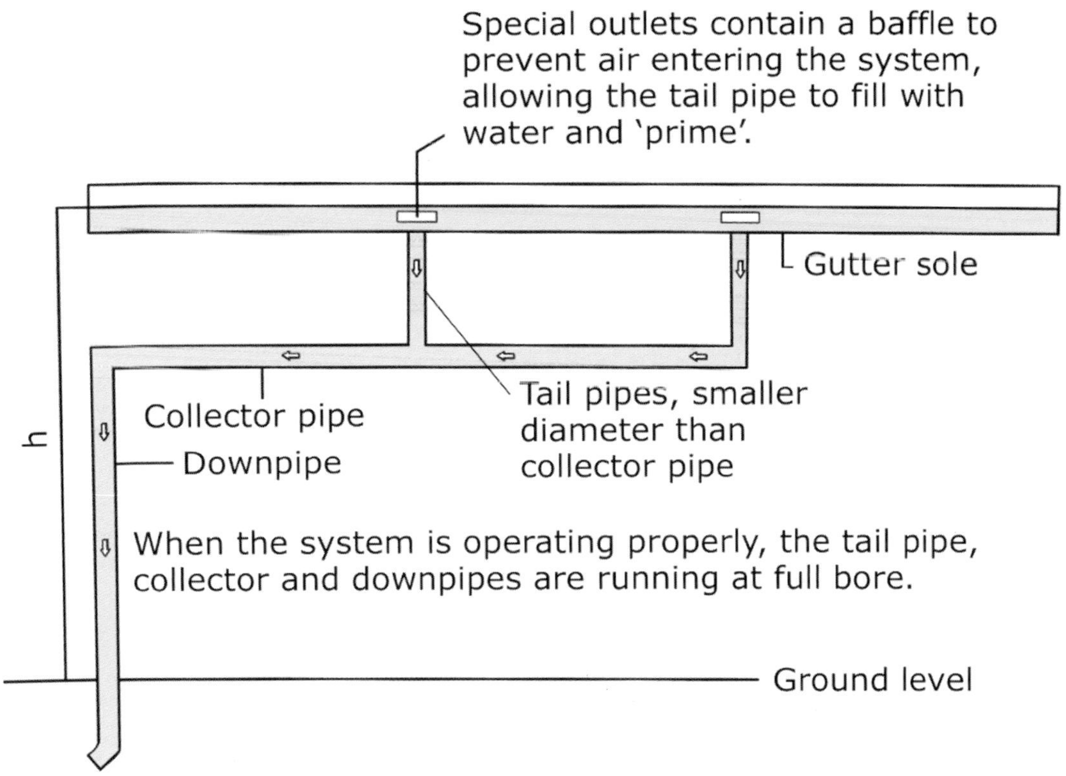

Figure 27 Essentials of a siphonic system. By comparison with a gravity system, the head of water, h, is equivalent to the height of the building.

the system to work properly, careful balancing of pipe diameters, fall and configuration is necessary to ensure that the pipes fill with water and do not continue to work under simple gravity flow. The system will take time to 'prime', i.e. develop proper siphonic action. If the system primes in two minutes there is every likelihood that the gutters will be overcome before the system works properly. Many systems work on a priming time of around 50 seconds, but in the past, longer priming times have led to problems.

Cavitation is a potential difficulty as this can cause serious problems and if severe can lead to the implosion of the pipework. The system must be designed so that the peak negative pressure in the system is within the working capability of the pipework. The overall height of the building contributes to the effectiveness of the system, but if the fall is too great and/or the pipework is not sized correctly, the negative pressures induced by the falling water in the downpipe could create an excessive pressure drop and resultant cavitation.

The siphonic system of drainage reduces the number and size of rainwater downpipes and therefore the extent of underground drainage. It follows therefore that care must be taken to ensure that the underground drainage is able to deal with the volume of water from the downpipe without surcharging as this could lead to problems with the effectiveness of the system. A vented manhole is usually provided so that if surcharge does occur, water can escape clear of the building. Design codes now dictate that possible surcharging must be factored into the design.

Although the basic theory is simple, when more than one roof drain is connected to a common collector main, the hydraulic calculations become very complex. Drainage design is therefore usually completed with the assistance of appropriate software packages. The final design is the product of specialist engineering; it will usually be the remit of a specialist subcontractor and so clear lines of communication and liability are essential. For example, difficulties could arise if the specialist retains responsibility for specifying the design rainfall intensity whereas design parameters ought to be the responsibility of the architect. (See case study 2 on page 63.)

It is important to remember that the siphonic system is merely a means of removing a large volume of water from a roof effectively; its ability to do so relies on outlet and pipe design. The system still relies upon adequate gutter design for which the traditional sizing calculations under BS EN 12056-3 apply. Furthermore, the provision of emergency overflows should still be made to safeguard against blockage events or a failure of the system to prime sufficiently rapidly.

8.4 MAINTENANCE OF THE SYSTEM

All gutter and outlet systems require routine maintenance regardless of type. Inspections are preferable at least twice a year, more so when gutters are likely to become congested with fallen leaves or other wind blown debris. Detritus should never be swept into gutters and allowed to wash away into the drainage system.

Siphonic systems, because of their reduced pipe sizes and more complex working arrangements are more vulnerable to blockage or malfunction due to lack of maintenance and so it is essential that maintenance staff are fully aware of the way in which the system works and the need to maintain it properly and in accordance with the manufacturer's recommendations.

In conventional gravity systems it is common practice to install leaf guards around outlets to reduce the chance of blockage. Leaf guards can silt up and reduce the flow rates into outlets, in some cases significantly. Under the design codes, the fitting of a leaf guard will mean a 50 per cent reduction in the capacity of the outlet. Leaf guards (or strainers) are usually required for siphonic systems. However, since most siphonic pipes are between 50 and 65mm in diameter they are likely to become congested very quickly with silt, organic matter, construction materials, dead birds and the like.[64] If the inlet to the system is restricted, the effectiveness of the siphonic system at large will be compromised. A traditional gravity system may perhaps be slightly more tolerant of silting, but regular maintenance is essential. Equally, safe access for regular maintenance must be provided.

8.5 SUMMARY OF POTENTIAL PITFALLS

Siphonic rainwater disposal systems can be very effective if they are designed and installed properly. However, failures can occur as a result of:

- inappropriate selection of rainfall intensity;
- gutters with insufficient capacity and no overflow;
- priming of pipework being too slow to deal with a peak rainfall event;
- negative pressure in system overcoming the capability of the pipework;
- the arrangement of tail pipes and collector pipes such that priming does not occur properly;
- insufficient drop between gutter sole and collection pipe to allow flow to be generated to fill the downpipe;
- a lack of provision for surcharging of underground drainage system;
- poor maintenance; and
- condensation on uninsulated gutters or downpipes.

Case study 1

Pearson Education Ltd v Charter Partnership Ltd [2007] EWCA Civ 130

The subject building was a book warehouse constructed in 1989 and situated in Leicestershire. The roof was drained via a syphonic drainage system designed to a capacity of 75mm/hr (0.0208l/s) under the then Building Regulations. Comparison with current standards would suggest that for this location and the sensitivity of the contents, a better choice would have been 0.072l/s. (See explanation on page 58.)

Two intense rainfall events, each with a return period of the order of 1 in 100 years, resulted in flooding as the system's design capacity was exceeded. The two events took place within 8 years of one another, on 24 July 1994 and 30 July 2002 respectively.

The court held that the defect in the roof drainage system was a latent defect enabling the limitation period for claims to be extended for 15 years from the date of the design error. The architects were found liable to the lessees of the warehouse for £2.1m plus interest.

Case study 2

Baxall Securities Ltd & Norbain SDC Ltd v Sheard Walshaw Partnership [2000] AB CLR 10/30

The subject building was one of a series of rectangular two bay warehouses completed in 1991 as a speculative development. The roof of the building was arranged in a pair of pitches running longitudinally, with a valley gutter between them. The outside slopes drained to eaves gutters. No overflows had been provided to the central valley.

A siphonic drainage system had been agreed in an attempt to reduce excavations for underground drainage in a somewhat difficult site. At the time of construction, siphonic systems were fairly rare within the UK. The specialist subcontractor selected a design rainfall rate of 75mm/hr, a figure that was not challenged by the architect. On 29 May 1995 and 4 September 1995 severe flooding occurred through the roof as a result of incapacity in the central gutter. The damage to stock within the warehouse amounted to some £740,000.

The court found that the first flood was due to a lack of maintenance in the system coupled with a lack of overflows; the storm intensity did not appear to have exceeded the design capacity of 75mm/hr. However the second event did exceed 75mm/hr and by that time a regular gutter cleaning regime had been put in place.

To enable the potential tenderers for the roof drainage to carry out their technical design, they ought to have been given information of the design requirement for the rate of rainfall. That requirement was not given. The tenderers provided their own design requirement and, left to their own devices, would have been tempted to put in the lowest and cheapest requirement. The judge agreed with expert evidence, namely that any shortfall in the capacity of the gutters to dispose of short duration rainfall appropriate to the location of the building would be as a direct result of the failure of the architect to incorporate the necessary information within the specification requirement.

Under the then code, a design rate of 75mm/hr would have been appropriate if some leakage could have been tolerated. Arguably any reasonably minded tenant would have a difficulty with that concept, and so the architect should have insisted on a rate of 150mm/hr, i.e. category 3 under the code. Arguments that the building was speculative did not cut much ice with the court who found the architect negligent in failing to insist upon a higher standard.

Furthermore, the architect had originally specified overflows, but these had not been fitted; he had failed to identify the omission when he ought reasonably to have done so and insisted upon their provision.

Although the architect did not have a direct contractual relationship with the claimant, the court found 'that it was reasonably, indeed easily, foreseeable by the defendants that a future occupier of the premises falling within a class of persons likely to occupy the premises would suffer damage to his goods as a result of the defects that the defendants caused or allowed to exist in the premises'.

9 Flat roofing

At a glance

- Broadly there are two variations of flat roof: cold roofs (in which the insulation is located at ceiling level) and warm roofs (in which the insulation is located either immediately below the roof membrane or above it (inverted roof). Earlier flat roofs tended to be of cold construction and were prone to condensation problems. However, because the roof deck could act as a heat sink, roof membranes such as asphalt performed well.
- With the gradual introduction of greater amounts of roof insulation, warm roofs became more popular. However, when insulation was located immediately below the membrane, the greater surface temperatures that arose would lead to quicker deterioration of the membrane. To avoid these problems inverted roofs became more popular.
- Asphalt roofing performs well when it is protected against extremes of temperature. The material is thermoplastic and when heated will tend to flow under gravity if unconstrained. At cold temperatures it becomes brittle and can be damaged under impact loads.
- Ponding on asphalt has not been shown to cause major problems. However, proper solar control measures are usually essential.
- Built-up felt roofs performed badly at one time, but the advent of polyester reinforcement brought a sea change in the industry with much greater flexibility and fatigue resistance than before. Attention to good detailing is needed to prevent damage to the covering from deck or substrate movement. Blistering due to entrapped air can be a difficulty but unless water-filled or greater than about 600mm, blisters are best left untouched.
- Single ply membranes are now much more popular than a few years ago. EPDM, TPO and PVC systems are common; EPDM needs adhesive bonded seams whilst TPO and PVC can be heat welded. Lap joints in these systems are all vulnerable if poorly executed. Loss of plasticiser can be a problem with some types, leading to embrittlement and cracking over time.
- Liquid membranes are also very common; hot applied systems are often used for new work but need solar protection. Cold applied systems often incorporate a polyester fleece for reinforcement and can be very effective, particularly where there is difficult detailing involved. However, adhesion failure arising from contamination of the substrate is one problem and the systems should never be used as a substitute for poor design and detailing. When used carefully and with compatible substrates, the use of liquid membranes for remedial work is often a good alternative to traditional roofing systems.

9.1 INTRODUCTION

In the late 1960s, aside from metal coverings, the choices facing a designer of a flat roof were generally fairly limited: either asphalt or built-up felt systems. Most roofs were of cold construction with minimal insulation whilst solar control was generally provided by chippings or promenade tiles.

Since then, much has changed, and whilst warm roof systems are now almost universal, there are a bewildering number of different membrane options, all of which, if properly detailed, are capable of delivering a good life expectancy. Defects in roof coverings are often due to inappropriate selection and use rather than a failure of the material itself (save for the expiration of normal life expectancy) with detailing accounting for a high proportion of failures. This chapter deals with the various types of covering in common use together with some of the defects to look out for. Common themes of workmanship and sensible design run throughout the various roof types.

9.2 DIFFERENT TYPES OF ROOF CONSTRUCTION

Roof type	Features
Cold roof	Insulation located at ceiling level; void or roof slab is insulated from heat loss from below and is therefore at a colder temperature than the room during cold weather. With this form of construction there is a risk of condensation forming within the roof void or on the underside of the slab. Good cross-ventilation is therefore essential. Vapour control layers at ceiling level are essential, but not 100% effective.
Quasi cold roof	Roof constructed as above, but without the insulation layer. Again, risk of condensation forming either on soffits of slab, interstitially or on reverse face of roof deck, i.e. the cold surface.
Warm roof	Insulation placed on top of the roof deck but beneath the waterproof membrane. Vapour control usually placed under the insulation. This form of roofing has a much reduced risk of condensation as the ceiling or roof void is at a similar temperature to the roof. However, the system prevents heat from infrared radiation from being dissipated.
Inverted roof	A roof in which the insulation is placed on top of the waterproof membrane. No vapour control needed. Roof void and or deck maintained at similar temperature to the room. The insulation protects the membrane from damage but must be anchored down. An inverted roof is also a warm roof.

9.3 SYSTEM FAILURES AND DETAIL FAILURES

Accurate diagnosis of a defect is an essential prerequisite to specifying a suitable repair. Understanding whether or not the failure is due to system or detail will assist in the selection of a suitable and cost effective solution without necessarily overspecifying or adopting an over-optimistic approach.

A system failure is simply a failure in the roofing material that could occur at any point over the roof as a whole. For example, a defective roofing specification (unsuitable insulation, missing vapour barrier, unsound deck) could promote multiple failures. If one defective area is repaired there is every likelihood that other defects could occur elsewhere for the same cause. In the case of a system failure, replacement of the entire roof is probably the only sensible option.

On the other hand a detail failure occurs at a specific point or points and is related in some way to the detailing of the roof covering at that position. Repair would be possible, with no reason to doubt that the remaining areas would perform adequately.

Examples of system faults include:

Basic cause	Materials at risk	Effects
Problems due to moisture	Affects organic materials such as chipboard, plywood, strawboard, etc. Results in swelling, sagging, degeneration due to biological effects, rot. Moisture damage to HAC roof decks – risk of corrosion to reinforcement or alkaline hydrolosis.	Swelling can cause compressive stress in the deck, relieved by bowing. If the membrane is fully bonded, tensile failure may occur. Degeneration of the deck will result in loss of support and rupture of the roofing system.
Problems due to temperature	Can affect certain types of rigid foam, resulting in shrinkage or expansion. Expansion movements in large concrete roof decks resulting in bowing.	Places stress on the membrane at board joint positions, rippling or rupture of membrane.
Problems due to a combination of moisture sensitive and temperature sensitive materials	Typically affecting woodwool slabs.	As above
Problems due to poor design	All	Inadequate design leading to deflection of the roof or risk of wind uplift; increased flexibility, risk of overloading, etc.

Examples of detail faults include:

Basic cause	Materials and locations at risk	Effects
Moisture	Localised faults in laps of BUR roofing. Missing or defective isolation joint at perimeter or upstand details – BUR and asphalt Defective upstand detailing, BUR and asphalt. Lack of provision for movement at movement joints or changes in structure.	Localised water ingress, blistering. Repairs can be executed locally; total renewal not required.
Temperature	Slumping of asphalt at changes in level, upstands, etc. Cracking due to brittle failure at low temperatures.	Localised water ingress, blistering. Repairs can be executed locally; total renewal not required.

9.4 MASTIC ASPHALT

Although a traditional material and with a proven life expectancy of up to 60 years in some cases, asphalt roof finishes, particularly in new-build commercial buildings, are gradually giving way to liquid membranes. It is now increasingly rare to find asphalt being specified as a primary membrane in new build.

When installed upon an uninsulated rigid concrete deck, asphalt performs very well (assuming that the detailing and workmanship are correct), but its long-term performance with insulated decks (warm roof construction) is less satisfactory. A properly installed mastic asphalt roof should have a life expectancy of at least 25 years, but if inadequate attention is paid to solar control, heat dissipation and maintenance it can be significantly less than this. The durability of asphalt is reflected not only in terms of high temperatures but also low temperatures where the same insulation layer can prevent or at least significantly reduce the amount of heat taken up from the structure.

BRE Information Paper IP8/91[65] considers the durability of different grades of asphalt. Following testing of various samples, BRE concluded that major failures in asphalt were more likely to be the result of contraction and embrittlement at low temperatures. Interestingly, the much vaunted problem of thermal shock (rapid change from high to low temperatures) was found not to be a significant factor in the performance of the samples.

The material

Asphalt is essentially a formless blend of asphaltic cement with graded fine and coarse aggregates. The material is liquid when heated but cools to a dense, cohesive and voidless mass on cooling.

Asphaltic cement is a mixture of bitumen or a mixture of refined lake asphalt (a natural product) with asphaltite (a material allied to bitumen but with a higher softening point and greater strength and stability). Bitumen is either taken from natural sources or distilled from petroleum; it is a thick viscous liquid, mainly non-volatile. Without aggregates, asphalt does not perform very well, so it is usual to combine the material with graded coarse and fine aggregates made from crushed limestone, siliceous, igneous or calcareous rock.

Whereas during the 1950s and 1960s the provision of asphalt roofing was common on flat uninsulated decks, the gradual increase in thermal insulation during the 1970s meant that whereas asphalt had hitherto performed well as a strong and durable material, its life was becoming much shorter. At that time warm roof construction was commonplace; it was not unusual to specify asphalt on top of cork, polystyrene or rigid foam board with a layer of chippings or solar reflective paint as protection. Whilst chippings of limestone, granite, feldspar, etc. offered a good level of protection to begin with, these materials could be subject to wind scour leaving larger areas of roof unprotected. Solar reflective paint systems were thought to be a better alternative, but in reality they lose effectiveness fairly quickly and offer very little benefit to flat horizontal surfaces. (Solar reflective paint is however the usual choice for upstands.)

Asphalt is relatively inflexible at normal daytime temperatures in the UK but inevitably the asphalt roof surface is subjected to heat from solar radiation. In older construction some of this heat is dissipated away into the structure, but if the asphalt is placed on top of insulation, the heat cannot be conducted away to the same degree. As a result the asphalt softens and is prone to elastic deformation. Slumping of upstands or the gradual sinking of plates or bearers into the surface of the asphalt are typical indicators of plastic deformation. Repeated cycles of cooling and heating lead to cracking, particularly if the asphalt is restrained in some way.

Solar reflective paint

Thermal cycling can cause problems with the solar reflective paint as well, and it is not unusual to find that the paint has applied significant stress to the surface of the asphalt leading to a pattern of regular cracking.

Various different methods of protection of roof surfaces have been attempted over the years, for example:

- silver reflective paint;
- white reflective paint;
- asbestos cement promenade tiles (and their non-asbestos counterparts);
- capping sheets;
- chippings; and
- inverted roof insulation systems.

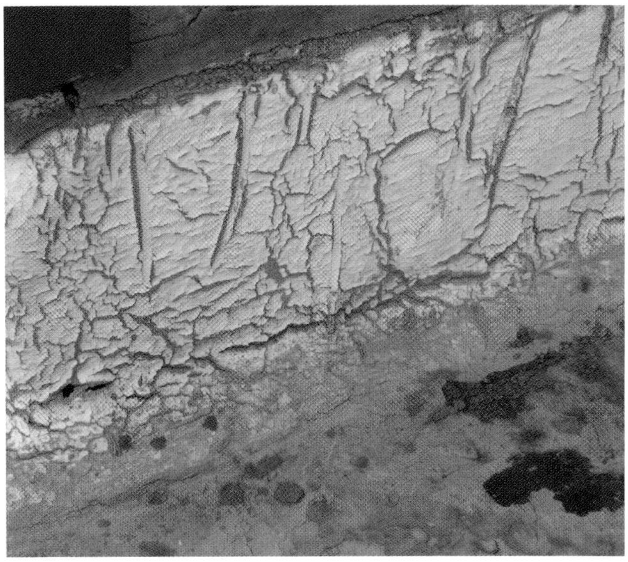

Figure 28 The solar reflective paint applied to this upstand has shrunk upon ageing, resulting in severe surface cracking of the asphalt. In turn the cracked areas absorb more radiation thus exacerbating the problem.

Until the late 1970s the predominant finish was chippings of one variety or another, often with asbestos based promenade tiles ('Spartan' tiles) in walkways or balcony areas. Towards the end of the decade a greater reliance was placed upon solar reflective coatings which gradually became more popular because of their ability to be added to vertical or sloping areas. However, solar reflective paints were not without problems and if the incorrect materials were used their performance could be disastrous.

A roof will absorb a measure of solar radiation, most of which will be converted into heat energy. Since solar radiation contains a variety of different waveforms in the spectrum, an amount of UV light can also be absorbed. UV light will cause organic materials to degrade; asphalt hardens following exposure to UV light, reducing its tensile strength and making it susceptible to movements in the deck or flashings, for example.

It is well known that a highly reflective surface will prevent or certainly reduce the amount of radiation absorbed, whilst a black or dark surface will absorb more. In terms of roof coatings, few, if any materials, will prevent absorption of radiation, although they will mitigate the effects. Conversely, a coating can serve to reduce (or increase) the amount of heat radiated from a roof at long wavelengths, thus affecting surface temperatures at night. During a cold, clear night, heat is radiated to the night sky (see section 7.8 on cold night sky condensation) resulting in a reduction in surface temperature several degrees below ambient air temperature. This can render the asphalt brittle, whilst the change back to warm temperature during the day can permit the formation of longitudinal cracking unless the effects are mitigated by a solar coating.

Figure 29 Asphalt hardens following exposure to UV light, rendering it less flexible. In this case, expansion and contraction of the lead flashing to this abutment detail has resulted in tensile failure of the asphalt. Ideally, joints in the flashing should be welted rather than lapped, and the lead securely fixed to a suitable timber edge batten bolted to the structure.[66]

Asphalt is thermoplastic, it softens on heating and exhibits creep characteristics which are not reversed upon cooling. Below a critical temperature (or glass transition temperature, Tg) the material changes from being relatively plastic to hard and brittle – and at greater risk of damage from impact loads. The provision of solar reflective paint can extend the time during which the asphalt is below its transition temperature.

The glass transition temperature (Tg)

The glass transition temperature provides a useful means of comparing the cold weather characteristics of a variety of polymers.

The chains of molecules that form a polymer can be likened to a plate of spaghetti. When warm, the individual strands slide over one another smoothly; this is described as an amorphous state. As the spaghetti cools it gets stiffer, the strands are less likely to move easily and eventually will reach a stage when no movement occurs at all. The temperature at which this occurs is called the glass transition temperature – at this point the spaghetti reaches its glassy state. The transition from amorphous to glassy state is critical in evaluating the performance of roofing materials.

The glass transition temperature can be lowered by introducing small molecules called plasticisers. Plasticisers increase the distance between the larger polymer molecules.

The beneficial effects of solar reflective paint can tail off quite quickly as a result of weathering, soiling or ponding on the roof. Initially, a high degree of visible light is reflected, but this gradually diminishes. Cleaning the surface in spring may serve to restore some of the benefits, but regular re-treatment, perhaps every five years or so, will be necessary to afford adequate long-term protection.

Some paints are not tolerant of surface contamination or moisture; application to an unclean surface or to one that is damp following overnight dew will probably result in failure and large areas of paint flaking off within a short time. In some cases performance is improved if the asphalt has been sanded prior to coating as this serves to give the paint a better mechanical key.

Research by BRE in the late 1970s sought to identify if there were any particular benefits in aluminium paints over white paints. Aluminium, which tended to be based upon a bitumen binder rather than a water emulsion, was a common finish at the time although less so now. BRE found that aluminium was successful in blocking infrared and UV radiation to about 65 per cent when new, but that reflectance dropped off after a short period of time. The paint served to reduce emissivity but not by enough to compensate for absorption. BRE were unable to identify clear advantages of aluminium over white paints although they did note that in some cases, presumably as a result of low permeability, blistering was more prevalent with aluminium paint.[67]

For a solar reflective paint to be effective, it must be compatible with the roof material and should be designed specifically for the purpose. Paints that are not specially formulated can cause more harm than good especially as they shrink with age. The shrinkage can cause the surface of the asphalt to crack and in turn the cracked area absorbs more radiation than the surrounding area. The localised increase in temperature results in localised softening and expansion of the asphalt – an effect that is not reversed upon cooling. Gradually, the problem increases in severity, the cracking is extended and the asphalt deteriorates.

Blistering

Blistering of asphalt is a common problem, often to poorly detailed upstands or as a result of a failure to provide effective vapour control or water ingress into a roof deck. The entrapped moisture vaporises and forms a blister in the asphalt. Eventually, repeated cycles of vaporisation result in the perforation of a blister and further water ingress. See also the material on blistering in section 9.5 on built-up felt roofing.

Upstands typically suffer from blistering as a result of poor detailing. Good construction practice dictates the use of raked joints in brickwork to provide a key to the asphalt and or the use of high bond primer to prevent the asphalt slumping. Ideally, the tuck at the top of a skirting should be protected by a lead or metal flashing; frequently this is neglected, and a mortar joint substituted instead. Such a detail leaves the tuck vulnerable to water penetration. If a lead flashing is used, it is imperative that the cavity tray is dressed over the top of the lead and not beneath it, otherwise water leaching from the cavity tray could be directed into the asphalt tuck and possibly behind the skirting. Similarly, the cavity tray must always project far enough to lap over the lead and not stop short of it. This problem is illustrated in figure 32.

For the above reasons, the use of asphalt in warm roof situations has become far less prevalent and is best avoided. As an alternative, inverted roof construction has proved far more successful in maintaining the long-term performance of the roof membrane. Here, the membrane is protected against mechanical damage as well as thermal cycling; being below the insulation, the asphalt is subjected to a much lower range of temperature variation.

Cavitation

Asbestos cement promenade tiles were a common finish to asphalt, offering good levels of solar as well as mechanical protection. However, smooth finish tiles tended to promote the growth of algae, in some conditions, making them very slippery and sometimes dangerous.

With the gradual restrictions on asbestos, cement fibre or GRC products became widely available. One early problem with these materials was there low permeability, meaning that water that became entrapped beneath the tiles could not necessarily escape.

Figure 30 Large vapour-filled blisters in an asphalt roof covering.

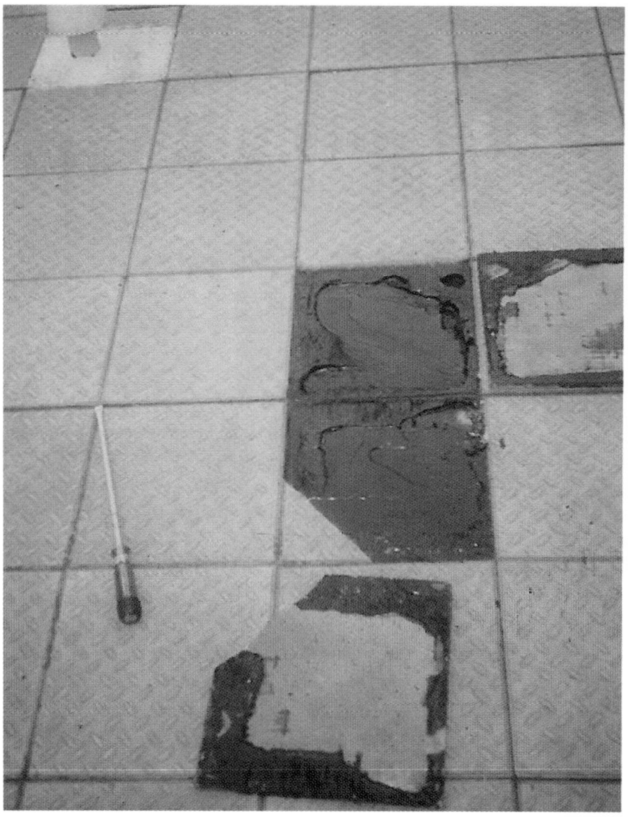

Figure 31 Cavitation damage below impervious promenade tiles.

During the day, the surface temperature of the tiles could cause the entrapped moisture to vaporise, a problem that was accompanied by the softening of the asphalt or bitumen bedding compound. Gradually, tiles could lift out of place, ultimately ratcheting upwards to present a trip hazard.

However, whilst ratcheting was a problem, the formation and subsequent collapse of vapour bubbles gives rise to a problem of the gradual deterioration of the asphalt – in effect a process of cavitation. Cavitation is defined as the process of formation of local cavities in a liquid, as a result of the reduction of pressure below a critical value. The problem frequently causes the pitting or wearing away of a solid surface.

To deal with the problem, promenade tiles were subsequently manufactured to be porous or vapour permeable.

Cracking

As noted above, asphalt is susceptible to deck movements, particularly following ageing and exposure to UV light. At low temperatures, the material is very brittle and can fracture under impact loads, or misguided attempts at opening up using a bolster. Similarly, embrittlement can be the product of overheating during laying.

Cracking will often be found around junctions with alternative construction, perimeter details and projections as a result of differential movements in the background materials. If asphalt is laid on a lightweight or flexible deck (such as woodwool), movement between the deck and the enclosing walls must be accommodated by isolated upstands, otherwise failure at the transition between horizontal and vertical work will occur. Cracking at upstands can sometimes be caused by failures not only in the provision for movement, but by poor workmanship during installation. For example, when the asphalter places the skirting, he/she will form a triangular fillet detail. If the asphalt is insufficiently cleaned prior to the formation of the fillet, poor adhesion will permit the formation of cracking along its edge.

Another reason for fillet failure could be a lack of support at the edge of the insulation in warm roof construction. If a gap exists between the insulation and the enclosing wall it is possible for the asphalt to slump into the void over time, thus compromising the asphalt at the perimeter.

Deflection of the roof structure or collapse of a roofing screed due to poor compaction could create similar tensile crack patterns in the base of an upstand.

Examination of the crack patterns in asphalt will give some indications as to their likely cause. For example, cyclic movements in the deck arising from thermal or moisture changes are likely to produce compressive rippling type defects, whilst tensile cracking can occur as a result of the structural arrangement of the deck or background materials. (See also section 9.5 on built-up felt roofing.)

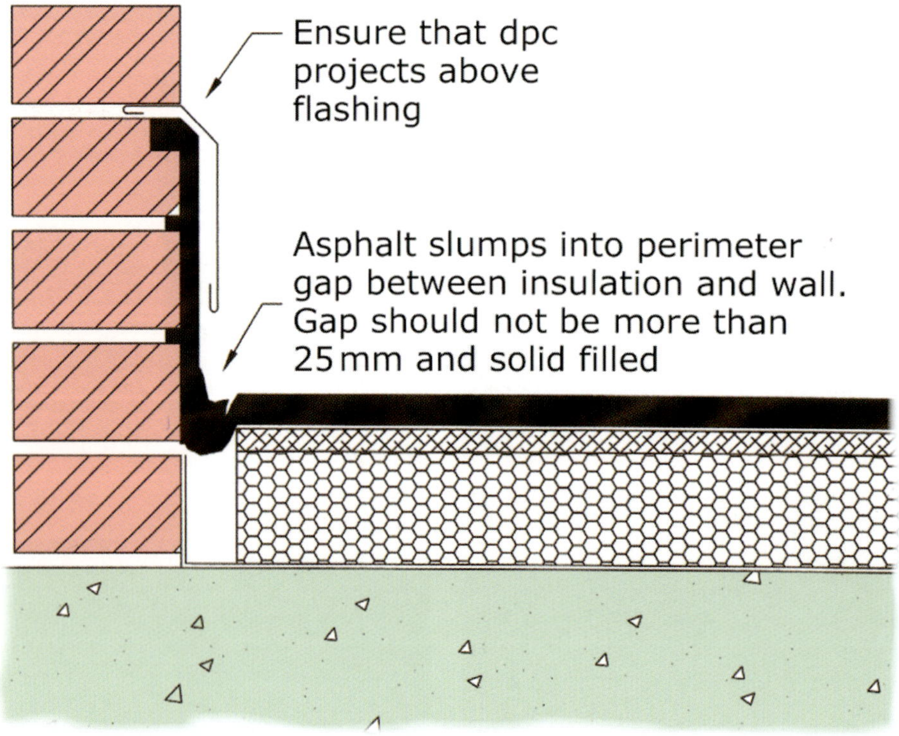

Figure 32 Gap between insulation and wall causes cold bridge and allows asphalt to slump.

Figure 33 Asphalt skirting has been taken across a parapet movement joint without provision for movement; it has cracked as a result.

Illustrations of deck movements arising from thermal movements can be seen in chapter 22 on multi-storey car parks.

Ponding

Ideally a flat roof should never be totally flat; CIRIA[68] recommends a minimum fall of 1:80 or greater, although this is often not attained. Deflection of the roof deck can lead to the reversal of falls away from outlets and to ponding.

There is some debate as to whether or not ponding is a bad thing for asphalt roofs; the general consensus is that as far as asphalt is concerned, ponding is not necessarily harmful, although of course if a leak does occur there is a greater reservoir of water available to penetrate the building.

Rigid foam insulation

Since polystyrene (EPS) is a thermoplastic material it is easily damaged during the placing of asphalt and so its use as an insulant in conjunction with asphalt is generally limited unless suitably protected (at least one manufacturer supplies a felt faced material for this purpose). Furthermore, it can be dimensionally unstable over changes in temperature and this can lead to the opening and closing of the boards and problems of rucking or splitting in the waterproof layer. Rigid foam boarding, usually polyisocyanurate (PIR) is a more popular choice because of its higher temperature resistance and good insulation properties. PIR is a thermosetting material, and rather than melting in the way EPS would, it tends to char. The material begins to degrade at temperatures above 150°C and so still requires protection in the form (usually) of bitumen impregnated fibreboard.

Until the early 1990s almost all commercial rigid foam products used CFC-11 as a blowing agent. CFC-11 was held to be an ozone depleting substance, and under various protocols was phased out of use and substituted with, for example, HCFC 141b. The transition was thought to have led to various manufacturing problems and product failures, although it is probably more the case that poor mixing of the component parts was to blame.

PIR foams are generally stable, but roofing failures involving these materials are not unheard of and can lead to expensive remedial work, often involving re-covering. Typical defects are as follows:

- formation of depressions in the roof after laying;
- shrinkage of boards creating board outlines in the asphalt;
- poor compressive strength; and
- cupping of boards giving rise to reflective undulations in the asphalt.

Almost all of the above problems can be attributed to manufacturing faults rather than issues of poor workmanship, save that improper storage of materials on site can give rise to performance difficulties.

The production of PIR and PUR foams involves the mixing of two component parts, part A and part B, to form a chemical reaction. Part A is usually Methylene Diphenyl di-Isocyanate (MDI), whilst the catalyst, part B, is a mix of polyol, various additives and blowing agents. The mixed components then react exothermally (producing heat) to form a rigid thermosetting polymer. The maximum strength of the product is determined by the cell structure of the foam; spherical or slightly egg shaped cells are needed and these are attained by mixing very precise quantities of A and B chemicals. If the ratio of chemicals is allowed to deviate during manufacture there is a risk that the cell structure will be more elongated rather than spherical. Elongated cells are much weaker, with the result that the crushing resistance of the foam is less; it becomes brittle and is easily damaged.

Incorrect mixing ratios therefore account for the subsequent deterioration of the insulation, particularly in areas that are regularly trafficked during installation or afterwards. Poor compressive strength leads to crushing damage under foot which may become manifest over a period of time after laying in the form of a series of depressions in the roof surface.

If severe enough, the deterioration of the boards will result in the delamination of the fibreglass outer covering.

Edge cavitation

Edge cavitation of the boards is a defect whereby the edges gradually become concave, creating gaps and reducing thermal insulation. The problem is usually associated with poor strength as well. Again a manufacturing problem, the root cause appears to be the introduction of small quantities of water into the B chemicals to produce CO_2 as a joint blowing agent. However, the water can dissolve some of the HCFC blowing agent resulting in an imbalance in the mixing ratio and subsequent weakness.

During mixing, the foam reaches temperatures of around 149°C. If CO_2 is present it will diffuse out of the cells at a rate that is much faster than air at ambient temperature can replace it. This leads to an imbalance of pressure which, coupled with the weakened cells, creates an elongated cell structure. The problem occurs most rapidly at the board's edges, resulting in poor foam structure and board-edge cavitation.

Cupping and bowing of the boards can result either from manufacturing faults or from poor storage of the boards on site. The latter is easily explained if the boards are maintained in damp conditions prior to laying. Since they are usually supplied in bundles it will be the top or bottom sheet that is at most risk if its outer surface is allowed to become saturated. If one face is wetter than the reverse face, differential expansion occurs causing the panel to cup. The problem is likely to be randomly spread over a roof and may be reflected in the asphalt layer.

Alternatively, poor mixing, or poor mixing ratios, can lead to an uneven distribution of cells (in terms of size and shape), which create dimensional instability and different properties throughout the thickness of the board. When placed under load, the different properties react in different ways, with the result that cupping or bowing can occur. Such problems result in the formation of regular ridges or depressions on the outline of the boards.

Case study 1 – Art deco building, South coast

The building comprised a large L shaped structure with an asphalt covered flat roof originally constructed in the 1930s. The roof had performed well in its time, but with no thermal insulation it gave rise to large heat losses. The designer of the re-roofing scheme concluded that the most appropriate form of repair would be to treat the existing asphalt as a vapour barrier and to provide thermal insulation in the form of polyisocyanurate boards, a layer of fibreboard and a new covering of polymer modified asphalt finished with solar reflective paint.

Within a few months of laying, depressions formed in the roof surface leading to ponding. The asphalt appeared to be performing well, but concerns were expressed by the freeholder over its long-term performance. The depressions were random in nature, but were more prevalent in areas of the roof where the asphalt boiler had been located – areas that would have received more foot traffic during installation.

A series of trial holes were cut into the asphalt; these revealed that whilst insulation was present it comprised two different types – a mixture of pink (PIR) and cream (PUR) types. Generally, the depressions occurred wherever the PUR had been used. In a few cases, moisture was found entrapped within the roof construction.

Since PIR boards had been specified, the contractor had clearly used the incorrect material and was in breach of contract. However, theoretically there was no reason why the insulation should not have performed properly as the fibreboard layer should have been of sufficient thickness to have protected the insulation (which is thermosetting) from heat damage during the laying of the asphalt. Further core tests revealed a manufacturing problem with the boards leading to a lower than expected compressive strength. The cells of the board were elongated rather than spherical which gave them a much weaker structure, easily damaged under load. Eventually the roof covering was replaced at the cost of the main contractor, the errant subcontractor having gone into receivership in the intervening period.

Case study 2 – An office building located in west London

The building was concrete framed with in-situ concrete roof decks constructed in the early 1980s. At top floor level were some residential flats, each with a large balcony enclosed by concrete parapet walls. The roofs were of asphalt laid upon an expanded clay insulation system, 'Insulscreed' – small balls of expanded clay coated with bitumen. The insulation layer was around 150mm thick, with the roof draining towards a shallow gutter along its front edge.

To protect the asphalt, precast concrete pavings had been laid on plastic pedestals; there was no separate insulation or geotextile membrane. At the perimeter of the roof, a shingle margin had been provided.

One of the residents placed a series of wooden planters around the roof, planted with bamboo for screening purposes. After five years or so a small leak occurred in the roof and following correspondence the roof was eventually exposed to reveal that the roots from the bamboo had broken through the lining to the planters and had entered the shingle margin. From here they had penetrated the asphalt and grown out again through the asphalt upstand as shown in the photographs below.

Figure 34 Shoots of bamboo sprouting through the asphalt upstand.

Figure 35 Bamboo roots had first penetrated the asphalt then spread through the expanded clay screed and exited out through the skirtings. Upon examination, the roots had spread through the screed to affect nearly the whole roof. The black 'pebbles' that are visible are the insulating screed, exposed after removal of the asphalt.

At first sight, the remedial work appeared simple: remove the planters, cut out the localised area of damage and make good the asphalt. However, upon exposure it became evident that the insulating screed had created an ideal growing medium for the roots, being both warm and damp (from the leaks), and the roots had developed into a tangled mass that spread throughout the screed. Given the propensity of types of bamboo to propagate from roots, the risk was that even if the damaged areas were to be repaired, further plant growth could occur again within a period of months or years. The only safe solution would be to remove the roof and screed in its entirety and replace with new.

Case study 3

The building comprised a terrace of three listed grade 2* large former houses in Kensington but converted for use as the headquarters for a large professional body. The building backed on to a typical London garden square. At basement and ground floor levels the building extended rearwards to form a large terrace at first floor level. The owner of the building used the terrace for corporate events and had for many years maintained a quarry tiled covering on its surface.

The terrace had fallen into disrepair and so it was decided to replace the asphalt covering with new and to provide a decorative finish of ceramic tiles. The specification included for the provision of foamed glass insulation cut to falls and laid over the original brick and filler joist vaulting. The roof was thus turned into a warm deck. The terrace extended over the rear of three houses, with a step at each party wall line. Asphalt had been dressed vertically up the steps to create a seamless membrane.

The tiles were laid using a flexible ceramic tile adhesive formulated for exterior use. Vertical tiles were placed at each upstand. However, within a few months of completion, tiles began to lift off and the asphalt began to slump at upstands and changes of level. Tiles covering the steps failed rapidly and became dangerous.

Eventually the problem became so bad that the entire finish was removed and replaced. The problems were due mainly to the tendency for plastic deformation within the asphalt. The layer of insulation beneath the membrane prevented heat that built up in the surface to dissipate away resulting in higher working temperatures than would have been the case previously. Thus, the asphalt would soften and under gravity gradually tend to 'flow' over steps in the roof. Movement in the asphalt was greater than that which could be accommodated by the tile adhesive. Rigid grouted joints created compression stresses in the tile layer which could only be relived by shear failure on the tile bed and 'popping' up of the tiles.

The difficulty for the designer of the remedial works was that in the intervening period the adhesive manufacturer withdrew its recommendation that the adhesives would be suitable for use on asphalt. In addition, if ceramic tiles were refixed, the problem would reoccur. After considerable debate the tiles were replaced with purpose made cement fibre tiles manufactured to resemble the original quarry tiles. Joints were not grouted but left open and compatible bedding adhesive was used. The solution was very much a compromise and whilst not matching the original in terms of quality did enable the terrace to be returned to its original use.

Mastic asphalt performs well as a roof covering if it is applied properly and if the design of the roofing system is such that the effects of temperature are reduced as far as possible.

However, annual inspections are a sensible precaution; probably best carried out in the autumn so that leaves and other detritus can be cleared at the same time. In areas that are heavily affected by trees or other sources of debris more frequent inspections may be necessary. The early signs of failure can then be identified and corrective action taken.

The areas most likely at risk are abutment details, penetrations, movement joints, gutters and outlets. Blistering and slumping of asphalt, particularly to upstands, will usually require fairly urgent work to arrest deterioration and more water ingress into the substrate.

Retraction of asphalt around rainwater outlet details or edge flashings can also lead to problems of water ingress, and thermal expansion of flashings will often lead to cracking at joint positions.

9.5 BUILT-UP FELT ROOFING (BUR)

At one time, the general rule of thumb was that a roof covered with felt to BS 747 would last about 15 years before re-covering was necessary. However, with the advent of high performance materials, life expectancies in excess of 25 years or more can be expected on a routine basis. Like most systems though, the success of a roof depends not only upon workmanship, but the design of the roof in the first instance.

Early roofing felts employed an asbestos base, although these performed poorly and had poor tensile and tear resistance. In the 1970s, asbestos was replaced with polyester giving rise to a significant improvement in performance. For example BRE fatigue tests on samples of BS 747 felt failed after around 32 cycles whilst polyester samples failed after 3,200 cycles. By using polyester based felt and modified bitumen over 30,000 cycles were attained.

Contemporary construction has adopted the nomenclature of RBMs (reinforced bitumen membranes); guidance for the selection and use of these materials is given in BS 8747:2007. BS747 was amended in 2000 but withdrawn in 2004 to be replaced by BSEN 13707:2004. Generally speaking the range of RBMs and their performance characteristics are now far superior to anything available in the early 1970s.

Whereas the norm for BUR roofs was at one time three-layer felt, the tendency in current practice is to utilise two or more layers or alternatively a single layer system using a specifically designed RBM. In a two-layer system the lower sheet provides the flexibility and the ability to accommodate movements in the deck, whilst the top sheet provides a protective function against static or impact loads. Many top sheets are self-finished with either mineral chippings or foil to provide solar reflectance. These sheets are generally referred to as cap sheets.

To prevent the effects of oxidisation embrittlement, modern felts are usually compensated by the addition of either:

- APP (atactic-polypropylene) – usually for torch-on felts. These plastomeric modified bitumen membranes are produced on polyester, glass fibre, and glass/polyester mixed bases. The base is saturated and/or coated with bitumen, modified with polyolefin or a polyolefin copolymer. Or
- SBS (styrene butadiene-styrene) – usually pour and roll, bonded or torch-on applications. These elastomeric modified bitumen membranes are produced on polyester, glass fibre, and glass/polyester mixed bases. The base or carrier is saturated and/or coated with bitumen which has been modified with thermoplastic rubbers.

As with asphalt coverings, the success of an installation is dependent upon the suitability of the deck to which it is laid and to the standard of detailing of abutments and roof penetrations. A deck that is likely to move and deflect under load or moisture changes is likely to impart significant stresses to the covering which, as it hardens on aging and exposure to UV light, offers less tensile strength.

Problems with polystyrene insulation

As noted earlier, polystyrene insulation is generally considered to be unsuitable for asphalt roofing, but it can equally cause problems with BS 747 type felts because of their poor flexural strength. Boards should preferably be adhesively bonded to the deck rather than mechanically fixed; they must be tightly butted and secured to prevent wind uplift. Some insulation boards are intended for use with BUR roofing systems and sheet sizes are limited. Large sheets will exhibit correspondingly greater movement capabilities at joints.

If the boards are mechanically fixed, their propensity to move is greater, and cyclical movements can stress

Figure 36 No matter how high the specification of RBM, it will be let down by poor attention to detail. Here, a membrane has been bonded over an older, asphalt membrane. The detailing of felt is almost impossible to achieve properly in these circumstances with the result that attempts have been made to seal the top edge of the felt using an incompatible mastic sealant.

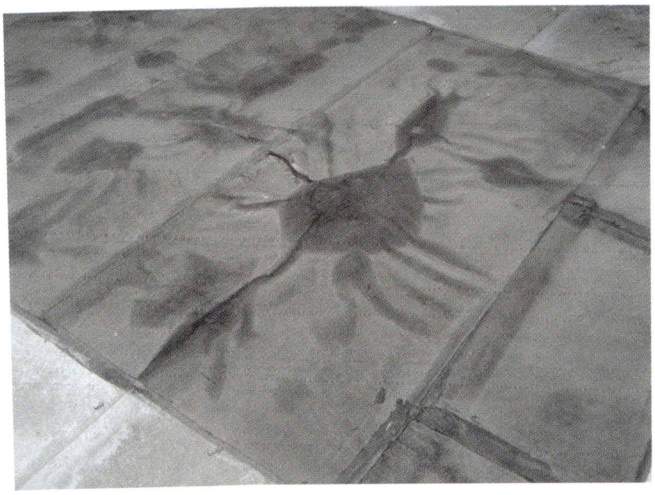

Figure 37 Felt used as a patch over existing failed asphalt. This treatment may be satisfactory as a short-term measure, but here entrapped moisture has formed severe blisters. The felt is also cockled and split.

the covering, possibly causing it to rupture. Similarly, changes in level at board joints brought about by inadequate or partial fixing can damage the roofing membrane. Modern high performance materials may be better able to withstand the amount of movement than older BS 747 felts.

Blistering

Cockling or blistering of felt roofing is usually a symptom of water entrapment within the interlayers or beneath the complete system. The problem can occur wherever there is a lack of adhesion between the cap sheet and the base layer. Blisters could be due to residual water from construction or other detail failures elsewhere on the roof. Similarly, application of a membrane over an already damp surface will cause problems as will work in cold weather if the material has been laid in unfavourable weather conditions.

A blister is an entrapped pocket of air and moisture (sometimes water filled) that can range over a few square centimetres to several square meters in area. Research by IRC-NRC suggests that the problem of blistering can be more prevalent with SBS type felts than APP types.[69] If the interface between sheets is voided, any moisture present within the void can vaporise during hot daytime temperatures. The volume change associated with the conversion of water to water vapour is in the region of 1,250 times greater, so even a minute quantity of moisture can be sufficient to exert considerable expansive forces on the felt. Given the felt will be warm and flexible, the expansion forces may be sufficient to overcome the

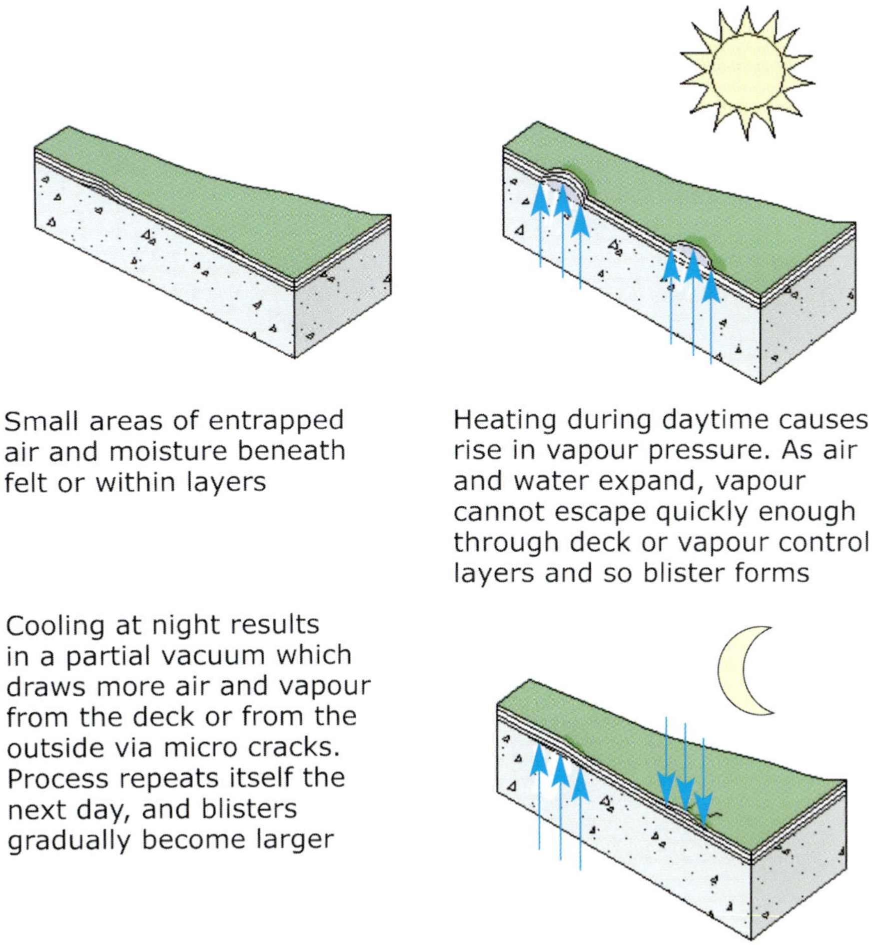

Figure 38 Blistering in built-up felt roofing.

bond strength of the membrane around the blister, thus tending to enlarge it.

At night the vapour condenses but since the void is now slightly larger, the felt is cooler and stiffer and may not return to its original shape. More air is drawn into the partial vacuum than was present at the start. The cycle is then ready to commence again.

Blisters do not generally affect the tensile strength of felt, but a rupture will make a large difference in tensile strength of the cap sheet. Small blisters can therefore be left alone so long as they are not in areas subject to foot traffic. Larger, particularly water-filled blisters should be cut out and repaired.[70]

Woodwool cement boards

Woodwool slabs were very common in the 1960s and 1970s but less so now, at least within the UK. Roof decks of (usually) pre-screeded and edge reinforced slabs are common in commercial buildings and schools – the latter being particularly popular when used in conjunction with 'Metsec' type lightweight steel beams.

Although a somewhat maligned material, woodwool could or can perform well if used in appropriate circumstances; in older buildings the felts used were generally unable to accommodate the cyclical changes in moisture and temperature which brought about dimensional strains. Commonly, woodwool slabs are believed to lose their integrity when wet, although this is by no means certain – indeed they have been used in external situations for sound attenuation purposes by the power industry. However, in common with many construction materials consistently damp conditions are unsatisfactory and can lead to softening and warping of the slabs. Warping of the slabs produces stress in the felt covering, with the risk of splitting along board joints.

Removal of old coverings during remedial works can be problematic, particularly where the felt has been fully bonded to the screeded finish of the slab. In these circumstances the screed can be damaged easily, leading to delamination problems and disintegration of the slab, necessitating a deck replacement in addition to the roof covering.

A further problem in 1960s and early 1970s buildings was a general lack of vapour control. If the woodwool was used in effect as a warm roof it would need a vapour barrier directly beneath it. Such a provision would be very rare, with the result that the dew point would occur at some point within the thickness of the slab giving rise to a risk of interstitial condensation. In severe cases, the slab could warp as a result.

One particularly vulnerable form of deck (now very rarely found due to life expectancy) is Stramit compressed strawboard – an insulation board material made from straw sandwiched between layers of building paper. Stramit boards had to be laid in a direction corresponding to the grain of the individual straws; a failure to do this could result in excessive deflection. Condensation occurring within the board could lead to rapid decay.

Fibreboard

Bitumen impregnated fibreboard is vulnerable to moisture movement and sagging over time, particularly when it gets wet. Whilst in contemporary construction it is unlikely to be used in isolation, it was a common form of roofing when used in conjunction with profiled metal decking supported by lattice beams – the fibreboard being laid over a felt vapour barrier on top of the decking and covered with three layer BS 747 felt. If condensation is allowed to form on the fibreboard it will sag between ribs, giving a characteristic multiple ridging of the covering above the profile ribs. Regular stripped pattern staining along the rib lines is a good visual indicator of this type of construction.[71]

Figure 39 Deck failure (chipboard). The outline of roof joists is clearly visible.

Figure 40 Effects of condensation on the chipboard deck in cold roof construction. The chipboard has lost its integrity.

Ponding

Ponding of roofs of BUR construction can be problematic as water can gradually penetrate at lap joints and find its way into the roof structure. BRE identified ponding due to undersized roof joists as being one of the main causes of failure in a survey of rehabilitation work on housing,[72] although in a separate study into the performance of roofing on Crown Estate buildings, no evidence was found that the life expectancy of roofing was compromised by the existence of ponding.[73] As with asphalt, the existence of ponding does mean that there is a larger reservoir of water available to enter the building should a leak occur.

Deflection due to undersizing (or bowing as a result of thermal expansion being constrained by the surrounding structure) need not necessarily be the only cause of ponding. Movement in decking materials (particularly organic materials such as plywood and chipboard) can also cause more localised depressions and any roof which exhibits a regular pattern of such depressions must be regarded with caution until the integrity of the decking material has been established. A frequent problem in older roofs is condensation occurring on or within the decking material causing it to expand, sag or decay.

Solar control

BUR roofs are demanding of solar protection in much the same way as asphalt (see section 9.4). Mineral flakes or chippings are often incorporated in the cap sheet to provide solar control – very light or white finishes are possible. A common alternative to reduce reflectance is the use of mineral chippings bedded in hot bitumen; however, as with asphalt, these can be subject to wind scour and can provide a useful base for the formation of moss and algae

Promenade tiles or a screed material called Paropa were sometimes used. Deterioration of the felt in lines between concrete tiles could occur due to their increased exposure and small cyclical temperature movements in the tiles.

Wind scour

Loose laid insulation in inverted roof systems must be secured in place with either loose ballast, paving slabs or by using composite sheets incorporating a cementitious finish. Some insulation slabs contain an edge interlock such that adjacent boards engage with each other and help to provide resistance to wind uplift. However, edge interlock by itself is unlikely to provide sufficient restraint, particularly when the size of the pressure waves travelling across a roof are at a similar dimension to the individual boards.[74]

Understanding wind uplift

There are two main influences upon a permeable covering to a roof:

- the global pressure field; and
- the local velocity field.

The global pressure field is the pressure that forms around the building or roof as a whole, influenced by topography, exposure, proximity of other buildings and of course wind strength.

The local velocity field is the zone of pressure around a subsidiary part of the building, for example the tiles on a pitched roof, insulation slabs, ballast and the like. Essentially, a rough projecting surface (such as individual tiles) disturbs the air flow and can make a large difference to the air pressure immediately adjoining the element. Conversely, a smooth finish permits uninterrupted airflow and so the local velocity field will be less significant.

The effects of either the local velocity field or the global pressure field depend upon the size of the gaps between elements and the volume of air beneath it. So, as pressure fluctuations affect the external air, small gaps and a large void will mean a time lag before the pressure can equalise. During this time the pressure beneath the element may be higher than above it, causing wind uplift.

A larger gap and smaller void will achieve pressure equalisation much faster and so the effects of uplift will be reduced. In effect this is why felt or timber sarking is placed immediately below tiles of slated roof coverings – it reduces the volume of the void and enables rapid pressure equalisation.

Roofs with loose laid insulation are vulnerable to wind uplift by this mechanism and so it is necessary to provide additional ballast or loading sufficient to overcome the forces generated by wind uplift.

In certain circumstances, wind scour of ballast systems can lead to the migration of the ballast into heaps and even blow individual stones off a roof – a possible risk to passers by or adjacent glazing systems, vehicles and the like. Much will depend upon exposure and roof profile, topography and location of adjoining buildings. Small chippings, for example those used as a solar reflective coating, are unlikely to be particularly harmful, but larger stones have the propensity to cause serious damage. Further information can be found in BRE Digest 311.[75]

9.6 THERMOPLASTIC POLYOLEFIN ROOFING

Single ply membranes have been popular for several years for both warm and inverted roof construction. Membranes usually comprise a reinforcement layer of polyester or glassfibre matting or fibres sandwiched between two layers of flexible polymer.

Broadly there are two generic types (aside from the bitumen based materials referred to earlier): thermosets (TS) and thermosplastics (TP). Thermosets are not modified by heating and therefore use bonded joints. A typical example of a thermoset single ply membrane would be EPDM (ethylene propylene diene monomer). Thermoplastic materials are modified by heat and can be welded. The family of thermoplastic sheets includes thermoplastic polyolefins (TPOs), now a particularly popular material. The terminology is confusing because 'olefins' is not specific, it simply describes any molecule containing carbon-carbon double bonds or 'alkenes'.

TPOs do not contain plasticisers, they exhibit similar UV resistance and heat resistance as EPDM but are weldable. Some forms are fairly difficult to install as they tend not to hold shape and do not relax quickly.

The durability and life expectancy of TPO roofing materials is very good, but in common with many roof systems the success of the system relies heavily upon good workmanship and the selection of materials that are appropriate for the particular circumstances in which they will be used.

According to a study conducted by NRC and others, membranes reinforced with continuous polyester fibre scrim had significantly higher tensile breaking strength than those reinforced with random short glass fibre mat.[76] In North America, experience with TPOs appears to have been mixed; the product was marketed as a cheaper alternative to PVC roofing, but its use has been marked by seam failures and age related embrittlement. However, the products used in North America tend to be thinner than their European counterparts, where there is greater experience in use.[77]

Some TPOs are reported to fade over time, but the most likely failure mode will probably lie in detailing or welded joints. Precise temperature control is needed to ensure proper fusion; too cold and an improper joint will be formed; too hot and the material may burn through. Since large scale welds are made by machine, the most likely problem areas will be at the start and end of a run.

In very cold weather conditions, the material may become brittle and suffer tensile failure. However, for the most part a review of manufacturer's data should enable selection to be made based upon the likely glass transition temperature (Tg) (see page 68).

9.7 PVC ROOFING

PVC roofing is another thermoplastic material that is capable of being welded in much the same way as TPOs. Performance of the material is generally good, with a design life of 30 years being expected. The material was first developed in the early 1970s but suffered from age embrittlement and loss of plasticisers. Today, these performance difficulties have been largely overcome by the provision of polyester or glass reinforcement and plasticisers that are more stable.

Hot air and solvent welding is possible; early membranes suffered lap failures as a result of problems with solvent welding, and although this method is still used, heat welding using an electric hot air gun is more popular.

As with all sheet materials, the success of the roof depends upon many factors, not just the durability of the membrane. PVC sheets can be fully adhered, partially attached, or loosely laid and ballasted. They are also available with a non-woven polyester fabric backing, which acts as a cushion to reduce the risk of perforation due to sharp ridges or material on the surface of the substrate.

PVC membranes are damaged by contact with bitumen or coal-tar pitch as plasticiser from the PVC migrates into the bitumen, reducing the flexibility of the sheet and promoting premature failure. For this reason, an isolating membrane is usually specified.

Figure 41 Misguided attempts have been made to seal the joints in this PVC roof with an unsuitable sealant material.

Defects in PVC roofing are likely to follow similar patterns to some of the defects listed under the EPDM chart below, with seam failure and contamination issues being reported by ASTM Committee 8.

9.8 EPDM MEMBRANES

EPDM systems are often referred to as rubber roofs or membrane roofs. Other rubber and rubberlike materials used are chlorinated polyethylene (CPE), chlorosulfonated polyethylene (CSPE, trade-named Hypalon), and polychloroprene (Neoprene).

EPDM is a thermosetting elastomeric polymer which must be bonded using appropriate adhesives. The material is made from:

- clay, which gives fire resistance;
- carbon black to increase tear strength and resistance to ultra violet light;
- process oils for flexibility; and
- curing agents for vulcanization.

Usually membranes are laminated together out of two plies. Like all single ply membranes, workmanship and proper detailing are essential for proper performance. Equally, EPDM roofs are not maintenance free and need to be inspected regularly to identify the early signs of system or detail failure. Manufacturers claim a life expectancy of at least 25 years for EPDM and despite early failures this does not appear optimistic.

Roof systems can be either fully bonded, mechanically fixed or loose laid and ballasted, the latter being unsatisfactory in conditions of high exposure because of the risk of wind scour. Performance data on EPDM roofing is not always easy to come by, manufacturers are, for commercial reasons, disinclined to publish information on membrane failures or system faults. Continual product development does mean that troublesome details and methods are gradually improved and so it is not necessarily appropriate to judge performance upon historic usage.

A study in 1990 of EPDM membranes between 6 and 13 years old revealed that a high proportion of roofs exhibited defects of one form or another. Of the sample roofs, 87 per cent had had one warranty claim, whilst 67 per cent had had multiple claims.[78] Many of the defects identified were reflective of workmanship and maintenance problems rather than inherent faults with the systems themselves. Judgment of the system based on 18 year old data would be unfair, since during the intervening period a number of advances have been made, such as improvements in bonding, elimination of neoprene flashings and their substitution with EPDM flashings, improvements to edge detailing and so on.

Typical defects with EPDM roof coverings, derived from ASTM Committee D-8 report, are summarised in the table below.

Typical defects with EPDM roof coverings

Fault	Comments
Puncture damage	Most often caused by a failure to drive home the mechanical fixings to insulation boards fully, leaving the fixing pin slightly proud of the stress plate. The pin can then gradually work its way through the membrane under normal loading or foot traffic.
Embrittlement of flashings (depends upon age of installation)	Often neoprene based detail flashings at wall abutments or other perforations, exhibiting signs of checking, splitting and cracking where exposed to south and west and where overworked during installation – forming by hand or with roller tending to thin the material.
Contamination	Contamination of flashings or field membranes with bitumen derived from previous roof coverings. Also contamination by grease or oil from roof mounted plant, extract equipment, etc. EPDM-based membranes are sensitive to hydrocarbons and should not be used in areas that are likely to become contaminated.
Delamination	Interphasal delamination of insulation facing sheets when bonded to fully adhered EPDM membranes. Possibly due to manufacturing faults with the insulation (see rigid foam insulation). If this occurs, the roof membrane may be at risk from wind uplift.
Seam defects	Birdsmouthing of seams, particularly at three-way junctions due to poor workmanship, adhesion failure, contamination of contact surfaces, etc. Moisture working its way into seams can disrupt the integrity of neoprene based splicing cement resulting in the gradual debonding of the entire seam. Since the 1980s, EPDM sheets have commonly employed factory applied adhesive seams rather than having to rely upon site applied adhesive.
Wrinkles in the membrane	Possibly due to inadequate attachment, curling of insulation or contraction of the membrane after installation.
Poor dimensional stability of insulation	Places undue levels of stress on the membrane
Defects in finishes	Risk of puncture from sharp debris left in ballast, or wind uplift due to missing paving slabs, wind scour of ballast, etc.
Defects in membranes	Due to poor or misguided attempts at repair using inappropriate or incompatible materials.

9.9 LIQUID WATERPROOFING SYSTEMS

Over the last few years numerous liquid applied roofing systems have been developed and specified for use in both new-build and repair situations. The performance of some coatings has not been satisfactory, but mainstream products are the result of considerable investment and research by producers and now offer considerable benefits over more traditional roofing systems or prefabricated single ply membranes.

Most of the new roofing materials are polymeric in nature meaning that they are made up of many ('poly') molecules ('mer') that are linked together in a chain-in effect forming very large molecules of different chemicals, for example poly(propylene) or poly(isoprene). Liquid applied membranes offer certain advantages over the traditional membrane materials such as asphalt and built-up felt systems – they are lightweight and flexile and can be worked around complicated details that would otherwise be difficult to achieve. However, their convenience can lead to a risk of poor application or selection with the inevitable result that the roof performs unsatisfactorily.

In broad terms there are two types of membrane – those that are applied cold and those that are applied hot. Generally, the market for cold applied membranes lies in repair and maintenance – they are particularly suitable for application over pre-existing roof coverings without having the need to strip the old roof. Hot applied membranes on the other hand are often used for new build applications and require some form of surface protection.

Like all roofing materials, care in selection of the product and proper attention to detailing and surface preparation are key for ensuring long-term performance. Ease of application is no substitute for proper application.

European Technical Approvals

As a requirement of the *Construction Products Directive*, individual member states were required to prepare harmonised standards for liquid systems; this has been achieved with the production of European Technical Approvals.

Within the UK, a parallel system of certification is provided by the British Board of Agrément (BBA), with many certificates now having been issued on a variety of liquid roofing products – some suggesting an effective life of 25 years.

European Technical Approvals (ETAs) exist for the following:

- polymer modified bitumen emulsions and solutions;
- glass reinforced resilient unsaturated resins;
- flexible unsaturated polyester resins;
- hot applied polymer modified bitumen;
- polyurethanes;
- bitumen emulsions and solutions; and
- water dispersible polymers.

In addition, solvent based acrylic roofing systems are available, these are not covered by ETAs and are often used for emergency repairs.

Most of the above products are cold-applied, but polymer modified bitumens are very popular for new build applications whereas cold applied materials are generally (but not exclusively) used for remedial works; both have good performance characteristics and if properly laid a life expectancy in excess of 30 years would not be an unreasonable expectation, although as with most materials this is highly subject to workmanship, protection, maintenance, etc.

9.10 HOT-APPLIED RUBBERISED BITUMEN MEMBRANES

By dispersing virgin or reclaimed rubber in asphalt together with limestone powder, oil and styrene butadiene-styrene (SBS) it is possible to manufacture a very flexible roofing membrane that can form a seamless membrane about 3–4.5mm thick. Some formulations include an additional spun bond polyester fabric interlayer with a further 3mm layer to form a roofing system. However, these materials are not usually resistant to UV light and so it is important to add a bonded membrane or alternatively a ballasted insulation layer to provide mechanical and solar protection.

The materials are usually supplied as cakes wrapped in polyethelene wrapping which are melted in a heater to around 200°C. The melt is discharged from the heater into a suitable container and applied to the roof using long-handled, rubber-bladed squeegees.

Prior to application, the roof surface must be free from contaminants. Liquid membranes depend upon a bond to the surface, so all traces of dust, grease and oil must be removed. Concrete surfaces are usually primed with a suitable asphaltic solution.

9.11 COLD-APPLIED LIQUID MEMBRANES

These membranes are polymeric compositions with one or more ingredients, such as modified asphalt or coal tar pitch, or various resins or elastomers, such as polyurethane, dibromobutadiene, silicone or acrylic. The bitumen based solutions are usually applied as emulsions, brushed or rolled in place.

Resin based systems can often be supplied in different formulations to give 'guaranteed' life expectancies of up to 25 years. These materials can also be dressed with sand or mineral granules to provide slip resistance, and for this reason are often used for waterproofing car park decks where a degree of friction is desirable and where various colour pigments can be used to advantage. (See also chapter 22 on multi-storey car parks.)

Type	Formulation	Modifiers (to provide durability, flexibility, etc.)	Application	Roofing system
Polymer modified bitumen emulsions and solutions	Aqueous emulsion or hydrocarbon solvent solution	Styrene butadiene styrene (SBS); styrene butadiene rubber (SBR); polybutylene (PB); rubber polyisoprene (IR); ethylene vinyl acetate (EVA); polychloroprene (CR); atactic polypropylene (APP)	Brush, spray or in some cases by spreader	Primer Emulsion or solution Internal reinforcement layer Solar reflective coating
Glass reinforced resistant polyunsaturated resins	Polyunsaturated resins with glass fibre (GRP)		Hand lay up process	Polyester resin basecoat/ matting/polyester resin second coat Chopped strand emulsion based glass fibre Polyester resin top coat Pigment and catalyst
Flexible unsaturated polyester resins	Unsaturated polyester resin reinforced with fleece		Spray or hand roller	Primer (type depends upon substrate) Fleece Catalyst Pigmented unsaturated polyester resin
Hot applied polymer modified bitumen	Reinforced or unreinforced	Rubber, SBR,SBS or IR	Spread	May contain reinforcement worked in between coats, but must be protected and laid below 15 degree pitch
Polyurethanes	Many types, e.g. air or moisture curing, isocyanate component mixed with polyol or prepolymer		Spray or roller	Primer (depending on substrate) polyurethane base coat, fleece, polyurethane second coat and or pigmented top coat depending upon sytem and durability requirements. May include non-slip grit in top coat for walkway applications
Bitumen emulsions and solutions	Aqueous emulsion or hydrocarbon solvent solution	Clay or resin soap	Brush or spray	Multi coat system Primer First coat Matting Second coat Solar reflective treatment
Water dispersible polymers	Many types, e.g. acrylic mastic, acrylic coating, co polymer, terpolymer, waterbourne rubber, SBR		Airless spray, roller or brush	Primer Reinforcement layer Polymer layer Non-slip grit

Figure 42 Cold applied membrane showing excessive adhesion failure; probably laid over poor felt roofing.

Figure 43 Incompatibility between substrate and coating has lead to extensive peeling and delamination.

Cold applied systems tend to be used for remedial operations, with the clear advantage that heat producing appliances are not required and the membrane can be sprayed over existing finishes without the need for a temporary roof. The ability to form complicated weathering details with the minimum effort places the materials at a distinct advantage over BUR systems. Like hot applied systems, it is usual to reinforce the membrane with a polyester fleece sandwiched between two layers of waterproofing. Some coatings require periodic refresher coats.

9.12 LIMITATIONS

Whilst both hot and cold applied systems are relatively easy to lay, this does not mean that they are tolerant of lower levels of workmanship. Most systems require good substrate preparation – clear of sawdust, swarf, dirt, hydrocarbons or organic materials, whilst the maintenance of the correct dry-film thickness is important to ensuring future durability. Similarly, many products have only a limited shelf-life before they deteriorate (six months or so).

From a designer's point of view, a failure to consider the usual principles of flat roofing design, rainwater disposal and the design of roof penetrations will lead to failure in much the same way as it would using more traditional methods. Similarly, because one of the great benefits of liquid membranes is their ability to bond to the substrate (and hence limit the propensity of water to travel through a sheathing layer) it is vital to ensure that the correct level of preparation is used and that the correct primers are employed. Compatibility with the substrate (and any likely contamination of the surface) must be established at the outset.

Figure 44 One of the benefits of liquid roofing is its ability to accommodate difficult detailing. However, it is no substitute for a failure to design details properly in the first place.

Liquid waterproofing systems – typical defects

Effect	Cause
Embrittlement of membrane due to loss of plasticiser. Regular pattern of fine cracks, material loses flexibility	Possibly due to lack of UV protection (hot applied system) or contamination.
Tearing, ruckling or splitting	Inability to deal with tensile stresses at places of concentrated movement, initiated, for example, by shrinkage in low temperature or creep during hot weather.
Splitting of membrane, lack of sufficient elasticity	Ability to bridge cracks – particularly those that open and close cyclically as a result of changes in temperature or structural movements. If a membrane is fully bonded, the percentage elongation over, for example, a 3mm crack is very high.
Puncturing	Debris or sharp materials dropped on roof, foot traffic, puncturing by plant growth.
Biological attack	Colouring/mottling, softening and degradation.
Loss of adhesion	Contamination of substrate with water, oil, sawdust, bitumen, etc. Incompatibility of background material.

10 Masonry facades

At a glance

- Since the early part of the 20th century, the advent of steel framed structures has resulted in the progressive reduction in weight of the external envelope. Masonry veneers are now much thinner (half brick instead of one or one and a half brick) or in some cases even thinner. The changes have resulted in facade design that is intolerant of movement or moisture and which requires high levels of workmanship to achieve effective performance.
- External brickwork falls into pervious or impervious categories, generally speaking impervious types are more prone to serious leakage as a result of water running down a facade than pervious types, which are able to absorb a certain amount of water and allow it to evaporate out over time. Walls constructed with dense, well burnt bricks demand a high attention to cavity tray detailing.
- In cavity work, cavity trays must be designed carefully and not left to chance. The design and installation must be treated with respect if water damage is to be avoided. Proper bedding, lap and stop end details must be provided, and trays installed at every point where the cavity is interrupted. Tray detailing should be as effective and meticulous as roof detailing.
- Movement in brickwork can be the result of moisture and or temperature changes, chemical problems (such as sulphate attack), structural movement (such as shrinkage of a concrete frame) or applied loadings.
- Cracking will usually occur in a direction that is perpendicular to the forces applied. Diagonal cracking is often (but not necessarily) indicative of structural movements whilst vertical cracking is often related to shrinkage or expansion.
- Clay brickwork undergoes an initial irreversible expansion after laying, followed by reversible temperature and moisture changes. Concrete and calcium silicate bricks have a certain irreversible shrinkage after laying.
- When used in conjunction with framed construction, care needs to be taken to ensure that adequate provision from frame shrinkage is allowed. Shrinkage of the frame due to long-term creep is a problem with reinforced concrete and timber structures; less so with steel.
- It has been estimated that the long-term creep of a reinforced concrete framed building can be as much as 3mm per storey height.

10.1 INTRODUCTION

Although masonry construction has been around for thousands of years and its performance is generally well understood, the gradual departure from heavy-weight construction to lightweight, framed structures with brick skins has brought about a range of problems that, until the twentieth century, were unknown.

The widespread use of steel in building from the early 1900s permitted a gradual shift away from the traditional heavyweight load bearing construction to one of thinner, lightweight methods – facades that were no longer load bearing (other than to support themselves) but which formed a weatherproof and decorative enclosure. In more recent years, with the advent of improved thermal insulations, brick skins have become even thinner, with brick slips or tiles fitted to insulated carriers becoming a popular alternative to conventional masonry.

These lightweight forms of construction suffer from various intolerances – intolerance to building movements, intolerance to water penetration and intolerance to poor workmanship. Many of the failure modes discussed in this book apply to domestic construction as well as commercial buildings; before investigating particular defects it is worth considering some of the properties of masonry and the structures to which it is attached. There are three principal considerations:

- water penetration;
- movement in the masonry; and
- movement in the structure.

10.2 WATER PENETRATION

The various mechanisms of water penetration are reviewed in section 15.3; they apply equally to water penetration through masonry. For example, water will not penetrate a masonry wall unless there is some mechanism to make it do so. Capillarity coupled with gravity and air pressure differentials will permit water to be drawn into the fabric, whilst surface tension allows water to flow over it and kinetic energy will drive shattered raindrops into any openings, pores or fissures. Broadly there are two construction principles that apply to brickwork:

- the overcoat principal; and
- the raincoat principal.

The overcoat principal

An overcoat works by permitting a certain amount of water to soak into its fabric until such time as it becomes fully saturated. At saturation point, water is rejected. In the case of a masonry wall, water gradually soaks into the brickwork as a result of kinetic energy, capillarity and gravity. If the overcoat is thick enough, water will not reach the inside face; it will gradually dry out again by evaporation. Bricks that have a high moisture storage capacity function in this way – the more the brick has been burnt the less absorptive it is.

The raincoat principal

By comparison with the overcoat principal, the raincoat provides a water resistant barrier on the external face. Glazed bricks or dense, well burnt engineering bricks act in this way, at least in theory. Depending upon the absorptivity of the surface and the rate of rainfall, a film of water can form and flow down the face of a wall. Surfaces of low moisture storage capacity readily become covered with a film of water that increases in thickness or volume flow toward the lower levels of multi-storey buildings. The flow of this film is influenced by surface texture, gravity and air movements along the wall face.

However, the downward migration of water tends to become concentrated at vertical irregularities in the wall surface. Surface depressions such as joints in brickwork are likely to attract flow rates many times greater than the average flow over the wall. Thus, while the fabric itself may be reasonably water resistant, the 'seams' in the fabric can allow water ingress, particularly if they have not been formed properly. Soldier courses of brickwork are particularly vulnerable because of the difficulty in battering the long faces of the brick with mortar and keeping it in place whilst the bricks are placed. Similarly, laying a damp-proof course (DPC) dry will create a multitude of water paths which can take water into the fabric by capillary action and pressure differentials.

In the author's experience, walls constructed from well burnt or engineering bricks, particularly with a smooth face, tend to suffer more severe leakage problems than brick of a more porous nature.

10.3 CAVITY CONSTRUCTION

In contemporary construction (and since the middle of the 20th century) external masonry has taken the form of cavity work, with a large air gap serving to break capillary action and movements of water due to gravity. The design of damp-proof courses or cavity trays had to be evolved in order to deal with the sometimes surprising amount of water that can and will penetrate a cavity – not just a small amount of leakage but in an exposed area many litres.

Regrettably, cavity tray detailing is often taken for granted, it being assumed that the contractor is familiar

Figure 45 Soft, porous brickwork in lime mortar. It will gradually absorb water and is therefore less likely to leak until the brickwork has become saturated.

Figure 46 Dense, well burnt engineering bricks; leakage into cavity is highly likely.

with the essential design requirements. Requirements such as cavity insulation and structural wind posts all serve to make careful detailing and workmanship essential. Construction drawings are often represented in two dimensions which makes an appreciation of the complex needs of a difficult joint hard to anticipate at design stage; the inevitable consequence is that the contractor muddles through during construction.

Most people would accept that a roof needs to be watertight, and that to be effective a gutter must have watertight joints and a means of draining water. Matters such as missing upstands, improperly formed joints and missing sections of roof membrane are all self-evident and would not usually be tolerated by a layman. Yet, when it comes to cavity tray detailing, the construction industry often fails miserably. Too often, one finds gaps, open laps or missing trays in what should be common sense construction. Water penetrating the cavity will move down the face of the outer leaf until it reaches an obstruction of some kind – a window opening, a floor or a shelf angle for example. To prevent that water travelling into the building the cavity tray has to serve as a watertight gutter and must have an effective means of disposing of that water. Cavity tray detailing is straightforward if one treats the tray as a watertight gutter – and treats it with respect.

BS 8215:1991[79] provides guidance as to cavity tray detailing, but there are so many permutations in building that it would be impossible to provide a library of standard details that will suit every eventuality. However, common sense has to apply – forget the outer leaf, if this was a roof how would you detail it?

Possible defects in cavity tray detailing include:

- **Sheet materials used instead of pre-formed components at corners:** Folding a sheet material to suit a cavity tray profile is easy when one is dealing with a straight run but forming a stop end or a corner is virtually impossible unless pre-formed trays are used. Often, stop ends are simply turned up from the same sheet material; however, this causes the tray to buckle with the risk that water can tip off the end of the tray.
- **Trays perforated by internal features such as windposts:** If one were detailing a roof, one would expect to see proper upstands around any features that passed through the covering – cavity tray detailing is exactly the same.
- **Lap joints unsealed or even unlapped:** Again, if one were making a roof watertight it would be usual to ensure that joints in the membrane are properly lapped and sealed to make them watertight, yet a failure to seal a cavity tray joint is a frequent occurrence.
- **No stop ends to cavity trays:** A gutter without a stop end will allow water to cascade down the face of a building in an uncontrolled yet concentrated manner. A cavity tray without a stop end is exactly the same, yet water is free to pass into the cavity in an uncontrolled manner. Stop ends should always be provided at the end of a tray – leaving water to cascade down inside the cavity until it meets a new tray is not a good idea.
- **Gable abutment trays incorrectly detailed; risk of water draining back into the roof void:** The detailing of trays at abutments is straightforward in conceptual terms but can be difficult to visualise properly on site. At a gable abutment the cavity tray will need to comprise a series of individual gutters each with a stop end and each overlapping the tray beneath. The stop ends must be arranged so as to occur at a point outside of the roof line. Alternatively, a fully jointed and sealed stepped arrangement could be constructed.
- **Cavity trays not bedded in mortar:** If the DPC material is laid dry onto brick or stone, the chances are it will create a myriad of capillary routes for water penetration, or larger passages if the material is not absolutely flat. However, failing to bed is a common fault, principally because it is difficult to provide a bed of fresh mortar, DPC and further layer of mortar within the confines of a standard 10mm joint.
- **Vertical DPCs and horizontal trays not lapped properly:** Vertical DPCs must lap into horizontal trays not behind them so that water can discharge into the tray.
- **Incorrect location of cover flashings at mono-pitched roof abutments:** The cover flashing should be positioned below the lip of the cavity tray or DPC and not above it to avoid water being discharged beneath the flashing.

Case study

The building is a purpose built office constructed during the 1980s. At top floor level there is a projecting balcony gutter and a masonry spandrel panel beneath half height ribbon glazing. The spandrel wall faces south-west in a moderately exposed location. A typical section is shown in the drawing below.

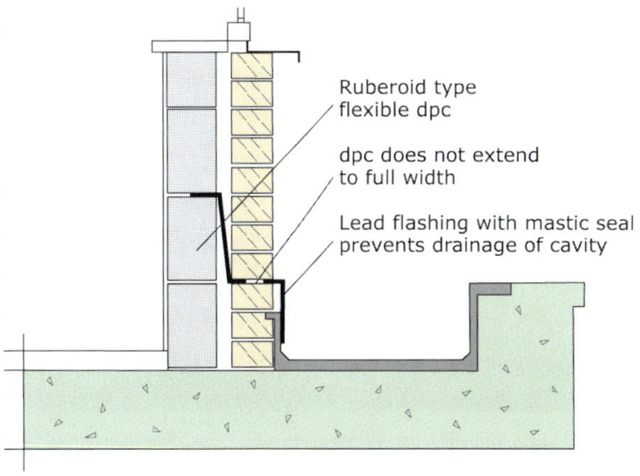

Figure 47 Sketch detail through spandrel wall.

Serious water penetration had occurred over the years since construction. The wall plaster had perished and the timber skirtings decayed. Given the proximity of the affected wall to the external gutter, the property manager had assumed that the asphalt lining to the gutter was defective and had arranged for the gutter to be lined with bituminous felt – the application of which was difficult owing to various projections and balustrade details. Attention was also given to the lead flashing and weep-holes, the latter being sealed with mastic following the assumption that water was being driven into the cavity during high wind. The joint between the flashing and the brickwork was sealed with copious amounts of mastic sealant.

None of the remedial measures had been effective which tended to suggest that it was the cavity tray and not the gutter that was at fault. The wall was therefore wetted over a period of three hours and then opened up so that the arrangement of the cavity tray could be determined.

The photograph below illustrates the findings of the investigation.

The well-intentioned remedial works had probably exacerbated the problem by restricting drainage, whilst the main problem was a poorly installed cavity tray. The joints between lengths of tray were not exposed, but given the standard of workmanship they must be highly suspect.

Several remedial options were considered including rebuilding the spandrel brickwork and installing a new tray or alternatively providing a rainscreen cladding system over the brickwork.

Figure 48 The bottom edge of the cavity tray stops short of the outer face of brickwork. The sealed weepholes and flashing prevent or at least restrict water drainage from the cavity which then is free to pass under the tray to the inside.

Figure 49 The lead flashing has been removed to expose the top of the upstand. Water trapped behind the flashing can only run down the face of the brick to be held by the poorly bonded felt against the bedding joint. The flashing had also been stuck to the upstand, further restricting drainage.

10.4 MOVEMENT IN MASONRY

Aside from movements created in masonry as a result of external influences such as subsidence, wind load, accidental damage, etc., brickwork is likely to suffer dimensional changes resulting from:

- changes in temperature;
- changes in moisture content (usually bringing about a volume change);
- changes in chemical composition; and
- changes in loading or structural movements.

Different types of brick display different movement characteristics; clay bricks tend to suffer an initial irreversible expansion after laying, whilst calcium silicate and concrete bricks tend to shrink. Understanding the movement characteristics of different masonry types is essential in diagnosing failure patterns.

10.5 CHANGES IN TEMPERATURE

With a few exceptions, most construction materials expand in response to an increase in temperature. Exposed surfaces are subject to the effects of solar radiation which tends to increase surface temperatures above ambient air temperatures.

Temperature ranges can be diverse, but for Europe a range of −25 to +80°C could be reflective of typical in-service surface temperatures. (UK temperature variations are less onerous – a range of 85 degrees is more appropriate for dark, heavyweight claddings).

BRE Digest 228 Part 2 (1979) deals with the estimation of thermal and moisture movements and stresses. Under that guide BRE advocate the use of a simple formula to determine linear changes of size (note that this does not simply apply to brickwork):

$R = \acute{\alpha} L t$

Where,

R = change of size

$\acute{\alpha}$ = coefficient of linear thermal expansion

L = length

t = temperature difference

The coefficient of linear thermal expansion of brickwork is typically in the range 5–8 $\times$ 10^{-6} per °C[80]. Thus, assuming that the brickwork was laid at a temperature of 10°C in a panel 6,000mm in length, theoretical expansion of 3.36 mm (70 $\times$ 6000 $\times$ (5 $\times$ 10^{-6})) would be possible along with a potential contraction of 1.68mm. Factors such as the texture and colour of the brick would affect the degree of movement. In practice, some measure of restraint is to be expected and it is probable that movements would also be accompanied with dimensional changes brought about by changes in moisture content.

Expansion in brickwork tends to occur uniformly in each axis (unlike some building stones which can display anisotropic tendencies).

Calcium silicate bricks exhibit thermal movement of about 1.5 times that of clay brickwork, the co-efficient of linear thermal expansion being about $8–14 \times 10^{-6}$ per °C. In restrained conditions, this movement may be less.

10.6 CHANGES IN MOISTURE CONTENT

An increase in moisture content of a material is usually accompanied by a volumetric change. There are two types of movement that need to be considered:

- irreversible or permanent changes (usually occurring during the early life of a component);
- reversible changes.

Irreversible changes usually affect clay brickwork because during the burning process moisture is removed from the brick. Once the brick cools the moisture content will gradually increase until it reaches equilibrium with its surroundings. Most movement of this kind occurs within the first three months of the life of the brick.[81]

Reversible changes are cyclical and take place as a result of daily or seasonal changes in moisture content.

Unlike the temperature based changes of most materials, movements due to moisture change are not necessarily the same in each axis. Furthermore, moisture changes are not necessarily uniform throughout the thickness of a material – the outer surface may be saturated whilst the inner surface remains dry, thus imparting a moisture and stress gradient through the material.

According to BRE Digest 228 Part 2 (1979) an estimation of the percentage increase (or decrease) is given by the expression:

$$\frac{\text{Factor} \times \text{dimension}}{100} = \text{R (change in size)}$$

Factors are listed in the Digest and for brickwork, irreversible expansion is shown as +0.02 to +0.07, whilst reversible moisture movements are shown as 0.02. So, for a brick panel of length 6,000mm, the possible change in length occurring between extreme wet and extreme dry conditions could be:

- $\frac{0.07 \times 6000}{100} = 4.2\text{mm}$ (irreversible expansion)
- $\frac{0.02 \times 6000}{100} = 1.2\text{mm}$ (reversible movement).

The example above assumes the highest figure given for brickwork in Digest 228; clearly if the lower factor (0.02) is used a smaller irreversible expansion will be estimated.

Calcium silicate brickwork, unlike clay, usually undergoes an initial irreversible shrinkage on laying (clay brickwork tends to expand) but so long as the propensity for movement is understood and catered for in the design, there is no reason why the brickwork should not perform adequately. Calcium silicate bricks often exhibit reversible movements of between 0.02–0.2 per cent and irreversible shrinkage of 0.02–0.06 per cent.

Calcium silicate bricks should not be used in solid work with clay facings or backings, this is because of the propensity of the bricks to shrink in contrast with the expansion of clay brickwork. If solid walling is to be contemplated, backings of concrete bricks or blocks should be used, as these have similar movement characteristics to calcium silicate bricks.

In cavity work, if differing materials are used, and providing the cavity is in excess of 50mm, there should be sufficient flexibility in the ties to permit the differing movements. However, care must be taken to ensure discontinuity around cavity closers to prevent cracking.

The estimation of dimensional changes in brickwork is rarely as straightforward as making the calculations described earlier. For one thing, expansion due to rise in temperature may be partially offset by a reduction in moisture content and subsequent shrinkage. Furthermore, movement in brickwork is often restrained, and considerable frictional restraint can be provided by mortar bed joints. Since masonry is weak in tension, cracking can occur instead of contraction of the whole panel. However, the calculations do produce a guide as to what is likely to be the dominant movement effect, and using this information can help to explain movement or crack patterns in brickwork. If movement is arising from distinctly separate causes, the total movement can be assumed to be the sum of the moisture and thermal movements.[82]

10.7 CHANGES IN CHEMICAL COMPOSITION

Arguably, a change in moisture content is a change in chemical composition; similarly the effects of ice forming in the pores of brickwork may be considered under this category. However, the agencies of concern are more likely to be:

- sulphate attack;
- corrosion of metals built into brickwork; and
- crystalline salt formation.

Expansion forces arising from these agencies cannot be calculated in the same way that moisture and temperature movements can; they are unpredictable and will continue for as long as the circumstances that bring about the change exist. The forces can be very large and can cause significant damage. Generally, one would not design movement control joints to deal with expansion due to chemical changes; it is far more appropriate to design out the risk of these changes occurring.

10.8 SULPHATE ATTACK

Sulphate attack is a chemical reaction between solutions of soluble salts such as potassium, sodium or magnesium and a component of lime or ordinary Portland cement. The reaction converts tricalcium aluminate into ettringite, which has a greater volume therefore resulting in the expansion and disintegration of mortars. Sulphate attack can also affect concrete floor slabs or foundations in certain circumstances.

Some bricks, for example Flettons, have a high sulphate content. If held in permanently wet conditions, the sulphates react with ordinary Portland cement used in mortar, causing the mortar to expand and disrupt the brickwork. The effects can be particularly damaging where brickwork has been rendered as the render restricts the evaporation of moisture.

Sulphate problems are likely to occur where:

- the brickwork can remain wet for long periods (for example a rendered parapet wall);
- permeable Portland cement mortars (i.e. weak mortars) have been used;
- no sulphate resisting cement mixes have been used in the mortar;
- bricks with a high sulphate content (> 5%) are used;
- calcium aluminate content of the cement is above eight per cent[83]; and
- bricks are not isolated from other sources of sulphates, for example groundwater.[84]

Typical risk areas could be:

- exposed parapets and brick on edge copings;
- brick on edge cills;
- retaining walls subject to groundwater; and
- brickwork below DPC level.

Sulphates can be produced as a result of the oxidisation of pyrites and other sulphides – the Mundic problem.

Mundic is the old Cornish name for pyrites, a mineral that is found in abundance in some of the geological deposits in Cornwall and some parts of Devon. In the early 20th century, waste products from the areas mining industries (tin, copper, lead) were used to provide a cheap and local source of fine and coarse aggregate for the manufacture of locally produced concrete blocks and in-situ concrete.

Some of this waste material contained unstable sulphides (pyrite being one of the main culprits), which deteriorated after the concrete became carbonated and also in damp conditions. There are two main mechanisms of deterioration, but both are based around the oxidisation of sulphide minerals in the aggregate resulting in the decomposition of the concrete and its expansion.

Parallel cracking along rendered block bed joints or map cracking of render finishes to in-situ concrete are a visual indicator of a problem. Rendered surfaces reduce the rate of carbonation and so it is possible for cavity walls to deteriorate from the inside out meaning that by the time cracking and other visible manifestations of a problem have occurred the concrete is badly deteriorated.[85]

Figure 50 Brick parapet wall suffering from sulphate attack; the mortar has expanded causing characteristic horizontal cracking. The render may have been applied to help repair the brickwork but has probably made matters worse. Reconstruction is the only practicable solution.

10.9 CARBONATION

Airborne carbon dioxide reacts with sodium, potassium and calcium hydroxide in cement based products gradually converting the product from a strongly alkaline material to a weakly acid state, a process known as carbonation. Whilst carbonation has more serious effects in reinforced concrete (see chapter 3), the loss of alkalinity in mortars for example can permit corrosion to occur in ferrous metals that may be embedded.

The effects of carbonation on concrete and calcium silicate bricks can be difficult to distinguish from normal shrinkages but nonetheless, carbonation induces volume changes in porous components such as lightweight blocks (autoclaved aerated products) within the first few years of the component's life.

Dense concrete blocks are affected by carbonation, but in these materials the rate of carbonation is much slower.

10.10 UNSTABLE AGGREGATES

Particles of calcium oxide and sometimes magnesium oxide are often formed during the burning process in clay brick manufacture. These particles can react with moisture to form calcium or magnesium hydroxide, a process that generates expansive forces sufficient to cause local spalling of the brick with characteristic 'lime popping' – a crater like depression with a white deposit at its centre. The effect is primarily cosmetic.

10.11 CRYSTALLISATION OF SOLUBLE SALTS

Efflorescence is the process of the dissolution of salts by moisture and their deposition on the surface of a material as crystals as it dries out. Severe efflorescence produces a white bloom on brickwork which is unsightly.

Figure 51 Example of crypto-efflorescence causing the brick faces to spall off (Photo courtesy of Steve Brewer.)

Where the efflorescence occurs below the surface of a material it is termed crypto-efflorescence; this is more damaging as it can cause face spalling of brickwork rather like damage arising from frost action. If the salts crystallise below the surface (within the pore structure) they exert a tensile force which is strong enough to disrupt the surface of the brick. The salts of sodium or calcium are relatively harmless, but potassium salts can be very damaging to brickwork when they occur in a crypto-efflorescent form. The problem is sometimes associated with rapid drying of water which causes the salts to crystallise. The source of salts could be from external influences such as sea spray, from the coal used to fire the bricks, contaminated mortar, etc.

10.12 CORROSION OF METALWORK

The corrosion of metal fittings, ties or other fixings can seriously disrupt brickwork or other porous materials in which they are embedded. The main problem is the formation of hydrated iron oxide (rust) which is accompanied by an increase in volume of around 400 per cent. One of the most common forms of failure is the corrosion of galvanised steel wall ties.

The earliest wall ties were fabricated from wrought iron, but various types have been used over the years including painted mild steel, bituminised mild steel, galvanised mild steel, stainless steel and plastic generally of the vertical twist, double triangle or butterfly type.

In 1945, BS 1243 was introduced. The standard covered cavity wall ties and required a minimum zinc coating weight for wire ties of 305g/m^2 and 610g/m^2 for vertical twisted types. In 1964, the standards were reduced to 275g/m^2 and 460g/m^2 respectively. The life expectancy of bitumen and zinc coatings on these ties is frequently well under the sixty years that was originally predicted. In fact, in 1981 BS 1243 tripled the minimum allowed zinc coating thickness on wire ties.

Whilst the reduction of galvanising standards under BS 1243 led to some failures, it is by no means certain that many failures were simply the result of non-compliant galvanising in the first place. Zinc coatings as low as 45g/m^2 are not unheard of. However, research work undertaken by the BRE suggests that average zinc loss is between 1.4–2.8 μm per year[86]. For pre-1978 ties this results in a predicted coating life of 13–26 years for wire ties and 23–46 years for vertical twist ties. Bitumen coatings have been found to last equally as well as some of the thicker zinc coatings. On the inner leaf, where the circumstances are less aggressive, the zinc coating can be expected to last much longer.

If wire ties have been used these have the unpleasant tendency of corroding away without any substantial physical disruption to the brickwork. Damage becomes manifest by the sudden collapse of an outer leaf particularly in conditions of high wind; this is why areas of particularly exposed wall (as described above) should be inspected as a priority.

BRE Information Paper 13/90 sets out useful guidance in terms of the buildings most likely to be a risk from wall tie corrosion. The paper, together with the later BRE Digest 461, paints a fairly pessimistic view of the life expectancy of galvanised ties generally.

Areas such as the south, north-west and north east coasts of England and the south coast of Wales are particularly vulnerable to corrosion problems in view of the harsh marine climate and the propensity of the walls to stay wet for long periods of time. Investigations have shown that in many cases the embedded faces of the tie will corrode first, with the outer leaf, followed by the inner leaf, showing signs of deterioration, sometimes before corrosion is evident on the portion within the cavity.

Also of particular risk are the following types:

- buildings constructed using black ash mortar;
- buildings located in areas of higher than normal exposure;
- buildings constructed in the period 1900–1940;
- post-war buildings constructed using poorly galvanised ties – especially during the early 1970s;
- buildings constructed before about 1955 with vertical twist ties;
- buildings constructed before about 1970 using galvanised wire ties; and
- timber framed buildings constructed before 1975 using galvanised frame ties.

Buildings constructed from black ash mortar are known to be particularly at risk from the corrosion of wall ties. Black ash mortar is generally composed of a mixture of hydrated lime and finely ground ash or coal dust. Black ash mortar is commonly found in the Midlands, industrial north and south Wales. Initially highly alkaline, the material soon loses its alkalinity as a result of exposure to atmospheric carbon dioxide and other acidic atmospheric gasses. Moist ash increases corrosion rates, particularly if it contains significant amounts of sulphides – these oxidise into sulphates. The presence of chlorides in the ash will accelerate corrosion rates yet further.

Walls most likely to be at risk will be in exposed situations, with frequent wetting or saturation, defective pointing or other similar defects – perhaps identifiable in dry weather by staining or moss or algae growth. Certain parts of walls may also be at risk from failure during high winds. In particular, large gable walls, large areas of cladding and narrow unreturned strips of brickwork should be examined as these can collapse given suitable exposure – as happened during the 1987 and 1990 gales in the UK.

Factors affecting the rate of corrosion

Moisture plays a significant role in the rate of corrosion although it is not so much the quantity as the duration of the moisture that will have an effect. In dry air, a film of zinc oxide will form on the surface of a zinc coated tie, mainly as a result of the reaction of zinc with atmospheric oxygen. The zinc oxide converts to a hydroxide in the presence of moisture and in turn this converts to zinc carbonate through reactions with atmospheric carbon dioxide. The zinc carbonate is very protective and reduces the rate of corrosion considerably. Up to about 70 per cent RH the corrosion of zinc is little affected by humidity but beyond 70 per cent there can be a marked change as a result of the formation of an invisibly thin layer of moisture on the surface. The moisture functions as an electrolyte and forms a corrosion cell.

Rainfall performs very much as humidity in terms of the formation of a surface layer; it also serves to wash away any surface deposition that might, in itself, have harmful consequences. However, rainfall tends to dry out and as long as the zinc is not held in contact with moisture the effects are not severe. It is in effect the degree and duration of wetness that is the governing factor.

Exposure to salt spray in coastal locations can lead to contamination by chlorides, although investigations have shown a marked drop in chlorides a kilometre or so back from the coast. Chloride ions are highly conductive and can form aggressive corrosion cells. However, except in cases where the metal is in regular contact with salt spray, the effects of chlorides on galvanised zinc are probably overestimated. High exposure probably carries with it the same risks as exposure to an industrial atmosphere, although opinions are mixed on this point.

Sulphur dioxide is one of the most harmful atmospheric pollutants and can alter corrosion rates drastically in test specimens. Sulphur dioxide is present as gaseous emissions in the atmosphere – primarily from the burning of fossil fuels. The gas is dissolved in rainwater and distributed as particulate matter. Contamination can have the effect of reducing the pH of the mortar and thus preventing the formation of a passive film.

Of course, mere depletion of the zinc does not mean that the component will fail immediately; merely that the base metal is no longer protected from corrosion. The rate of corrosion will depend more or less on the factors listed above and the amount of oxygen, moisture and electrolytes present.

Visual indicators

Unless wall ties have failed completely, there is often very little visual evidence as to the possible failure of individual ties or a patch of ties. However, indicators such as bowed or cracked brickwork (possibly horizontal cracking at around 450mm centres) and random bulges can be indicative of a problem. Corroded steel occupies about four times its normal volume and so if vertical twist ties are used, corrosion of these usually becomes manifest in the form of cracking, either through brick joints or reflected in rendered surfaces. Cracking may be more pronounced towards the top of a wall.

Expansion of an outer leaf as a result of cavity tie corrosion can sometimes be restrained by the roof construction, in which case bulging of the leaf is likely to occur.

Twisted wire ties are much thinner in section; corrosion of these is unlikely to lead to the same volumetric expansion as vertical twisted types. In these circumstances, cracking may not be apparent.

Detection

Wall tie detectors can be used to attempt to locate the position of ties. Tie detectors comprise what is in essence a hand held metal detector with interchangeable heads to scan for ferrous or stainless steel. Care in the use of this equipment is necessary as signals can sometimes be very weak and it is important to try several scans at different intensities before drawing any conclusions. Sometimes, ferromagnetic or paramagnetic particles can give misleading results and scanners with large coils can make it difficult to locate the tie accurately. A great deal of effort can be expended attempting to determine reliable locations with little practical result.

Cover meters can sometimes be used to give a guide but they are calibrated to find steel running parallel with the face of an element and may not give satisfactory results.

A more reliable assessment can be achieved by using portable pulsed radar equipment, although such a form of diagnosis involves specialist equipment that is not normally hired. Specialist testing consultancy practices are best able to provide this type of service. A hand held scanner is passed over the wall; this identifies any major discontinuity of dielectric constant in relation to the antenna surface and thus enables the position of the cavity and the position of the ties to be located with reasonable certainty.

Experiments with infrared photography have met with some measure of success, although reviews of this form of technology are mixed and the picture can become confused. Equipment can be hired reasonably easily and utilised with a minimum of training. The system enables a fairly rapid assessment to be made; this can then be validated by means of physical opening up or metal detection.

A visual examination using a fibre optic probe or a physical opening in the brickwork will give a better indication. The problem with fibre optic probes is that they do not necessarily reveal whether the tie has corroded – the area most at risk is that section buried in the wall.

The placing of ties is sometimes erratic, but will depend upon the thickness of the cavity. For an outer brick leaf of standard thickness, and a cavity of 50–75mm, ties should be placed at 900mm centres horizontally and 450mm centres vertically in a staggered pattern. If the cavity width is increased to 100mm, horizontal centres should be reduced to 750mm. For a cavity width between 100–150mm, ties should be at 450mm horizontal centres.

Replacement strategies

BRE Digest 329 (2000) sets out very useful data in terms of the selection of suitable replacement strategies for wall ties and includes a decision making tree to assist the process.

There are many techniques for the replacement of defective ties, involving epoxy, grouted or expanding type bolt systems. Generically they can be summarised as follows:

- mechanical expansion types;
- resin/grouted anchors;
- resin/grouted and mechanical combination ties;
- friction fixed ties; and
- inflated sock ties.

Care must be taken however to select the correct remedial method, as some of the expanding types cannot be used in lightweight inner block leaves. Pull out tests should invariably be carried out to ascertain the capacity of the inner leaf. It is also important to provide replacement ties with a suitable level of flexibility in situations where long term creep of the frame is likely to occur. For example, fitting rigid ties to a timber framed building could have serious effects.

It is important also to remove the remains of the old ties as further corrosion in these could create further disruption to the brickwork.

Polyurethane foam

One method of repair identified by Digest 329 is the use of polyurethane foam injected into the cavity. A suitable polyurethane foam system may be injected into a masonry cavity wall by injecting liquid polyurethane through properly spaced holes in the outer leaf. The foam expands in the cavity and sets to become substantially closed-celled, cross-linked, rigid polyurethane foam which strongly adheres to the inner and outer leaves, so that they become bonded together.

The foam also serves the dual purpose of increasing thermal insulation values and because it is substantially closed cell, does not take up water.

The foam stabilises a wall by adhering to the surfaces of the cavity and providing a continuous structural connection between the two leaves. It has the ability to form a good bond with surfaces that are unprepared and in a raw state. The adhesive bond strength is better than $60kN/m^2$ which is generally in excess of the wind suction likely to occur on the wall.[87]

Stabilisation and thermal insulation of cavity walls is covered by British Standard Code of Practice, BS 7456:1991 *Code of practice for stabilisation and thermal insulation of cavity walls (with masonry or concrete inner and outer leaves) by filling with polyurethane (PUR) foam systems.*

Foam filling is not suitable for timber framed or metal stud construction.

10.13 CHANGES IN LOADING OR STRUCTURAL EFFECTS

Forces applied to a material create stress or pressure within that material; the amount of stress depending upon the magnitude of the loads. Stress is expressed as force/area and measured in N/mm^2.

Under load, most building materials deform to some degree, although the magnitude of that deformation can be very small. The physical extension or shortening of a material under stress is called strain. Strain (e) does not have units as it is a ratio of change of length to original length, but it is sometimes expressed as a percentage or in units such as mm/mm. Because these movements can be very small the term 'microstrain' (me) is sometimes used, which is $e \times 10^{-6}$.

Within a range known as the elastic limit, materials will return to their original size once the loading is removed. Within this limit, the deformation due to load is proportional to the stress induced by the load. Thus, it is possible to express the properties of a material according to its stress and strain ratio or 'modulus of elasticity' (E):

$$E = \frac{\text{stress}}{\text{strain}} \quad \text{i.e.} \quad \frac{\text{load per unit area}}{\text{deformation per unit dimension}}$$

$$\text{Strain} = \acute{\alpha}\, t$$
$$\text{Stress} = E \times \acute{\alpha}\, t$$

Where,

E = modulus of elasticity
$\acute{\alpha}$ = coefficient of linear thermal expansion
t = temperature range

So, knowing the properties of the materials in question enables not only the change in length to be calculated for temperature and moisture changes, but also the stresses that will be generated by those changes. If the calculated stresses exceed the ultimate tensile strength of the brickwork, cracking will result.

Earlier in this chapter we considered the effects of dimensional changes resulting from temperature, moisture and chemical changes. However, long term loadings on materials can produce additional deformations, i.e. movements that do not return elastically. These deformations are termed creep (or sometimes plastic flow) deformations and characterise many standard building components, particularly concrete.

Creep can be defined as the gradual increase in strain with time. All columns and walls, irrespective of the material from which they are constructed, will shorten due to loading from above. In multi-storey construction this can be a significant consideration. Again, creep will increase the shortening of concrete columns. For most buildings the elastic shortening of steel columns is sufficiently small to be ignored. In timber columns axial creep is negligible and is not normally considered in design.[88] The actual level of creep depends upon many factors, including relative humidity, the ratio of surface area to volume, the type and quality of cement, the amount of water present, the stress in the concrete and the age of loading.[89] If loading is applied before it has reached its 28-day strength, creep deflections can be appreciably more.

Early framed buildings tended to be of heavy beam or girder construction which, with heavy brick or stone claddings, were able to withstand the effects of gradual shrinkages in the frame without damaging the fabric. However, advances in design lead to progressively stronger forms of construction with the ability to sustain higher stresses. Thinner, more highly loaded members were the result, often with increased flexibility. Such forms of construction revealed the effects of the combined problems of creep and shrinkage both axially in columns and horizontally in the form of increased deflections.

Buildings are not static; they move constantly. Concrete frames in particular do not stop moving once they have been completed, but continue to move for many years as a result of the effects of drying shrinkage and creep. In horizontal members the total deflections will be the sum of creep deflections for permanent loads, elastic deflection for transient loads together with the effects of temperature change and shrinkage.

Creep is not peculiar to concrete frames; masonry can suffer as well, although the effects are not so well understood. Research by CIRIA suggests that creep in walls seems to be somewhat greater than in columns.[90]

From the above it can be seen that unless measures are taken to permit movements, the combined effects of all of these forces may generate failures. If a component is free to move without restraint, all well and good, but in building terms there is usually some form of restraint whether this be intentional or not. In an unrestrained condition, deformations will not result in cracking, but where restraint is present, deformations may exceed critical levels for stress or strain with the result that cracking or other forms of movement occur.

In addition to the natural properties of materials, the question of building inaccuracy must not be ignored. In building terms there is no such thing as perfection; it is simply not possible to construct something to the degree of precision that eliminates the need for constructional tolerance. For one reason, many materials are natural products that change shape and size according to the factors listed in the table on page 1; for another, the practical difficulties of assembly and achieving exactness cannot be underestimated. However, whilst a designer might consider the need to provide for elastic and long-term movements in a building, he/she must also take into account reasonable allowances for building tolerance. A designed gap of 15mm to allow for expansion could turn out to be an actual gap of 20mm or 2mm depending upon the materials used. Thus, the effects of expansion in a clay brick facing could become pronounced if a movement joint did not have the required capability.

Creep in steel framed structures is usually negligible; the same would apply to concrete. However, for reinforced concrete the movements can be significant.

Depending upon the aggregate used, the shrinkage of concrete can be estimated to be between 0.066 per cent and 0.085 per cent (660–850 microstrain). Loading of around 10N/mm^2 produces creep of around 300 microstrain therefore over a 3m high storey height the combined shrinkage could be a shade under 2.9–3.5mm. The effects of temperature and moisture must also be considered. A 3m high panel will exhibit around 1.5mm of vertical movement due to expansion plus around 2.7mm for irreversible and reversible moisture expansion. Thus, expansion is probably greater than contraction of the frame and

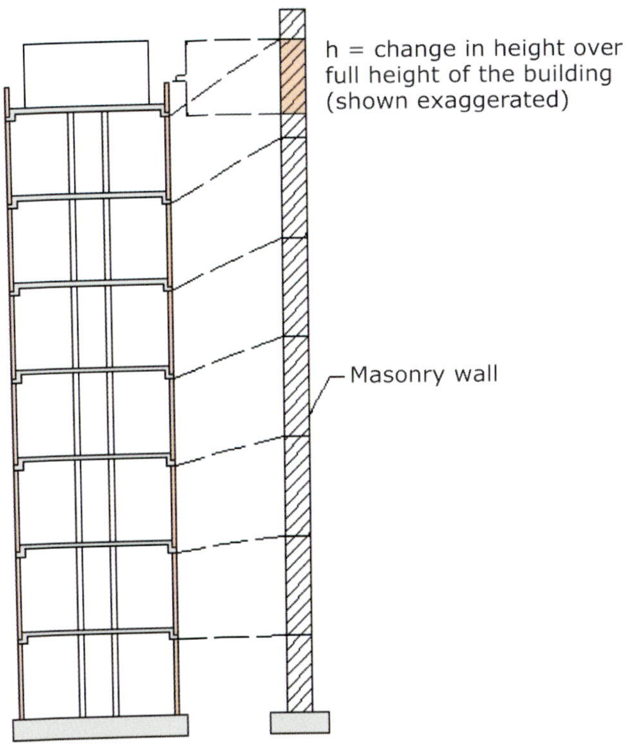

Figure 52 Drawing illustrating the cumulative effects of frame shrinkage over the height of a building

whilst a bitumen DPC has some 'give' it will not in itself be able to take up the difference. Estimating movement for joint width design is notoriously unreliable and usually results in overestimation.

Little data are available on actual in-service measurements of building shrinkages. According to BDA Technical Note 9,[91] one American source quotes shrinkage of 1 inch per 100 feet of height – 25mm per 30.4m. As a rule of thumb, 1mm per metre of height would be a reasonable estimation.

BDA Design Note 10[92] provides a list of useful rules of thumb that will cover a majority of situations, whilst various recommendations are also given in CIRIA Practice Note 44[93]; unfortunately the two documents provide conflicting advice. BS 5628 Part 3[94] also gives guidance as to the location of movement joints. Taken together a fairly conservative set of rules of thumb are shown below:

- For restrained walls of clay brick, allow a minimum width of movement joint of 15mm; joint spacing should never exceed 12m.
- For clay brick walls that are lightly restrained or unrestrained (for example spandrel panels or parapets) the distance between joints should not exceed 7–8m.
- For calcium silicate bricks, allow shrinkage control joints at 7.5–9m centres; the aspect ratio of panels must never exceed three. If greater than three provide joints to form smaller panels.

The expansion and contraction of brickwork is similar in both the vertical and horizontal dimensions. Horizontal movements can be reduced by providing bed joint reinforcement.

In relation to cladding (whether this be brick or other forms) the effects of long term creep must be taken into account during design. Typical problem areas can be illustrated by an examination of:

- brick support knibs or shelf angles;
- brick slips; and
- masonry infill panels.

Brick support knibs

In multi-storey construction it is usual to support the external leaf of a cavity wall at each storey usually by means of a shelf angle or (less common in contemporary construction) a concrete knib designed to carry loads back to the structural frame. Without this support, the weight of each storey would be carried by the masonry below it, inducing excessive compressive loads.

Although simple in theory, the design and installation of support angles has, over the years, provided fertile ground for failures and disputes.

From earlier in this chapter, we know that the supported brick panels will be subjected to certain movements – expansion or contraction due to moisture and temperature as well as long term creep. The loads so generated must also be taken by the support. However, the structure to which the support is fitted will, (assuming it is of concrete) also suffer from long term creep as well as deflections. Expansion of brickwork and contraction of the frame may well work against each other; the forces so generated can be sufficient to shear concrete knibs in the manner illustrated below.

In the drawing, a lack of provision for movement at the top of each masonry panel has also led to load transfer between panels.

The alternative to an in-situ concrete knib is a steel angle, usually of stainless steel to prevent corrosion problems. Whilst it may seem obvious, the angle must be properly supported to reduce deflection and to transfer loads to the frame. However, it is not unknown to find deficiencies with support, and seemingly innocuous evidence of local disturbance in brickwork can be indicative of serious problems. The adjacent case study illustrates a particular set of defects that had significant repercussions – the need to rebuild the entire brick skin.

Careful attention must be given to shelf angles, they must be capable of adjustment to make up for construction tolerances. In the case study described above, the combination of large radius to the angle and deviations in the line of the ring beam meant that the bottom course of brickwork could not fit properly.

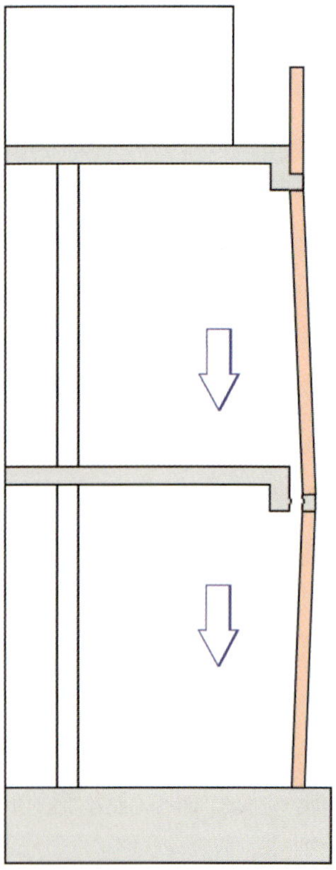

Figure 53 Shear forces resulting in knib failure.

Case study

The building is a 1980s office in the south-west of England, comprising around 7,000m² of accommodation arranged on a ground and six upper floors. The frame is of in-situ reinforced concrete. The external walls are clad with clay brickwork constructed in faceted panels and with a corbelled parapet at roof level.

An inspection from ground level revealed isolated cases of brick spalling below horizontal movement joints at each floor level. The movement joints appeared to be pointed with sealant in the usual way and so the managing agent's initial conclusion was that there may be localised bridging of the soft joint with mortar. Although not an unreasonable suggestion, further investigations were undertaken; these revealed a series of potential problems best illustrated in the drawing shown below.

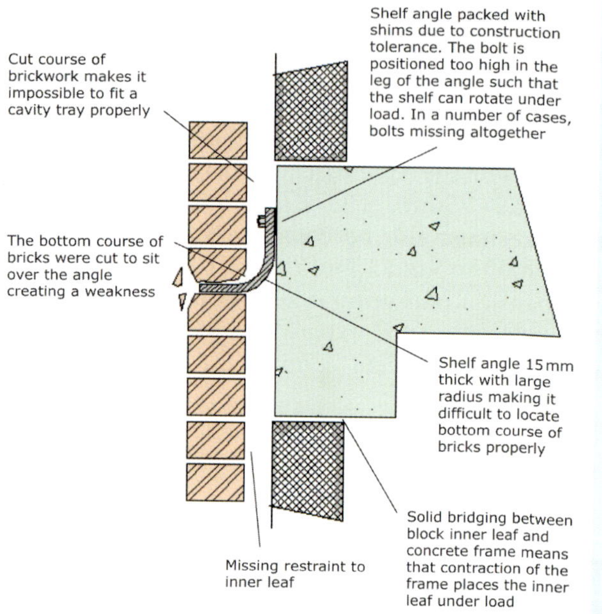

Figure 54 Steel shelf angle detail.

The construction exhibited several faults, some of which were the accidental consequence of poor design, whilst others were more wilful. First, the soft joints were found to be nothing more than a layer of mastic sealant applied over a mortar joint, the arrangement permitting load transfer to adjoining panels of brickwork and the immediate cause of the spalling edges of individual bricks. However, and more significantly, the angle supports themselves had not been attached to the frame properly. The angle was designed to be bolted back to the concrete at regular centres, but intermediate 'bolts' were found to be nothing more than the head of a bolt carefully cut off and stuck onto the steel with epoxy resin. Deflections in the angle could therefore transfer loads to the masonry below causing overstressing. In one location the brickwork was effectively unsupported to virtually full height.

The inner leaf of the enclosing cavity wall was of lightweight concrete blockwork built off a ring beam at the perimeter of the building. The blockwork was rigidly pointed to the underside of the ring beam so that any shortening of the frame would place the blockwork under compression, potentially causing the inner leaf to bow.

The designer selected a thickness of stainless steel for the angle support that was of unusual thickness. The angle had been roll formed with the radius of the roll being greater than the 50mm cavity width. As a result the bottom course of brickwork would not sit on the angle properly, so the bricks had been hand cut with the edge of a trowel leading to an irregular bedding width. The rough cutting weakened the brickwork and made the bottom course vulnerable to rotation and spalling. Poor positioning of the support meant that the bricks also had to be cut to achieve proper coursing.

Given the severity of the support problems, the only practicable solution was the removal and replacement of the entire outer skin of brickwork.

Figure 55 A prism of cracked brickwork at an external corner resulting from support angles being stopped short of the corner (see figure 56).

Problems can occur if the shelf angles do not extend to the full width of the masonry panel and are cut short of the corners of the building. If this happens, the masonry then bears onto both the angle and the masonry below. The forces generated by expansion of the brickwork are then carried partly by the support (and thence back to the frame) and partly by the masonry. This condition creates a shear force which, if greater than the strength of the masonry will cause vertical shear cracks at the corner – a problem that could be confused with expansion movement. Given that the positioning of ties at corners is undesirable, the shear cracks could create a tall, unbonded prism of brickwork to become unstable.

10.14 BRICK SLIPS

The alternative to support by brick nibs or support angles is to continue the floor beam through to the face of the brickwork or to stop it short of the face by 50mm or so and to tile it with slips of cut bricks. Such construction was very popular in the 1970s.

Failures in the adhesion of brick slips had lead to an expression of caution over the use of this material. For several years, brick slips were included in lists of deleterious and hazardous materials simply for the fact that slips could become detached from a building and cause a safety hazard.

The use of 20mm brick slips became common in the 1970s as a means of disguising the edge of a floor slab or downstand beam in situations where an expressed concrete frame was considered to be undesirable. The slips were often bedded on polymer modified mortar and attached as tiles or in some cases had additional mechanical fixings (although to begin with this was rare).

A failure to accommodate movement in the frame and brick cladding can lead to serious failures. The mechanism of failure mainly involves shrinkage of the concrete frame and expansion of the clay brickwork. A failure to provide a soft joint between the normal courses of brick and the brick slips could lead to the transfer of forces to the thin slips which in turn are then forced off the building. Even the provision of soft joints could be problematic if they are locally of insufficient thickness or have become bridged by mortar.

The bonding of slips to the concrete frame could be problematic if there was insufficient key, or if the face of the concrete was contaminated with mould oil or other contaminants. Similarly, if the slips were cut on site, it was very important to remove brick dust from the bonded face to ensure that a suitable bond was achieved. The use of PVA building adhesives was unsatisfactory since PVA is water soluble; SBR adhesives may have given a better performance. Epoxy based mortars were expensive and so during the 1970s were unpopular. Furthermore, variability of the line of the concrete frame could mean that the backing adhesive

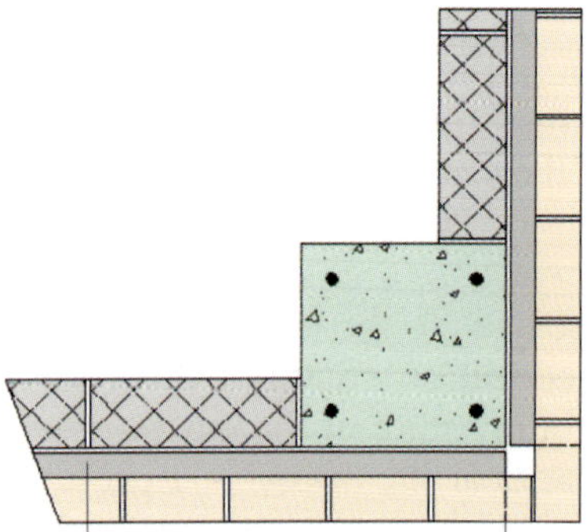

Figure 56 Illustration of potential cracking due to shelf angles being too short.

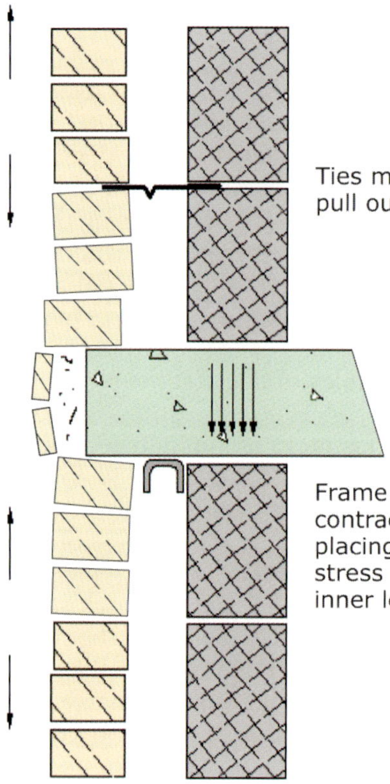

Figure 57 Brick slip failure due to contraction of frame and expansion of brickwork.

Figure 58 Brick slips spalling off; note outward bowing of remaining slips to the left of the photograph.

would need to be of varying thickness, meaning that some slips would need to be supported temporarily whilst the adhesive set. More importantly, drying shrinkage resulting from varying thickness could create planes of weakness.

If the expected life of the external fabric could reasonably have been predicted at 60 years it is doubtful whether any adhesive based systems could have given satisfactory performance for that period of time. Indeed, one current manufacturer of polymer modified adhesives predicts a life of around 20 years.

Given uncertainty over the life expectancy or effectiveness of adhesive based systems, some form of mechanical fastener could have been expected to give better performance. One type of fastener comprised a stainless steel clip which fitted into a slot in the top edge of a floor slab and then hooked into a small slot in the bottom edge of a slip. Another alternative was to provide a flat metal plate beneath the slips to deal with the effects of gravity and to use a mortar or epoxy modified mortar although this is subject to the limitations expressed above.

10.15 BRICK PANELS

Whilst problems with brick slips led to their decline from favour, a failure to accommodate movement also affected brick panel construction leading to some significant failures. In extreme cases, expansion of the brick coupled with contraction of the frame has lead to fracturing of brick support knibs or, and more commonly, distortion in the brick outer leaf leading to bowing and in some cases loosening of the brickwork.

A common problem is the excessive overhanging of bricks at the bottom edge of the panel. The maximum overhang of the brick over the panel should be no more than one third of the thickness of the brick – for example, with 102mm bricks, a maximum of 34mm. However, inaccuracy of construction can lead to parts being rather more than 34mm – perhaps up to 50mm or more. This eccentrically loads the bricks and coupled with the effects of expansion can lead to the bowing and loosening of the bottom courses of brickwork.

The effects of expansion in a brick panel can be significant, particularly if the joint between the frame

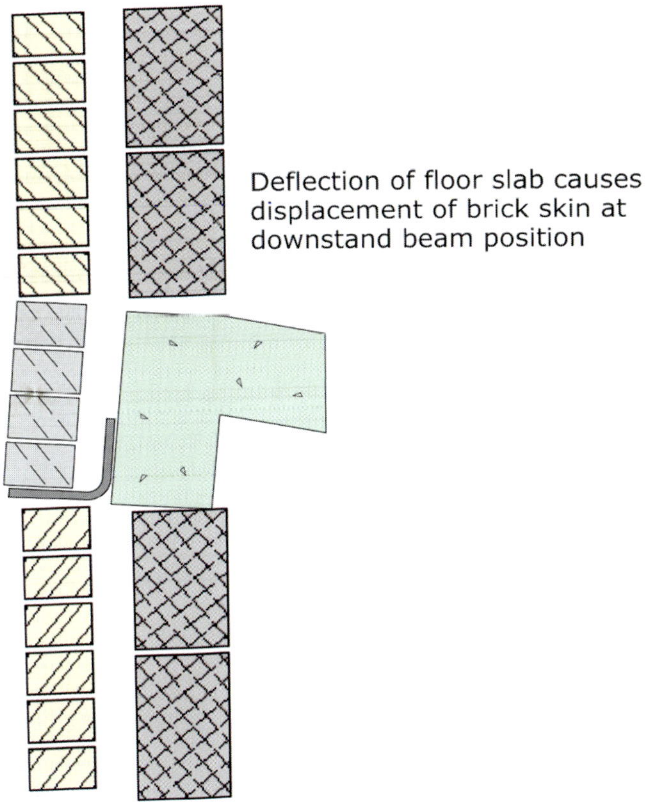

Figure 59 Deflection of floor slab rotates edge beam and disrupts brick cladding.

and the panel is unable to accommodate it. Movement of as little as 0.5mm in a 3,000mm panel can cause the panel to bow out of plane by as much as 25mm at mid span if the ends are restrained.[95]

10.16 CRACK PATTERNS AND SYMPTOMS

The previous sections have covered some of the mechanisms by which movement can occur in a masonry facade. It is now appropriate to consider the effects of those movements and the bowing or crack patterns that they create. The diagnosis of building movements is not necessarily straightforward, and sometimes cannot be fully understood until a pattern of movement has been established after a period of monitoring.

In theory, brickwork that is totally unrestrained ought to be free to expand and contract according to moisture and temperature changes without harm. In historic construction, lime based mortars created a degree of flexibility that would allow brickwork to move without necessarily cracking; this is not the case in contemporary construction where brittle Portland cement mortars are used. However, brickwork is not necessarily homogenous anyway and since it is generally very weak in tension, cracking can occur as a result of shrinkages, or, according to location, expansion.

Putting aside cracks that occur as a result of structural or chemical changes, cracking in brickwork of 0.5mm to 5mm or so is often symptomatic of expansion or shrinkages due to temperature or moisture effects. Vertical cracks usually occur at right angles to the forces that cause them, whilst horizontal cracks (usually finer) indicate the slip plane. It is usually possible to distinguish between shrinkage cracking and thermal expansion, because with expansion there is usually some form of displacement, for example brickwork sliding off a damp-proof course, cracking at corners or evident disruption. By contrast, shrinkage cracking does not generally produce these manifestations.

Assuming a brick panel expands say by 2mm as a result of temperature or moisture, the compressive forces generated may be sufficient to overcome friction and create a slip plane, for example along a damp-proof course. When the temperature drops, or moisture content reduces, tensile forces will be set up as the wall tries to return to its original size. This time the frictional forces along the slip plane are serving to place the wall in tension. It is probable that the strength in tension of the panel is less than the frictional restraint and so the only way in which the stress can be relieved is by the development of a crack.

This type of problem can often be seen in contrasting brick string courses, particularly if these are of a darker colour than the adjoining brickwork. The darker band absorbs more heat than the adjoining brick and as a result expands more. Friction prevents or reduces the ability of the band to return to its original length and therefore cracking ensues.

If the brickwork were free to move along the slip plane and take up the width of the crack as it expands again, all well and good. However, in practice, the crack will probably become filled with small pieces of

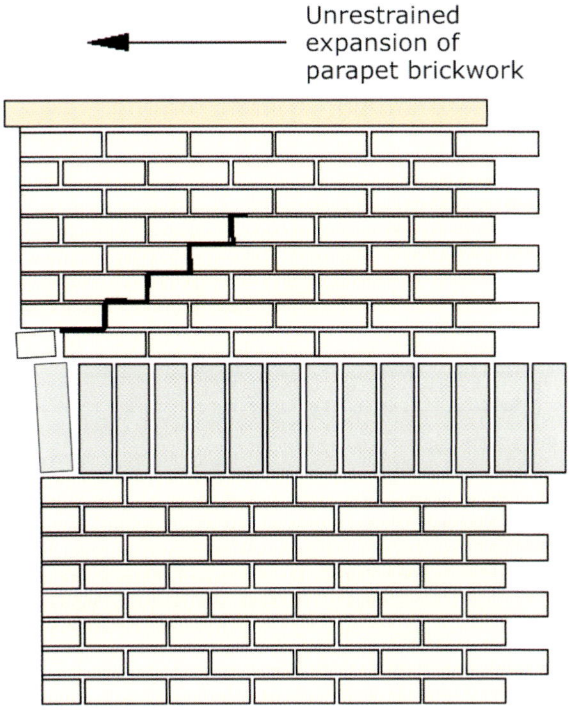

Figure 60 Damage arising from unrestrained expansion of a parapet wall.

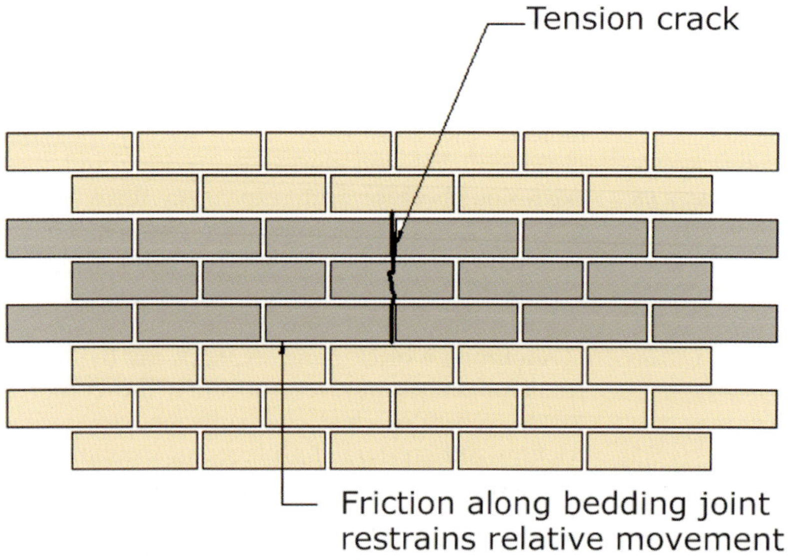

Figure 61 Tension cracks in feature courses.

debris, mortar, etc. which will prevent this from occurring. The result will be progressive movement as the cracks gradually 'ratchet' up.

Vertical cracking can often be seen at the vertical end of a cavity wall where it returns. Commonly, brick spandrel panels beneath ribbon glazing exhibit this type of problem – usually caused by the restraint provided by wall ties placed close to the return and by the effects of expansion in the longer section of the spandrel panel.

Moisture and temperature effects can also become manifest in brickwork that is arranged in panels of different aspect ratios, such as the section between windows and the longer spandrel section (as shown in figures 65 and 66). The relative expansion and

Figure 63 Example of cracking arising from rotation of short return (see figure 64).

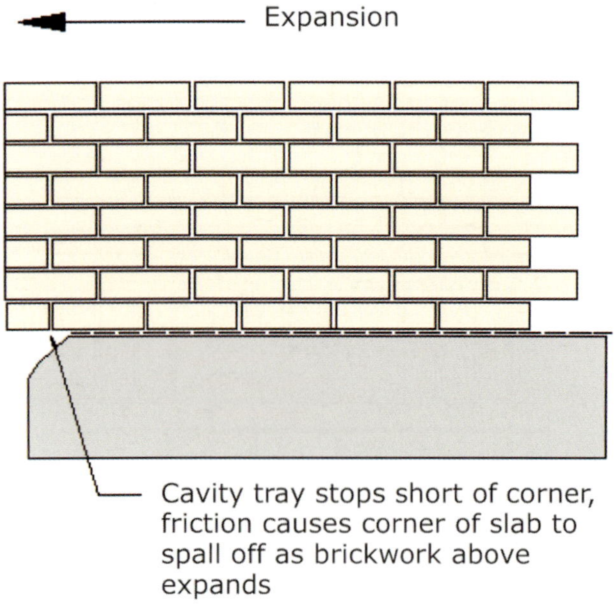

Figure 62 Spalling of concrete slab due to expansion of brickwork and missing slip plane where DPC stops short of the corner.

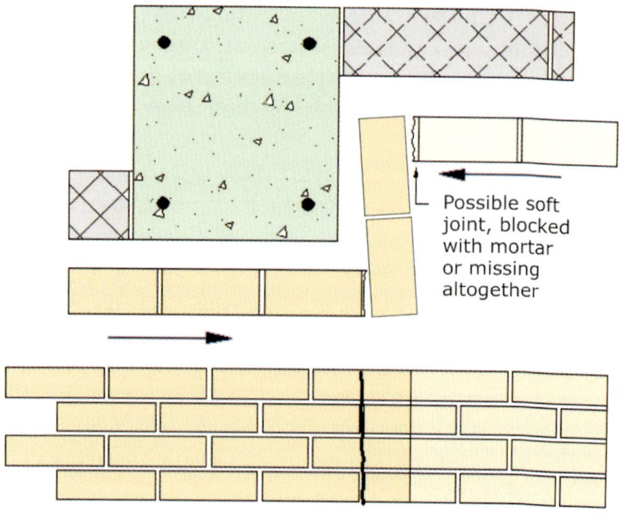

Figure 64 Cracking on line of short return due to expansion of long panel of brickwork.

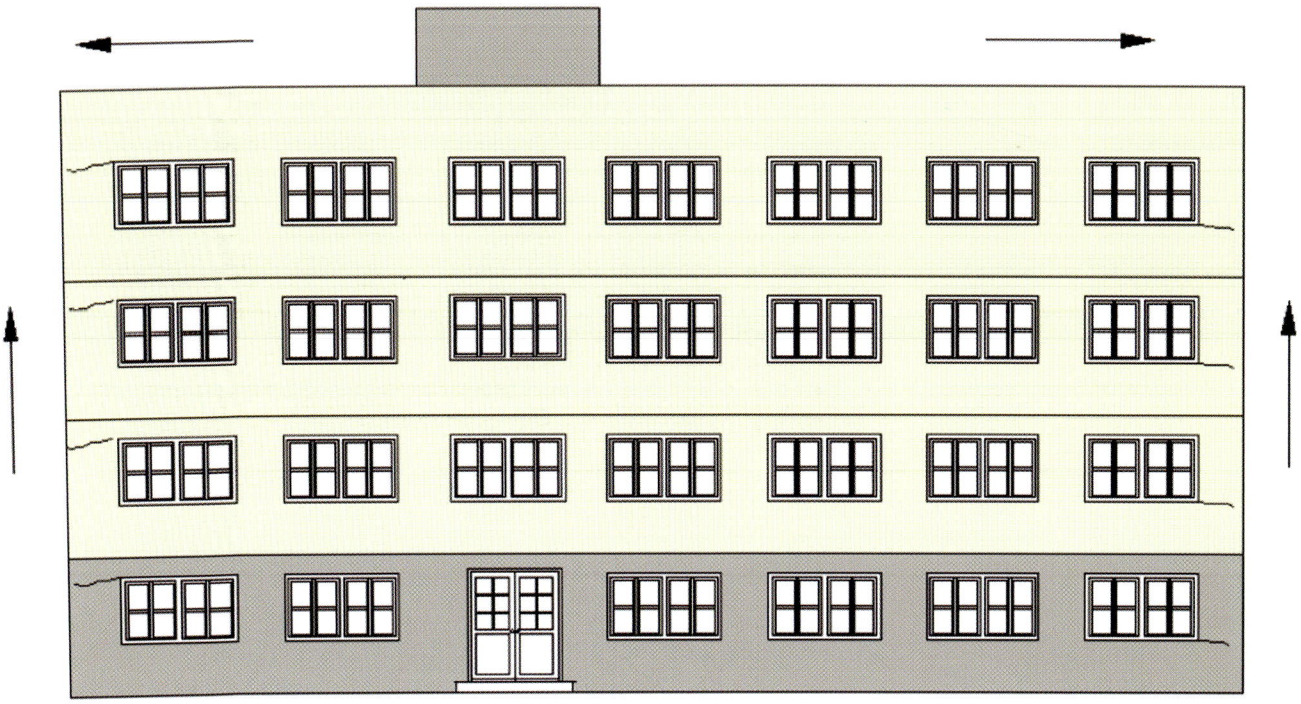

Figure 65 Cracking due to expansion of long brick panels and shorter panels between windows. Expansion occurs vertically as well as horizontally.

Figure 66 Alternative crack pattern due to expansion of brickwork in long run.

contraction of the two sections can result in horizontal cracking.

Occasionally, cracking and displacement of brickwork can occur at ground level, particularly above and below DPC level. The problem is often associated with the brickwork oversailing the DPC. Such movement could be indicative of thermal and moisture movement, but could also be an indication that the floor slab is expanding, possibly as a result of expansive reactions such as sulphate attack.

By comparison with thermal and moisture movements, damage resulting from structural problems such as lack of support, foundation failure, impact damage, etc. can be up to 25mm or so in width. Cracks created by structural movements may be vertical (tension) or diagonal, being reflective of shear (slicing) forces. Such movements often affect both leaves of a cavity wall rather than simply the outer leaf.

Movement control in walling is not the only issue; thought needs to be given to the use of brickwork in conjunction with elements of structure that could provide a restraining influence. Any form of construction that will prevent free movement will allow tensile forces to build up and cracking to form; for example, concrete columns or walls cast up against brickwork will create a restraining influence unless separated from the brickwork by a slip membrane. The propensity of concrete frames to shrink over time is well known, and provision for this needs to be allowed in construction – a requirement that is probably more important with clay brickwork than with calcium silicate bricks.

In all of the above cases, adequate restraint needs to be maintained and so the use of appropriate sleeved anchors and ties must be considered.

In 1986, CIRIA published the results of an investigation into cracking problems experienced on a sample survey of 85 buildings constructed of both clay brickwork and concrete blockwork.[96] The study examined aspects of the structure and fabric that were most vulnerable to cracking and concluded that for unbroken wall lengths exceeding the limits given in BS 5628 moderate level cracking was no more likely to occur than no or slight cracking. The study went on to consider the influence of structural form and details, with the following features being vulnerable to cracking:

Feature	Effect
Short returns to a long straight wall	Returns that are less than 900mm are vulnerable owing to expansion of longer length of wall. The shorter the length of the return, the greater the risk of cracking.
Spandrel walls	Spandrel walls are, or can be, relatively unrestrained permitting circumstances where the amount of free movement is greater.
Long parapets	Parapet walls are also relatively unrestrained, although expansion of all four sides of a square or rectangular roof can lead to bowing or displacement of the parapet.
Stronger mortars	Strong mortars are brittle and weak in tension. Lime mortars are able to accommodate movement by forming many tiny cracks rather than several larger ones. Mortars with a lime or low cement content possess some plasticity.
Discontinuous movement joints	A movement joint reduces the tendency of the masonry to crack in other locations, it effectively permits what would otherwise be a crack in a specified location. If the movement joint is discontinuous, the notional crack will continue as an actual crack through the masonry.
Discontinuous DPC	The DPC acts as a slip plane, permitting a small degree of movement. If the DPC is discontinuous, frictional resistance is increased in a localised position.
Brick slips	See section 10.14 relating to brick slips and the propensity for damage resulting from creep of concrete frames.
Incompressible joint fillers	Fillers such as bitumen impregnated fibreboard do not compress particularly well and can act as a restraint to expansion.
Changes of vertical load	The abrupt change of vertical load, for example at the junction of a low rise and high rise building, will give rise to cracking unless the structures are isolated.
Changes in shape, thickness and height of wall	Changes in the aspect ratio of brick panels mean that the amounts of thermal and moisture movement will be dissimilar.
Eccentrically confined walls	For example, projecting brick panels confined within a concrete frame.
Bonding to dissimilar materials, for example clay bricks with calcium silicate backing	Different propensity for moisture or thermal movement, for example expansion on one face coupled with contraction on the reverse.

Case study

The building was a late 1970s office located within the conservation area of a small provincial town. The accommodation was arranged on two floors with a pitched roof terminating in a series of rendered gable walls. The roof structure was constructed of standard fink trusses.

Horizontal cracking (0.5–1mm) had occurred along the base of the gable and a dispute arose between landlord and tenant as to the cause. Concern had been expressed by the tenant that the cracking was symptomatic of a lack of wall ties to the gable, but investigations revealed that whilst the positioning of ties was not strictly in accordance with BS 5628 they were sufficient in number not to be of concern.

Figure 67 Horizontal cracking to base of gable wall.

Measurements with a plumb bob revealed a slight inward lean of the gable wall on the south-west elevation and a corresponding slight outward lean on the north east gable. Internally, an inspection of the roof void revealed that whilst galvanised steel restraint straps had been installed between the roof trusses and the end gables, the work was defective because:

- the straps should have been taken over at least three sets of rafters (they had been fixed to one or in some cases two);
- solid bridging had been placed between the two end rafters;
- no bridging had been provided between the inside of the gable wall and the nearest rafter; and
- diagonal bracing to the fink trusses was insufficient.

Figure 68 Lack of solid bridging to restraint strap between gable wall and nearest adjacent trussed rafter. The metal strap is also laid on top of the rafter rather than on the underside, and hooks over a cut block rather than a full block, thus reducing its effectiveness.

The lack of solid packing between the trusses and the gable walls meant that there was no immediate restraint against inward movement of the wall. The gable walls needed to be firmly anchored to a braced roof structure; they were not. The effects of positive wind pressure on the south west gable, which was acting as an unrestrained cantilever, were sufficient to cause a small amount of movement in the wall and overcome the tensile strength of the mortar bedding and render. Similarly, negative pressure was sufficient to influence the slight outward movement of the north east gable.

Whilst fink trusses are virtually standard in housing, they are somewhat less common in commercial buildings since spans tend to be greater. Up to a span of around 12m detailing can be compared with domestic construction but beyond this special measures will be needed. An additional problem with the subject building was that conventional lay-in grid suspended ceilings had been installed at the first floor, meaning that the stiffness that would normally be provided by a plasterboard ceiling was missing. (Further details of bracing may be found in Timber Engineering Guidance Document 8 published by TRADA.)[97]

The restraint problem could be resolved by providing new straps and solid bridging, ensuring that the rigidity of the three trusses nearest the gable wall were sufficient to provide proper lateral restraint.

10.17 EFFECTS OF FIRE ON MASONRY

Whilst a small fire of low intensity will have little effect on a wall other than cosmetic damage, a more intense and prolonged incident will cause expansion and probable cracking, spalling and bulging. The effects will depend upon the materials used, their thickness and other factors such as the degree of restraint. Loss of mortar bond is quite possible.[98]

Clay bricks are burnt during their manufacture so will usually respond reasonably well to fire. If temperatures are very high, some vitrification of the clay can occur. Concrete and calcium silicate bricks are not exposed to the same temperatures during manufacture and can suffer more harmful effects such as spalling and delamination.

The main risks to masonry arise as a result of structural components that are built into or bear upon it. Steel frames can distort significantly and disrupt masonry, whilst other components such as galvanised plates, wall ties and the like can lose their corrosion protection. Plastic wall ties, although relatively unusual, can be destroyed.

11 Applied exterior finishes

At a glance

- With the widespread use of framed construction and the move away from loadbearing stone and masonry, applied finishes became popular. Finishes were usually applied as a finish to reinforced concrete, and in most but not all cases relied upon the application of a suitable render background to take out any undulations or inaccuracies in the concrete structure.
- Typical finishes include ceramic tiles, terrazzo, plain render or Mineralite render. In contemporary construction, applied external insulation systems are becoming increasingly popular.
- External finishes rely not only upon good adhesion to the background, but also good adhesion of the background to the substrate, be this concrete or masonry. An interphasal bond failure could have serious consequences as applied finishes are very heavy and can collapse if unbonded.
- Typical causes of failure include traces of contamination on the substrate, a lack of mechanical key, unsuitable bonding materials or a lack of provision for building movements.

11.1 INTRODUCTION

External finishes to concrete structural elements are many and varied, offering different performance characteristics, appearances and costs. During the post-war period, several products were popular:

- tesserae;
- ceramic tiles; and
- render, in particular Mineralite render

All three finishes share a common thread in that they rely upon adhesion firstly to a suitable substrate (usually render) and secondly good adhesion between the substrate and the background (often concrete). Typically, points of failure are as follows:

- tile adhesive adhesion failure;
- background adhesion failure;
- cracking;
- background failure;
- water penetration and associated frost or salt damage;
- poorly positioned movement control joints or lack of continuity of joints; and
- unsuitable movement joint materials.

Failures in the adhesion of the background render can give rise to serious health and safety risks.

In order to overcome the difficulties of achieving a plain, regular background, it was often necessary to introduce an intermediate layer between the background material and the finish. Usually this was of cement render, perhaps 10–15mm in thickness and preferably no more than 20mm. Normal rendering techniques and design rules applied in limiting shrinkage and suction. It is however the interface between the rendering and the background material which is the critical part of the whole finish and indeed when serious failure of surface finishes occurs it is more often than not the failure of the adhesion of the render rather than the tile adhesive.

From the late 1950s various experiments were tried with bonding agents – PVA to begin with and later acrylic resins and SBRs. Both PVA and acrylics were water soluble and thus vulnerable to water ingress, some of which was bound to occur. Weakening of the bond between the render and the background could create plates of delaminated materials that would suffer from frost action and fail suddenly.

Other failures could result from traces of mould oil left on the surface of the concrete background, or alternatively failure to remove laitance or, if a bonding agent was not used, a failure to provide adequate mechanical key or suction.

The combined weight of finish and rendering is significant (probably around 50kgm^{-2}) and the consequences of failure are therefore much higher than with localised failures of individual tiles.

11.2 CRACKING

Cracking within the surface finish can result from a number of causes and aside from the obvious structural faults, the effects of moisture and temperature movements in the background could induce stresses within the finish which can only be relieved by cracking or delamination. Long term creep within a concrete structure can induce stresses within a tile finish unless adequate provision for movement is made.

Movement joints should be at at least 3–4m vertical centres and be at least 6mm in width. They should also pass through the tiles, the intermediate background and where required to the background itself. Joints should also occur at changes in background material and coincide with main movement joints.

Dark coloured finishes absorb heat and exacerbate thermal movements that must be accommodated by the provision of joints. A 4m wide concrete panel with a dark finish could have as much as a 4mm range of movement.

The formation of cracks permits water ingress with subsequent risk of frost attack, chemical reactions and also localised carbonation of concrete backgrounds (which could then permit rebar corrosion).

11.3 BACKGROUND FAILURE

Background failure is most likely to be as a result of chemical problems such as sulphate attack within brickwork. If unsuitable materials are used, soluble sulphates can be dissolved by water ingress behind the finish. These sulphates may then react with water to form ettringite at the interface of the background and the rendering finish, with subsequent spalling together with expansion of the brick joints.

Similarly, the effects of reinforcement corrosion can cause spalling of render and tile finishes, the expansive forces generated being sufficient to push off applied finishes.

11.4 TESSERAE

Mosaic tesserae were a common external finishing treatment to many buildings during the 1960s and 1970s. Comprising small marble or ceramic tiles bonded to a background material such as concrete or sometimes brick, the finish has been found to be durable, but problems of adhesion failure are common, leading to the loss of individual tiles or more serious delamination of large areas of backing material.

To correct a common error, the term 'mosaic' strictly relates to an artistic design that has been commissioned and not to a uniform finish of small tiles or 'tesserae', often applied to entire elevations or to exposed structural components.

There are no British Standards by which to select and compare tesserae suitable for external applications although guidance as to the application can be found within BS 8000-11.1:1989. The subject is also covered in more detail within BS 5385 Part 2, 1991 (formerly CP 212, Part 2). Neither of these documents was of course available for designs executed during the 1960s but they do now give us a useful standard by which to comment upon the earlier levels of work.

Tesserae come in a variety of styles and sizes – for external applications commonly around 20 × 20mm

Figure 69 Defective tesserae applied to a concrete parapet. Both the background render as well as the tile finish has failed, possibly as a result of lack of key and water ingress via the parapet capping.

in glazed or unglazed ceramics, marble or glass. Some rough finished stone materials were also used and these required special care in fixing.

The individual tiles were bedded face down upon a paper backing or alternatively nylon strips or filaments to facilitate handling and fixing. In the case of paper-faced tiles, these were pre-grouted before fixing to prevent adhesive from squeezing up between joints. The paper facing was soaked off before final setting to enable minor adjustments to be made.

11.5 CERAMIC TILES

Ceramic tiles are now classified in BS EN 14411 and fall into two main categories according to their method of manufacture.

- Extruded tiles which are shaped in the plastic state in an extruder and then cut into tiles of predetermined thickness and size.
- Dry-pressed tiles which are formed of powder or small grains and shaped in moulds under high pressure before firing.

Ceramic tiles are further subdivided according to their water absorption (see BS EN 14411).

In themselves, ceramic tiles are virtually impervious although the finish necessitates a larger proportion of joints than other materials, with the result that the weak point in the system is often the quality of the joint grouting. Tiles are generally very durable, so long as first quality materials were selected and used. If the quality was not specified during construction, there is a risk that second quality or 'commercial – run of kiln' tiles were used[99]. These 'warts and all' tiles may not be of the same consistent quality.

Most external tiles are glazed, that is they are finished with an impervious glassy layer that can be either matt or glossy in appearance. Tiles that have water absorption of less than three per cent are described as vitrified; porous body tiles are manufactured, but these are not usually installed in external situations.

Ceramic tiles are inflexible; they cannot accommodate movements in the background and will crack or spall if subjected to stress arising from expansion and contraction of the background or pressure arising from salt growth or frost action. Movement joints in the structure must be carried through to the outside face of tiling, and flexible joints are preferable along the line of dissimilar background materials where differential movements are possible.

Crazing of glazed tiles can occur, particularly where the tiles are incorrectly fixed, as a result of expansion, frost and soluble salts. Vitrified tiles will not usually deteriorate as a result of crazing, but in areas that are very exposed, and where the background material has a high sulphate content, failures can occur. Industrial areas with high atmospheric sulphur content can degrade pointing and grouting materials.

Panels of tesserae and ceramic tile finishes could be fixed in a variety of ways using either tile adhesive or sand and cement. In areas where a smooth uniform background was available, thin bed adhesives could be used (3mm thickness) but there was a risk of voids forming which could collect water and freeze, thus disrupting individual tiles. Furthermore, if the bed thickness were exceeded there was a serious risk that excessive shrinkage stresses could be created with consequent cracking of the surface finish.

If the background was less accurately formed, thick bed adhesives could be used (3mm plus), but again care was needed to avoid voids. In both cases, the late fixing of tiles on to adhesive could also create adhesion failures that may not necessarily become manifest for several years.

Problems have been encountered with nylon backed tesserae panels used in external situations due to rotting of the nylon or difficulties with the adhesive spread on the back of the tiles. For these reasons these types of tiles, when used in external situations, must be treated with caution.

The results of adhesive failure can be spotted easily and invariably by numbers of individual tiles scattered around the ground at the base of a building. Although individual tiles are light, when they fall from a great height the effect can be dangerous.

11.6 REMEDIAL WORK

Provided failure is confined to the tile adhesive, tile replacement is relatively easily effected.

Failure of the rendering is more of a problem. The extent of failure can be detected by a hammer survey but this is time consuming and close access is needed.

Figure 70 Severe spalling of glazed finish to porous body glazed brickwork. The pattern of spalling suggests expansive forces, possibly frost or crypto efflorescence behind the glaze. Note white staining in centre of photograph.

Final initial appraisal, cradles, hydraulic platforms or abseiling techniques will need to be used. For large areas, where a fairly rapid assessment is required, thermographic surveys can help to pinpoint areas at risk provided delamination has occurred and that a suitable temperature differential can be created.

Apart from removal and replacement it can be possible to re-bond areas of delamination using vacuum grouting techniques. (Pressure grouting will only force the finish further away). Specialist firms will undertake this work, however, care in the selection of resin adhesives is vital; some early epoxy resins tended to absorb water and expand on wetting, thus perpetuating the very problem that it was hoped to address. Some controversy remains as to the effectiveness of vacuum grouting.

Cracks in finishes can be grouted but care must be taken to determine the reason for cracking and to select a rigid or flexible sealing system as appropriate. The bonding of a reversible movement crack will only induce further cracking elsewhere.

In instances where the substrate has failed, particularly where sulphate attack is the cause, it will be necessary in all probability to reconstruct the affected area.

In extreme cases the use of overcladding may be considered desirable provided the existing finishes are removed, stabilised or restrained before installation of the new cladding system. An alternative method may be the use of a stainless steel render carrier and reinforced render system.

11.7 SAND AND CEMENT RENDER

Sand and cement rendering, whether used as a finish coat in its own right or as a background to tiling or other finishes, can be problematic for several reasons. The choice of mix will influence the performance of the coating and indeed the substrate to which it is attached. For example, a dense cement rich render will absorb little water, but it is also inflexible so that when the inevitable cracks occur, water may penetrate behind the render to promote deterioration either by frost action, chemical attack or dampness.

By contrast, a weaker, more flexible render, such as a mix of 1:1:6 cement:lime:sand is able to absorb more moisture and will crack less. Such materials are more able to promote drying by evaporation so that damp penetration and resultant chemical or frost problems are less prevalent.

With a cement rich render, any moisture that does penetrate to the interface between the render and the background will take longer to dry out. Certain masonry, such as Fletton or common brickwork contains high levels of soluble sulphates. In persistently damp conditions the sulphates react with the alumina in Portland cement to form ettringite: a process which, causes expansion, disintegration and loss of strength, with the render often developing a series of parallel cracks along bed joint lines.

Cement render shrinks on drying and imparts a force upon the background. The adhesion of the render to the background must be good enough to overcome the shrinkage force otherwise delamination will occur. When render is applied to smooth dense surfaces, such as concrete and certain types of well burnt brick, it is essential to provide an additional mechanical key otherwise adhesion failure is likely.

Moderately porous surfaces such as clay and calcium silicate bricks usually provide a better key, although some wetting and possibly a spatter dash priming coat would be beneficial. For very soft or otherwise unsuitable backgrounds, stainless steel render carriers are essential – indeed these can be configured so that the render coat functions in the

Figure 71 This panel of tesserae was repaired about ten years ago; vacuum grouting was used to secure the background render to the concrete substrate; repair appears to be performing satisfactorily.

same way as a rainscreen cladding. Some carriers have a building paper lining to facilitate this purpose.

To avoid unnecessarily high shrinkage stresses, rendering should not have a total thickness in excess of 20mm. Each coat of rendering should be no less than 8mm thick, nor greater than 16mm thick.[100]

In the UK, cement based renders were often left smooth faced and painted in contrast to European practice which often employed Tyrolean render to give a textured surface. Generally, textured surfaces will absorb more water than smooth faced surfaces and so help to protect walls against dampness. Modern practice has drifted away from traditional render systems however, and external insulation systems incorporating polymer modified thin coat renders are becoming common in both commercial and residential construction.

11.8 MINERALITE RENDER

First developed in 1923, Mineralite render became a popular finish in the 1960s and 1970s and although less popular now, is still marketed.

The product is intended to be the finish coat to sand and cement render and was usually applied to a thickness of around 5mm by specialist applicators. Mineralite render is a cementitious based product manufactured from white cement, graded sand and granite and glass aggregates. Various pigments can be added to create a variety of different appearances.

Mineralite was intended to be applied as a finishing coat to a 12mm thick layer of sand and cement render (1:0.5:4 cement:lime:sand or stronger) although it was commonly applied direct to visible concrete elements such as columns, beams and spandrel walls.

Whilst the product is generally durable, non-combustible and of low permeability, it was susceptible to poor workmanship or delamination of the background render from the structure. Such problems could be due to lack of adequate key, laitance on the surface, mould oil, PVA adhesives and the like (see also section 11.4 on tesserae). It is therefore not unusual to see sheets of Mineralite peeling away from the surface and creating a hazard for passers by.

Figure 72 Brown painted Mineralite render applied to the columns and edge beams of this building. The material has been applied direct to the concrete structure without a backing render. Sheets of the material are delaminating.

12 Thin stone cladding

At a glance

- Changes in cutting technology over the last 50 years or so have led to the ability to saw thin stone veneers for cladding purposes. Whereas older, heavyweight claddings were able to deal to some extent with the effects of thermal and moisture movements, thinner claddings can be more susceptible to damage.
- Proper design of the fixings is critical to the long term durability of the cladding. The design of the fixing is also dependent upon the nature of the stone cladding – whether it is hand placed stack or half bonded or whether it is to act as a rainscreen. Panels used as rainscreens must be supported individually, whereas hand placed finishes can derive some support from adjacent units as long as the bottom unit is properly supported.
- Typical problem areas include detachment, cracking around fixings, cracking within the body of the panel, spalling around fixings, misalignment and bowing of panels.
- Bowing of thin white marbles, especially calcitic marbles is a known problem and can be very harmful, usually leading to a need to re-clad. Marbles demonstrate anisotropic tendencies – they expand and contract at different rates in different axes. Summer temperatures in Europe are sufficient to initialise bowing or warping of panels.
- Anisotropic movement also loosens the individual grains of marble leading to sugaring of the surface and gradual disintegration of the stone.
- Although limestones are durable they can be seriously eroded by the build up of calcium sulphate, the product of the chemical reaction between calcium carbonate and rainwater containing dissolved sulphur. The chemical is deposited on the surface of the stone to form a crust which eventually spalls off taking some of the limestone surface with it.
- Some concerns have been expressed over the long term durability of flamed granite, and in particular whether the flaming process extends any pre-existing microcracking in the stone.

12.1 INTRODUCTION

Traditionally, building stone was seen as a durable and robust material; its properties were understood and building practices evolved to deal with particular characteristics of the material. It was used as a heavyweight form of construction – a structure in its own right – and this meant that large stones were used; these were able to accommodate the effects of thermal and moisture movements and were more tolerant of the use of incompatible materials, for example iron cramps, which could corrode and cause localised damage.

However, stone was (and is) an expensive commodity and gradually it began to be used as a facing material with often a brick backing behind. Even these forms of construction were still fairly heavyweight, with stone of at least 100mm thickness being common.

Once framing was introduced in the early part of the twentieth century, designers lost little time in reducing the thickness of stone claddings yet further; this coupled with advances in cutting technology enabled stone to be cut to as little as 25–30mm for use as a facing material. This trend towards lighter weight buildings resulted in buildings that were inherently more flexible. Stone, recognised by its durability, was being expected to perform in ways that could result in significantly less durability than was first envisaged.

When it is in its natural form, stone will usually be in compression, simply because of the weight of material bearing down upon it. When quarried, the resultant release of stress may lead to dimensional changes in the stone. Buildings constructed from loadbearing stone replicated some of the natural forces that the unquarried stone endured, but the current practice of using stone as a facing material, hung from the building, often creates tension forces – the opposite of the natural condition.

12.2 FIXINGS

Whilst there are very many different forms of fixings for natural stone there are essentially two fundamental types:

- support or gravity fixings; and
- restraint fixings.

Support fixings are designed to take the dead loads of the stone back to the support structure whilst restraint fixings prevent the stone from rotating away from it. A standard form of fixing is a combined support and restraint fixing – effectively a stainless steel angle inclined at 15° to the horizontal and inserted into a kerf towards the bottom of the stone.

Proper design of the fixings is critical to the long term durability of the cladding. Not only will the fixings be concealed from view and inaccessible for the life of the cladding, if they are wrongly positioned they can influence the performance of the stone; for example, by failing to allow for thermal or structural movements additional stresses may be transferred to the stone resulting in its premature failure. The design of the fixing is also dependant upon the nature of the stone cladding – whether it is hand-placed stack or half bonded or whether it is to act as a rainscreen. Panels used as rainscreens must be supported individually, whereas hand placed finishes can derive some support from adjacent units as long as the bottom unit is properly supported.

To avoid unequal stresses building up in the stone, it is important to try and aim for fixings that are equally positioned and arranged so as to avoid eccentricities. For this reason, it has been usual to restrict the number of support points to two for each stone panel. The exception to this is for thinner stones where there is insufficient thickness to install conventional pin fixings without fear of breakage. In these circumstances, the trend is to use a continuous angle fixing that is set into an angled slot (or kerf) in the back of the panel

CIRIA Good Practice Guidance C524[101] gives recommendations for the positioning of fixings. In the case of stack bonding, fixings should be located along the panel at points 20 per cent of the length of the panel whereas for half-bonded, they should be positioned at 25 per cent points. Damage can occur if the fixings are located too far away from the top or bottom edge; for example, a fixing point that is located at more than the limits given above would create a long lever arm of unsupported stone, making it more vulnerable to mechanical damage from ladders or window cleaning cradles. No fixings should be positioned closer to the corner than 75mm.

The thickness of the stone panel dictates to a point the nature of the fixing – although the essential physical principles are the same. For internal applications where the stone may be even thinner, wire ties and adhesive dab fixing may be permissible, but externally, and above first floor level, such techniques are unsatisfactory.

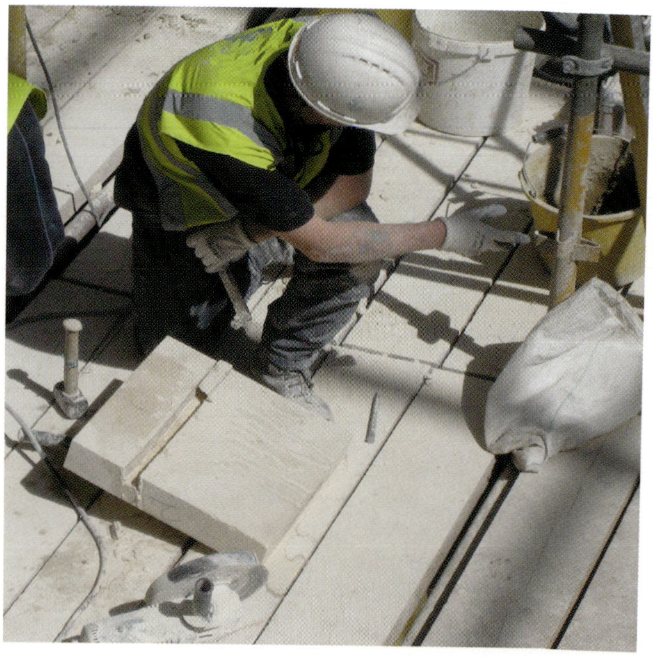

Figure 73 Site fixing of sandstone cladding. The rear face of the stone has been cut out with an angle grinder with no proper depth control. The stone has then been broken with a bolster and mallet leading to the possibility of overcutting and or microcracking of the stone. The slot extends virtually to the middle of the panel.

Until recently, guidance within British Standards recommended the following thicknesses:

- for thin stone (marble, slate, granite, hard limestone (travertine)), 20–30mm for low level work and 40mm above; and
- for Portland limestone and sandstones, 50mm for low level work and 75–100 mm above.

However, these thicknesses are now no longer the rule and it is common to find thin stone panels of 30–35mm thickness on high rise buildings. Thin stone panels – those 20–30mm thick – require a different fixing regime, usually being continuously supported.

Modes of failure

According to CIRIA, 80 per cent of cladding failures are due to defects in fixings. The table overleaf illustrates the key issues.

The following modes of failure are typically found in thin stone panels:

Detachment of panel

Stone spalls around all fixing and restraint points, or fixings are omitted or become defective through corrosion or other fault.

Cracking around fixings

The formation of cracks around a fixing may be due to the formation of microcracks during the initial

Type of failure	Causes
Design failures	Incorrect selection of fixings for particular service conditions. Fixings cannot be installed or adjusted properly or inappropriate materials for exposure conditions. Poor specification of fittings leading to incorrect ordering. Lack of understanding of buildability issues.
Location failures	Incorrect positioning of the fixings or a failure to pre-drill in the correct places. For shared restraints or support fixings, poor alignment of holes or kerfs. Holes oversized to resolve alignment problems.
Component failure	Possibly overstressed as a result of packing, or by installer trying to overcome the misalignment of holes – due to construction inadequacies, lack of tolerance. Failures in bending, shear or pull-out resulting from unintentional loading or miscalculation.
In-service failure	Due to corrosion or inability to accommodate natural building movements such as thermal expansion and contraction, long-term creep, deflection, etc.

drilling process (especially in thin stone panels). More likely is that a fixing has moved as a result of inadequacy (perhaps it was improperly attached to the structure) or that the structure itself has moved. The problem may also be due to insufficient thickness of stone around the fixing point, or holes being drilled too close to the surface.

Cracking within the body of the panel

Given the propensity of concrete framed buildings to shrink over their lifetime it is possible that panels could end up taking loads that they were never expected to withstand. If the panels are held in too rigid a fashion, they may warp as a result of thermal or moisture movements. Restraining this movement can cause cracking to occur and possible disruption of the stone around the fixings.

Spalling

Localised stress within the stone caused by corrosion of the fixings or compression due to a movement joint being bridged, possibly coupled with shrinkage or deflection of the frame.

Misalignment

Possibly due to initial construction faults, lack of tolerance in fixing positions, building movements or the effects of expansion and contraction.

Curvature

The modes of failure described above are symptomatic of problems with fixings and or movement of the background. However, a further mode of failure affects marbles in thin panels – curvature of the marble itself. The phenomenon is described in detail in *Investigating Hazardous and Deleterious Building Materials* but a brief resume is given below.

Marbles are usually made up of the minerals dolomite and or calcite. These minerals have anisotropic tendencies – thermal expansion and contraction of the material is not uniform – which can occur at different rates and in different directions. Whilst a block of steel will expand at the same rate in all directions, calcitic marbles tend to expand in two directions but contract in the third – the same effect produced by rolling out a slab of pastry. On the other hand, dolomitic marbles tend to expand in all directions but at different rates.

Figure 74 Cracking around a stainless steel fixing bracket indicating a loose fixing.

Figure 75 Spalling of stone cladding adjoining each fixing tie. The elevation faced a football field and was thought to be struck regularly by balls. The pins may have been fixed rather too close to the surface of the stone. (Photo courtesy of Garth Ball.)

Such forms of movement create internal stresses in the stone, but because of the crystalline matrix of the stone itself, the individual particles cannot revert back to their original positions and so expansion tends to be progressive rather than reversible. If left to its own devices, a large sheet of marble would try and revert to something approaching a spherical shape, in real terms causing marble cladding panels to warp. The movement is irreversible and can be sufficient to crack the panels and to tear them away from fixing positions.

Slates and granites do not suffer from the same problems and are generally more stable materials, although if subjected to unintentional loads bowing and cracking can still occur.

Some concerns have been expressed that flame texturing of granite can cause durability problems although there seems to be little published evidence of this. The theory is that the aggressive treatment of the stone (the same could also be said for water blasted surfaces) increases the severity of the natural (and existing) microcracking of the stone, making it more vulnerable to the agencies of deterioration.[102]

Similarly, there is some evidence to suggest that repeated thermal cycling, particularly if the stone is restrained, can have an impact upon the flexural strength of certain limestones, gradually reducing its resistance to flexural movement.

12.3 WORKMANSHIP PROBLEMS

It is all very well constructing stone cladding using concealed, corrosion resistant fixings of high quality, but unless the method of attachment is properly considered, there will be locations where it is impossible to fit the intended fixing. Consider for example a series of stone panels that are being fixed within an opening of set width – each panel is secured by two support fixings and two peg type restraint fixings. Panels can be fixed progressively, but unless the fixings to the last top corner panel are of a different type it will be impossible to restrain it – there would not be enough room to slide the panel into place. Faced with these circumstances one would like to think that a competent mason would make provision for an alternative restraint – perhaps a dog cramp to the last but one panel. Sadly, we know that common sense is not always common enough, and one may well find that the difficult fixing has been left off altogether, giving rise to a potential local defect.

Similarly, if the fixings are such that have no tolerance for construction inaccuracy, any that are found in the wrong position may simply be removed or bent over, leaving a reduced number of fixings and again a potential unreliability.

For peg type restraints it is usual to attach these to dovetail slots cast into the concrete panel (if the building is concrete framed) or bolted to a secondary steel frame. Dovetail slots allow a degree of flexibility in the vertical dimension although they need accurate positioning in the horizontal dimension. Sometimes, site drilled plates may be attached, but there is a risk that the fixing may be less reliable and unless the plates are slotted, it may prove very difficult to ensure correct alignment if, for example, the fixing hole coincides with a reinforcement bar.

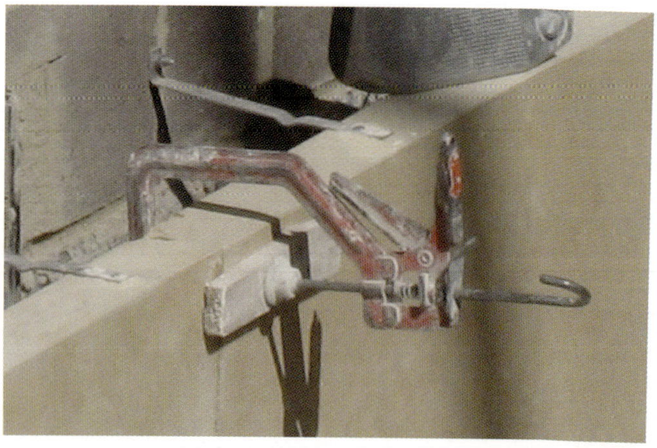

Figure 76 Restraint cramp fixed to block inner leaf using an expansion bolt fitted into a mortar joint. A better fixing would have been made into the body of the blockwork.

CIRIA Guidance C524 sets out a useful table of workmanship faults leading to failure or potential for failure as summarised below:

- use of incorrect fixings for the location;
- dome fixings omitted for convenience or to accelerate the programme;
- incorrect hole size formed;
- incorrect torque on fixing bolts;
- incorrectly located kerfs or dowel holes;
- poorly mixed or cured resin for fixings or inadequate preparation of holes – failure to remove dust prior to placing of resin;
- plastic shims substituted for stainless steel shims or different metals used for shims leading to galvanic corrosion;
- dovetail anchor slots filled or contaminated with cement grout making installation of anchors difficult; and
- ad hoc amendments without proper design consideration.

12.4 DURABILITY OF STONE

The chemistry of stone is complex and a detailed review of the various mechanisms of deterioration is well beyond the scope of this book. However, a basic understanding of the principles is useful in understanding the processes that are taking place.

There are three primary types of stone:

- igneous;
- sedimentary; and
- metamorphic.

Figure 77 Plastic shims used to correct errors in alignment – note poor fixing to top cramp and poorly installed cavity tray below.

Figure 78 Expanding bolt fitted too close to edge of block causing spalling and fixing failure.

Figure 79 Beware of explosive (shot fixed) pins into structural steel. In this photograph the head of the pin has sheared off giving only minimal restraint to the fixing cramp.

The table overleaf provides a summary of how the types of stone are formed and their features. Both sandstones and limestones are vulnerable to deterioration through atmospheric agencies, chiefly sulphur in urban and industrial areas (with reference to limestone, see figure 80 overleaf).

In sandstones, dissolved sulphur gas also attacks the calcium carbonate that bonds the particles of calcareous sandstones leading to severe weakening of the stone. Salt laden air can also affect the performance of both limestones and sandstones, resulting in powdering of the surface and expansive damage to surface layers.

12.5 CONTOUR SCALING OF SANDSTONE

In this condition calcium sulphate is deposited within the pores of the sandstone causing internal stresses to occur within the surface layers. BRE Digest 177 suggests that the stresses occur as a result of differential thermal expansion between the contaminated stone and the body of the sandstone or alternatively to differences in moisture movement. Eventually the contaminated crust spalls away, with the damage following the line of the original stone.

12.6 SUGARING OF MARBLE

As noted above, marble is essentially limestone that has been modified by extreme heat and pressure. Both calcitic and dolomitic marbles exhibit strong anisotropic tendencies that gradually cause the individual grains of the stone to loosen. In turn, microcracking around the grains increases the porosity of the marble, making it more vulnerable to atmospheric pollutants and freeze-thaw cycles. As the process develops, the surface of the stone becomes sugary, losing its highly polished finish.

Features of common building stones

Type	Formation	Examples	Features
Igneous	Solidification of molten magma.	Granite (most common as a building stone), pumice, basalt (may be present in aggregates for concrete).	Granite is highly resistant to the normal agencies of decay; has very low moisture movement and low coefficient of thermal expansion.
Sedimentary	Limestone is mainly composed of calcium carbonate with some varieties containing magnesium carbonate. The main component of sandstone is quartz, combined with silica (siliceous sandstone), calcium carbonate (calcareous sandstone), calcium carbonate and magnesium carbonate (dolomitic sandstone). The chemical components of limestone are derived from the skeletons of dead sea animals or other organic particles, fine particles of other stone, and/or chemical deposition.	Limestone, sandstone.	Very popular building material, easily worked and cut. Has a natural bed, although not always apparent.
Igneous	Derived from sedimentary rocks, but converted due to heat and pressure. Marbles may be calcitic or dolomitic or a combination of both.	Marble (from limestone). Slate (from clay).	Usually very durable although contamination from sulphur dioxide can lead to surface deterioration (sugaring) of marble. When cut into thin sections, marble has a tendency to curl. Impurities in slate, such as iron pyrites, can cause deterioration.

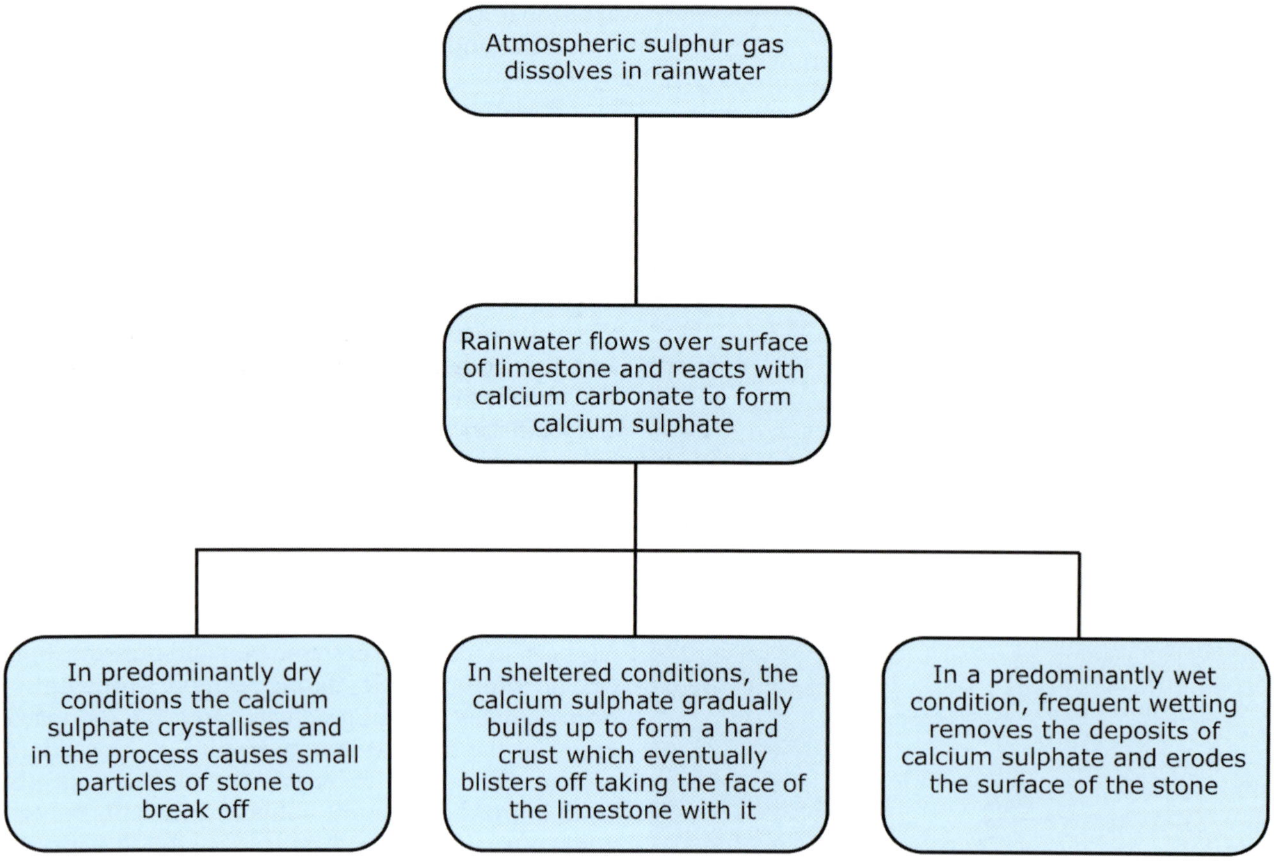

Figure 80 Atmospheric deterioration of limestone.

12.7 STAINING OF GRANITE

Flame finished and polish finished granite cladding panels are a popular and durable cladding material. However, the appearance of the panels can be marred by water absorption at joints and by the migration of plasticisers from joint sealants. The effects are largely cosmetic, although moisture absorption could theoretically lead to freeze-thaw problems.

Flame finishing and polishing involve heat, which can fuse the grains in the surface of a building stone rendering it reasonably impervious. However, the cut edges of the stone may not necessarily be treated in this way, leaving the unprotected stone vulnerable to water entry.

Certain types of sealant backing rod are known to absorb water; these materials can then hold water against the face of the stone, allowing it to soak into the stone over a period of time to cause dark staining at joints. Replacement of the backing rod and sealant with non-absorptive materials will usually cure the problem.

However, a more difficult problem to resolve is fluid migration from the sealant into the stone. The problem is often associated with low quality silicone sealants, although problems with other types of sealant are not unknown when they are placed in contact with porous surfaces.

If low viscosity fluids do leak into the stone, they can cause a number of aesthetic problems such as:

- darkening of the stone in a margin around the joints;
- a patchy appearance caused by the material leaching out onto the surface and making the stone more water repellent in areas; and
- permitting airborne dust and other contaminants from rainfall to stick to the surface of the stone.

These effects are usually confined to a zone 20mm or so wide each side of a joint but larger areas can be affected if the fluids are not absorbed into the stone but instead flow down its surface.[103]

Problems of fluid migration are usually dealt with by poultice treatments, although this can be an expensive remedy.

Figure 81 Example of the effects of fluid migration into granite cladding, note dust and dirt to horizontal joint and staining to sides of vertical joint.

13 Concrete cladding

At a glance

- Heavyweight concrete cladding panels can function well in service, but are prone to the usual forms of deterioration arising from carbonation, chloride attack, low cover, etc.
- Total failures are rare, but isolated failures are much more common. Older types of fixing can suffer from corrosion problems although it is often the concrete that fails before the fixing. Copper alloys were popular before the 1960s, but before the widespread use of stainless steel, some of the bronze alloys including manganese bronze, copper zinc alloy and high tensile brass suffered from stress corrosion.
- Panels can suffer from the effects of frame shrinkages or deflections, and if fixings are unable to cope with the resultant movements, failures can occur. Panels will usually have two support points unless fitted to a continuous floor edge, although some means of adjustment and levelling is usually needed.
- More than two support points can lead to misalignment and overloading of adjacent fixings.
- Thermal stresses can be high and must be accommodated by the fixing system.
- Some forms of large panel contain insulation; these function as sandwich panels, but corrosion of (particularly) phosphor bronze ties between the inner and outer leaf can lead to the detachment of the entire outer face. For this reason the nature of the ties should always be ascertained.

13.1 INTRODUCTION

Relatively heavyweight claddings made from precast concrete are common in many commercial buildings, especially those constructed in the late 1950s and during the 1970s. In contemporary construction concrete panels tend to be faced with brick or stone as part of a pannelised system manufactured in storey height panels in the factory and craned into place. Concrete cladding systems can be very durable if correctly designed and manufactured, although durability problems can mar the performance of some claddings and create a need for expensive remediation.

Total failures of cladding systems are rare, although isolated failures in individual members are much more common and can be attributed to:

- excessive deflections in the structure;
- carbonation coupled with poor cover to reinforcement resulting in cracking and spalling;
- chloride additives or exposure leading to corrosion or spalling. (Because of the desire to strike panels from the mould on a 24-hour production line, chlorides were often added during manufacture – this was often true of corner panels and short end returns because of the manufacturing process involved. However, seaward facing panels or those within salt-laden atmospheres are also at risk);
- poor installation, e.g. grout not applied to fixing dowels, inadequate fixings;
- cracking of corners and arises caused during handling or erection which could lead to more advanced deterioration as a result of corrosion, carbonation, etc;
- weathering and water ingress leading to corrosion of fixings;
- detachment of exposed aggregate or stone finishes;
- temperature induced strains;
- elastic shortening of the structure leading to compression failures, crushing and distortion where movement provisions are inadequate;
- bowing as a result of differential temperature and moisture effects; and
- user damage.

The above list is of course not exhaustive and indeed many individual failures can be the product of more than one problem.

13.2 FIXINGS

Given that fixings, once installed, are generally inaccessible, it makes sense to ensure that the materials used will have a design life at least equal to that of the cladding installation. For many years, fixings have been selected according to established principles of corrosion protection, but this does not mean that corrosion problems do not exist. Furthermore, failures could also occur as a result of the incorrect choice of fastener, for example an expanding bolt fitted into a soft substrate.

Typical failings may include:

- corrosion occurring as a result of a combination of dissimilar metals and a failure to provide a suitable barrier between them (see section 2.3 on bimetallic corrosion);
- incorrect choice of fixing metal, for example the wrong grade of stainless steel or the use of galvanised steel as opposed to other more durable metals;
- poor installation of fixing, for example fixing secured to a weak part of the structure or too close to an edge or not set correctly; and
- failure to use correct equipment, leading to over-torquing and resultant overstressing of fixing.

Typically, non-structural concrete panels will have two loadbearing (gravity) fixing points which are designed to take the dead load of the panel together with a further four restraint fixings designed to deal with live loads.[104] Live loads (wind pressure or suction) act horizontally on the panel, whilst dead loads act vertically downwards.

For design purposes, loadbearing fixings are usually considered to be pin jointed and take the form of concrete nibs or stainless steel brackets.

Panels may be either hung or bottom supported, the latter being the most usual. Panels must never be fixed using both methods together as this can lead to serious distortion problems (see the case study on page 121). Given the construction tolerances that can normally be achieved, it is usual to provide a levelling mechanism of some kind on the gravity fixings.

Restraint fixings are usually positioned as close as possible to the corners of the cladding panel and should be so designed to facilitate the truing up of the panel. Movements in the structure to which the cladding is fitted must be accommodated by the fixings and by the joint design between adjacent panels. The stresses that can build up if these issues are neglected are considerable and can be sufficient to fracture a horizontal support or the units themselves, break ties and create bowing or detachment of the panels. The calculation of the forces is complex and whilst general guidance on thermal and moisture movements is readily available, the accurate prediction of creep coupled with other forces acting in combination is very complex and can only be estimated approximately.[105]

The movement patterns that are of most significance to the design of the cladding are:

- **Thermal movement** where this is relative to the panel and the frame to which it is attached.
- **Differential settlement** between columns (usually limited by design to 1/500 of the distance between columns).
- **Elastic deformation under load:** In steel-framed concrete-encased buildings this should rarely exceed 1mm over a storey height. If the frame is not protected with concrete deformation can be greater, particularly given fast-track construction, and movements of 3–4mm per storey may be needed over a 25–35 storey building. With concrete framed buildings, the movements need to be calculated, particularly in respect of the elastic deflection of beams and floor edges and where spans are large.
- **Drying shrinkage, creep and moisture movement:** The effects of initial irreversible shrinkages can be minimised by delaying the installation of concrete cladding as long as possible after the removal of formwork to the main structure. Typically the long term shrinkage value for structural concrete may be taken as 0.03 per cent. Creep of the main structure will cause permanent shortening of the structure, with the amount of reduction being influenced by the age of the concrete when it was loaded, ambient relative humidity (RH) and stress within the concrete. According to BS 8297:2000, concrete under load and constant RH develops about 40 per cent, 60 per cent and 80 per cent of the final creep within 1 month, 6 months and 30 months under load respectively.

13.3 SANDWICH PANELS

Some early concrete cladding systems were also designed to be loadbearing. These 'large panel' systems were common in residential construction but developed a bad reputation following the partial collapse of the Ronan Point Tower block in May 1968.

Large sandwich panels comprise two layers of concrete separated by an insulating core of polystyrene or machine made mineral fibre. The panels would have been constructed either in face up or face down construction but would commonly contain a number of metal tie bars holding the two leaves together. The leaves would usually contain mesh reinforcement.

Once erected, sandwich panels would usually be so configured that vertical loads would be taken by the inner leaf.

> In face up construction the inner leaf is cast in a mould, with the insulation then placed on top of this. Tie bars are pushed through the insulation into the concrete behind and the outer layer is then placed. The external finish could be exposed aggregate or other worked finishes – usually exposing the aggregate by washing off before final setting has taken place. Face down is not surprisingly the reverse process and was used most often for mosaic, tile or bush-hammered finishes.

If the ties are incorrectly positioned or composed of unsuitable materials, the risk exists that the outer leaf could move relative to the inner leaf or in extreme circumstances, become detached.

The table below shows the materials used for ties (derived from the IStructE Guidance note[106]).

Problems with this type of panel are not confined to the detachment of the outer leaf. In many cases, the connection between the floor structure and the wall panels was defective, having been either poorly designed or, more likely, poorly executed. Frequently, the joint relied upon a structural connection being made with exposed hoops of reinforcement linked together and protected with in-situ grout or make-up. If the grout was poorly placed the reinforcement would be at risk of corrosion and deterioration.

Insulated sandwich construction was not used solely for loadbearing panels. Many non-loadbearing configurations were also supplied (and indeed are still available); these tend to be strengthened with concrete ribs or stiffeners on the reverse face.

Materials used for ties

Material	Comments
Mild steel	Can be found particularly in older panels. May have been used without authority as the metal has poor resistance to corrosion in carbonated or chloride contaminated concrete.
Phosphor bronze	Acceptable use, should perform satisfactorily.
High-tensile brass	An alloy of copper, zinc and manganese. The base metals can be mixed in various proportions and may also contain small amounts of iron or nickel (which increases tensile strength) and aluminium or silicon (which enhances corrosion protection). High-tensile brass can suffer from corrosion and stress corrosion cracking but this is dependent upon a number of factors such as applied stress, metallurgical condition and exposure to contaminants. Ties made from high-tensile brass should be regarded as suspect. If discovered, IStuctE recommends that additional retro-fit ties be installed as quickly as possible.
Stainless steel	Can be subject to stress cracking when held in warm, chloride-rich environments or in a highly alkaline condition; however, the conditions needed to stimulate this form of corrosion in stainless steel can, for all practical purposes, be ignored for cladding panels. Austentic stainless steel possesses good resistance to pitting corrosion whilst ferritic stainless steels are more resistant to stress corrosion.

Case study

The building comprised a regional distribution warehouse located in southern Europe. The building is Z-shaped with a projecting central wing and has a total plan area of around 52,000m^2 being subdivided into 12 open plan warehouse areas of 50,000m^2 with a small two-storey area for ancillary office accommodation.

The structural frame comprised a grid of precast concrete columns and beams supporting a flat roof formed with prestressed concrete T-beams. The walls were formed of non-loadbearing precast concrete panels each being 2.4m wide and 11.2m high. The panels were hung from the beams around the perimeter of the roof and rested on a cast in-situ reinforced concrete plinth/ground beam. Adjacent panels were interlocking via male to female edge details and the joints were sealed with silicone.

Each panel was 200mm thick with a composite construction comprising an external and an internal skin of concrete, separated by a 100mm thick polystyrene slab core. The panels were framed with a solid perimeter of approximately 130mm wide 'stiles' to the sides and 160mm to the top and bottom 'rails'.

The concrete skins were reinforced with a 200mm × 200mm welded mesh of 5mm diameter steel bars lapped by 200mm. Dowel fixings were also cast into the bottom of the panel to connect to fixings cast into the top of the ground beam.

Within a year of construction, cracks started to appear in the concrete panels. The cracks were sealed with mastic as a rather poor cosmetic repair, but they continued to form with increasing regularity over the following five

Figure 82 Concrete panels (11.2m high) showing extensive cracking. The panels are top hung but also bear onto the in-situ plinth at ground level. The cracks shown in this photograph tend to be diagonal but a majority within the building were horizontal as shown in the panel above the fire point.

years or so. A majority of the cracks were hairline in nature, although others suggested slightly more significant movements. In all cases the concrete appeared to be suffering from very low carbonation, although the water:cement ratio of the mix was somewhat higher than would have been desirable.

Some of the cracking had clearly been caused by early-handling damage (and a suspicion that the panels may have been installed too soon after casting) but of more significance was the combination of the panels being top hung and supported at the bottom by the plinth beam. This, coupled with deflections brought about by wind loading and temperature differentials between the internal and external surfaces, was sufficient to overcome the tensile strength of the concrete and this led to cracking.

Construction drawings were unclear as to the design intent; it is possible that the bottom fixings were intended simply as a restraint fixing, but that inadvertent errors led to a transfer of load, placing the panels in compression. Normally, concrete panels work better with bottom gravity fixings as concrete will perform better in compression than tension.

14 GRC cladding

At a glance

- Fibres have been used to reinforce concrete products for many years. The excellent tensile properties of asbestos were unmatched by man-made products although the introduction of alkali resistant glass fibres saw some benefits and enabled lightweight, moulded cladding panels, balustrade panels, planters and the like to be manufactured at a fraction of the weight of conventional reinforced concrete products.
- Normal glass fibres degrade in alkaline environments such as concrete, but alkaline resistant varieties have been found to survive much better through the addition of zirconium dioxide.
- The inclusion of chopped strands of glass fibre in to the cement matrix creates concrete with high impact and flexural strength, and enables thinner and therefore lighter components to be made.
- Over time, the strength of GRC degrades, but provided that this is taken into account at design stage, this need not be a problem.
- GRC is susceptible to moisture and temperature movements and so proper care in the design of fixings is very important.
- Early GRC cladding panels tended to be large units, sometimes of sandwich construction – these exhibited a tendency to bow, particularly if curved units were used. Larger units are now considered to be less satisfactory.

14.1 INTRODUCTION

Without any form of reinforcement, concrete and cement mortars are brittle materials with little tensile strength. The inclusion of fibres creates small pockets of very high strength which resist the propagation of cracks from small voids and thus increases tensile strength and impact characteristics.

GRC is a mixture of hydraulic cement, fine aggregate and glass strands, sometimes with additional pozzolanic materials and fillers. The fibres usually make up about five per cent of the weight of the concrete. The inclusion of fibres in the concrete mix enables increases in the flexural and tensile strengths of the material and enables thinner (and therefore lighter) structures to be manufactured.

GRC loses some of its strength and ductility over time and for this reason its use for large, especially curved, cladding panels is now discouraged.

14.2 THE MECHANISM OF DETERIORATION

One of the properties that give GRC its strength and toughness is the ability of the glass fibres to slip within the concrete mix. Over time, this ability to slip is reduced by the formation of crystals within the reinforcement.

The individual strands of glass fibre are made up of about 200 individual filaments, and thus contain numerous voids. When it is first mixed, the concrete matrix surrounds the strands, but enables them to move relative to the concrete – this creates high impact strength.

Essentially what happens is that when the concrete is mixed, calcium hydroxide (lime) is present in aqueous solution. The calcium hydroxide is gradually drawn into the voids between the filaments by capillary action and the voids are gradually filled. Over time the calcium hydroxide crystallises, filling the pores and locking the fibres into the cement matrix and preventing the fibres from slipping. The changes are moisture and temperature dependent but over the long term the properties of the hardened GRC become stable.

GRC is not in itself satisfactory for structural applications such as loadbearing elements, floor slabs, large cladding panels or indeed any application where the results of failure could place life at risk. This is not to say that the material is unsuitable for quasi-loadbearing applications such as smaller cladding panels, balustrade panels and the like where resistance to overturning or wind loads is appropriate. So long as the aged strength is taken into account during design, there is no reason to be cautious over the use of GRC.

One of the early uses of GRC was for the production of cladding panels – these were either single skin, or quite often double skin, including a thin layer of insulation. The opportunities afforded by GRC enabled designers to be adventurous with the shape and profile of the panels. What was not fully appreciated was the moisture and thermal performance characteristics of the panels and the need to provide for free movement at panel fixings.

Some of the early panels did not perform well; panels were too large or mechanical restraints created large tensile stresses that the GRC was unable to accommodate, particularly as its strength reduced with ageing.

Like all building materials, GRC is affected by changes in moisture content and temperature and so fixings must be designed to accommodate movements resulting from these changes. Resilient pads are usually placed at gravity fixing points, but restraint fixings are also needed at the bottom as well as the top. To prevent the symptoms of bulging caused by restraint, the top restraint fastenings must be designed to move in four directions – this can be achieved by the use of special fixings incorporating PTFE washers.

The bottom restraint should have movement in two directions, but slotted fixings are vulnerable to over-tightening and should not be relied upon in themselves. The locating point can be fixed or two-way but all fixings need to offer some form of rotational freedom. Actual support is provided by the two bearing pads.[107]

Fluctuations in these factors give rise to dimensional changes, and if restrained in some way, the product will crack.

The coefficient of thermal expansion of GRC is around 20×10^{-6}. In the UK, light-coloured cladding panels on the face of a building may be subjected to temperature variations from around −10°C to +60°C. The range for darker colours, which absorb more heat, could extend to around 80°C. Therefore, where the temperature, for example, is 70°C and the panel width is 1,500mm, the change in length (Δ) would equate to $1500 \times 70 \times 20 \times 10^{-6} = 2.1$mm.

Irreversible and reversible movements are present at all times; irreversible movement is slower if the GRC already has a high moisture content. Reversible movement relies only upon moisture content. Assuming that the GRC was wet cured, the irreversible movement would normally be expected to be around 0.05 per cent increasing up to 0.15 per cent reversible. The greatest amount of shrinkage will occur shortly after curing.

Within the UK, and indeed most locations, it would be most unlikely to find that the concrete product would be subjected to the full range of moisture movements. For example, the UK climate is such that moisture will always be present within the product. In service conditions, a good rule of thumb would be to allow for 0.15 per cent, i.e. 1.5mm per metre run. In the case of the 1,500mm wide panel given above, this allowance would mean designing for movement due to moisture of 2.25mm.

Panel fixings must be designed to take into account the range of moisture and temperature movements identified above; they must also take into account the effects of long term creep of the structural frame (in the case of concrete) or loading deflections.

Aside from the propensity to cracking, restraint of panels can also produce the effect of bowing. Because GRC panels are generally fairly thin, the temperature gradient through the thickness of a GRC sheet is

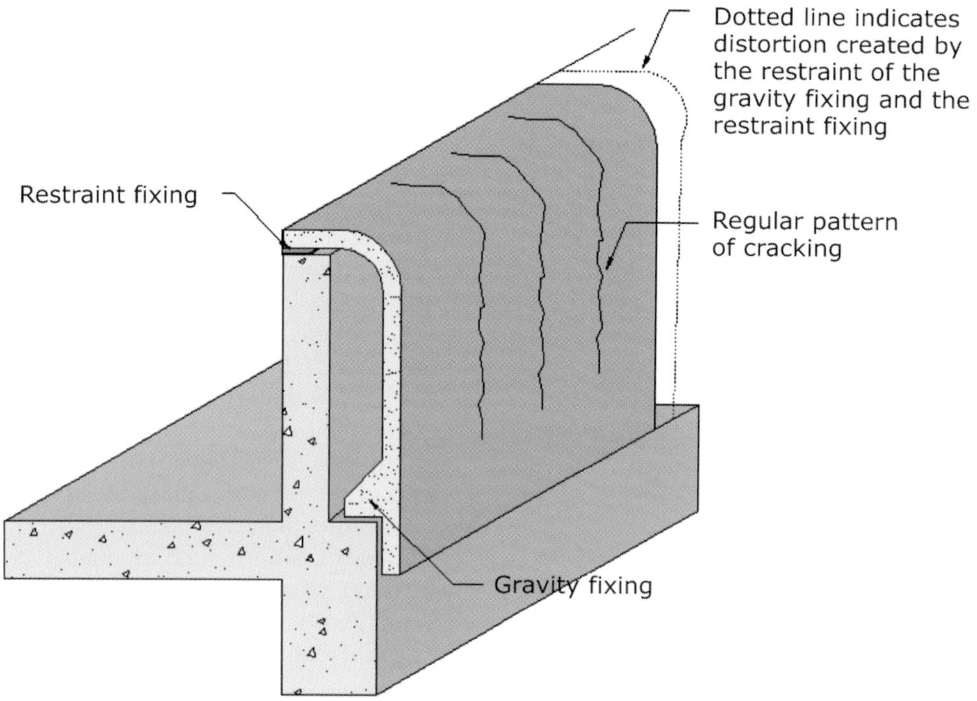

Figure 83 Curvature and cracking of typical curved GRC cladding panels. Dark colours increase range of movement.

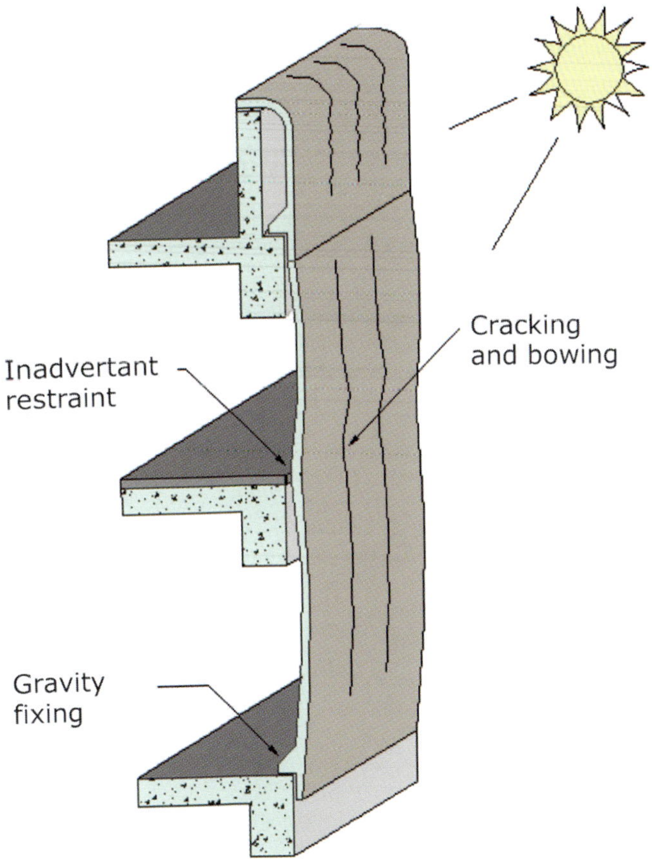

Figure 84 Double storey height panels. Inadvertent restraint, for example from a metal bracket at the edge of a floor screed, restrains the tensile forces due to expansion and creep of the frame; these forces are relieved by bowing.

negligible. However, the same cannot be said of moisture content. If the panel has become saturated, it is to be expected that the surface exposed to the sun or to the drying effects of the wind will dry out first, meaning that the dry face will contract before the rear face. When this happens, the panel can bow. The effects of such movement are usually temporary, with the panel returning to shape as the moisture content stabilises through the thickness of the panel.

Difficulties could occur in cases where uneven drying was occurring across the face of a panel and this could create difficult stress patterns within the body of the panel, possibly leading to cracking. The answer therefore is to avoid intermediate fixings and to ensure that the fixings provided allow free movement. Unintentional restraint may be provided by floor screeds abutting panels on their inside faces. The screed could restrain movements of the panel or alternatively lock the free movement in fixings that are designed to move.

Whilst the above description applies to simple, flat panels, for many years it has been customary to construct sandwich panels containing an insulating core of polystyrene or mineral wool. In this case, the thermal and moisture gradients through the thickness of the panel are likely to be more demanding, and the effects of bowing need to be considered carefully in terms of the fixings. Irrespective of this, curved or corner shaped panels will develop tensile stresses irrespective of the fixings and are therefore best avoided as sandwich panels.

15 Curtain walling

At a glance

- Curtain walling may be defined as a weatherproof, non-loadbearing enclosure of windows or metal panels in a light metal framework which is fixed back to the structure of a building, usually at intermediate floor levels. There are various systems in common use ranging from simple, site assembled 'stick systems' to larger, pre-assembled panels or unitised systems.
- Many glass curtain walls comprise a system of mullions and transoms with the insulating glass units held in place with clamping plates. In more recent years, silicone bonded systems have become popular; in these systems either two or four edges of the glass are held in place with silicone adhesive.
- Very few, if any curtain walling systems can be made totally weathertight; it is usual to accept that some leakage will take place from the outer, weather facing seals, but to drain this water away in a controlled fashion. There are several mechanisms of water penetration into a system: force of gravity, kinetic energy, capillarity, pressure assisted capillarity, surface tension and air pressure differentials. Of these, air pressure differentials are possibly the most significant.
- Wind pressure against a facade creates localised areas of high pressure. If the air pressure inside a building is lower, air will leak through, taking moisture with it. If the air in the cladding system can equalise quickly with the external air pressure, the propensity for leakage will be much reduced. Drained and pressure equalised cladding systems (which are generally most common) utilise this principle, and divide the ventilated void in the cladding into small fields to permit rapid equalisation.
- Problems of leakage are possibly the most common form of defect, although defective fixing details are not unknown. Fixings must be robust and durable; they are inaccessible and cannot be inspected or repaired easily. The fixing system must be capable of accommodating building movements due to loading, creep, thermal and moisture movements, etc.
- Most systems rely to a large extent upon the performance of sealants. Broadly, there are several basic types of sealant failure: adhesion failure, cohesion failure, delamination failure and reversion.

15.1 INTRODUCTION

Whilst in historical terms a curtain wall formed the enclosure to a fortification, the current use of the term describes a weatherproof, non-loadbearing enclosure of windows or metal panels in a light metal framework which is fixed back to the structure of a building, usually at intermediate floor levels. Curtain walling is a form of cladding, providing a weatherproof enclosure and increasingly serving as a means of controlling the internal environment of a building.

The science and technology of cladding is fairly complex and consequently the cladding industry is often plagued with design or workmanship problems. Over the last few years, dozens of cladding manufacturers and installers have failed, often as a result of difficulties with particular projects. One major cladding manufacturer is reputed to have purchased an entire office building from the employer as a means of resolving a dispute. It is not surprising therefore that many surveyors shy away from cladding inspections, preferring to leave the inspection and resolution of defects to those with appropriate skills.

The cladding supply chain is also complicated – populated with a large number of specialist contractors of varying degrees of competence and a smaller number of correspondingly larger manufacturing organisations. Frequently (and especially with window systems) the components of the system are supplied by the manufacturer to a franchised fabricator who may decide to introduce non-franchised components into the system, and produce an installed product that is not fully in accordance with the manufacturer's specification or design.

Whilst there may be an infinite number of variations of cladding designs, there are some fairly basic concepts which are worth examining as these will help develop an understanding of how cladding works and why it can go wrong.

15.2 TYPES OF CURTAIN WALLING SYSTEM

Stick system

Stick construction is the traditional form of curtain walling, comprising a grid of mullions and transoms into which various types of glass and/or insulated panels can be fitted. The systems are usually fairly low cost, using extruded components supplied by the system manufacturer. Mullions are usually assembled first, with transoms then fitted in-situ.

A variety of jointing systems are employed but most stick systems will have some form of stub connection between individual lengths or transoms. Mullions are usually fixed back to the structure at floor levels, although longer spans are possible according to the section properties of the component.

Aluminium is the usual material for the framework, although some older buildings experimented with steel. Most of the grid assembly work is done on site which permits some measure of dimensional adjustment and tolerance, but it is highly workmanship-sensitive and will rely upon accurate cutting of components and the correct use of sealants at joints.

Assembly on site is fraught with risk as many systems are superficially similar so that a contractor who is familiar with one system may omit crucial components such as seals at junctions when assembling components provided by another manufacturer.[108] Similarly, if shop drawings are hard to read (perhaps in a different language) or do not detail items such as drainage or ventilation slots there is a risk that these items are neglected or installed incorrectly.

Unitised system

Unitised systems comprise narrow width storey height units of aluminium framework containing glazed and/or opaque insulated panels. The entire system is pre-assembled under factory controlled conditions thus avoiding some of the problems of assembly encountered on site. The unitised panels are usually craned into position, with pre-positioned brackets attached to the floor slab or the structural frame. Modern installation techniques increase the speed of erection and often minimise the requirement for scaffolding. Unitised systems do have higher direct costs but nowadays the curtain walling to most prestige buildings is of this type.

Panellised systems

Like unitised systems, panellised curtain walling is constructed in the factory to improve quality control and speed of erection. However, the systems comprise large prefabricated panels of bay width and storey height which are connected back to the primary structural columns or to the floor slabs. Panels may be of precast concrete or comprise a structural steel framework and are often clad with a variety of stone, metal and masonry cladding materials.

The advantages of these systems are improved workmanship as a consequence of factory prefabrication, allowing improved control of quality and rapid installation with the minimum number of site sealed joints. Panellised systems are less common and more expensive than unitised construction.

Structural silicone

Structural sealant glazing is a form of glazing that can be applied to stick or unitised curtain walling systems. With structural sealant glazing the insulating glass units (IGUs) are attached to the grid framework with factory applied structural silicone sealant rather than by pressure plates and gaskets in a more traditional system.

Structural glazing

Structural glazing typically comprises large thick single panes of toughened glass assembled with special bolts and brackets that are supported by a secondary steel structure. This form of glazing is often referred to as 'Planar' glazing and is commonly used to form the enclosure to atriums and entrances.[109] Planar is actually a trade name rather than a generic type of glazing system.

Rainscreen systems

Rainscreens are not curtain walls as such as they provide a screening function rather than an enclosure in its own right. A rainscreen is used to shield a wall, whether this be of masonry, metal studwork or in some cases glass. Usually a rainscreen is designed to permit some controlled leakage with the main functions of resistance to air and water being provided by the shielded wall behind.

15.3 WEATHERTIGHTNESS

Patterns of wind and rain exposure

Whilst all openings in a facade are a potential source of water penetration, the geometry of the building, its height and orientation all affect the performance of the facade in terms of air and water leakage.

Walls facing the wind will be subjected to greater driving rain, whilst leeward walls are more sheltered. However, as discussed later in this section, variations in pressure occur between the sides, centre and top of a wall. As an air current meets a building, it splits to travel

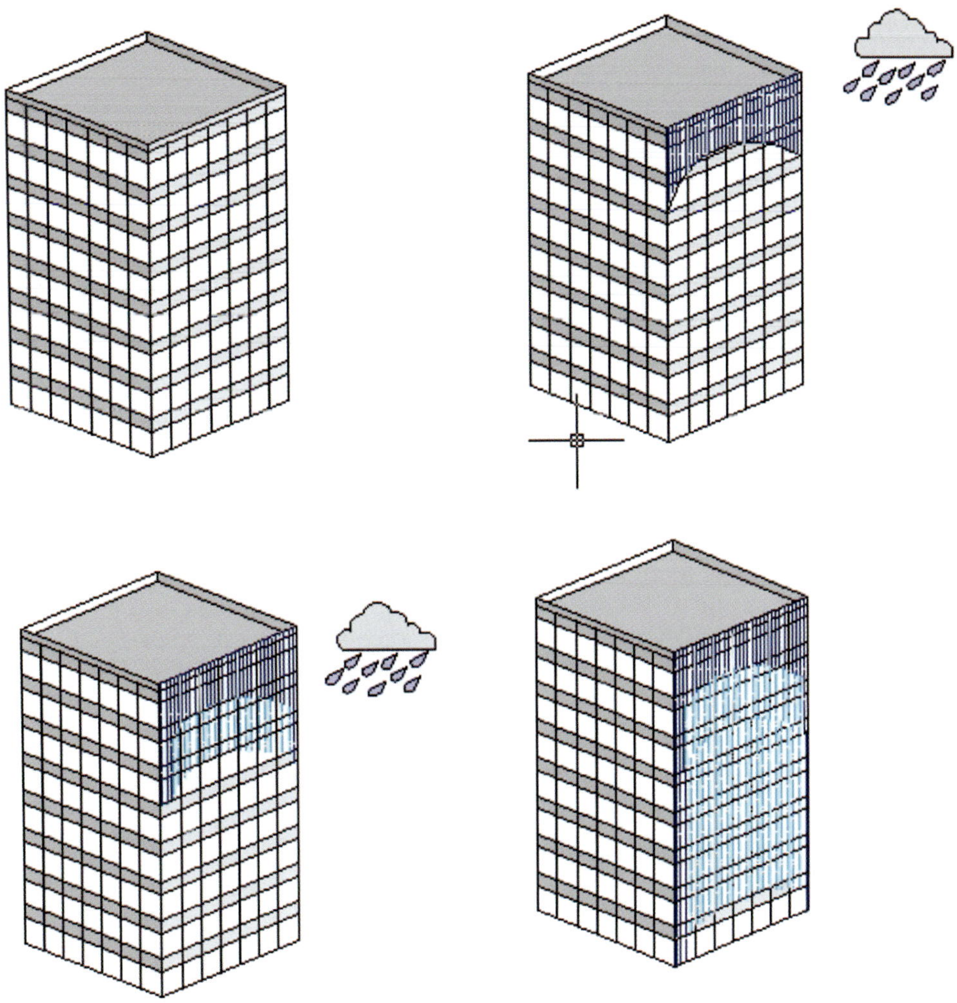

Figure 85 Illustration of rainfall effect on a facade.

round and over it. This parting of the air creates a cushion of high pressure but relatively still air at the centre of the wall, which tends to reduce the amount of rain hitting it at that location. By contrast, the amount of rain hitting the wall at the corners or top of the building can be as much as 20–50 times that at the centre, since the movement of air accelerates around these features. A tall, narrow building will suffer more from this type of problem than a wider, more squat building.

The diagram above shows the effects of rainfall on the facade of a tall building. After about ten minutes the migration of wetting commences, gradually extending down the building and down the corners. With impervious claddings, such as metal and glass curtain walls, water simply flows down the wall surface and the accumulated flow can be significant by the time it reaches the bottom of a tall building. Wind flow around corners and parapets can also draw water laterally and even upwards.

Aside from appearance and climate control, the key function of a curtain walling system is to exclude air and water. Indeed, most failures are related to water leakage in one way or another and a proper understanding of the mechanisms of water entry and the main ways of preventing this will assist in the diagnosis of leakage problems.

There are five mechanisms of water ingress:

- kinetic energy;
- surface tension;
- capillarity;
- gravity; and
- air pressure differentials.

Essentially, the methods of prevention of water ingress are as follows:

- to use sealants;
- to design joints in such a way as to prevent water leakage;
- to use mechanical seals (gaskets); and
- to use a combination of two or more of the above.

Whilst it is theoretically possible to construct a curtain wall using only one of the above methods, in practical terms it is usual to employ more than one; different options are possible:

- fully sealed or front sealed systems (usually where a flat plane is required);
- drained and back ventilated systems; and
- pressure equalised rainscreen systems.

Kinetic energy *(see figure 86)*

A rain drop possesses translational kinetic energy: a function of the mass of the raindrop and its speed. When driven by wind, drops of rain may have sufficient forward momentum to be carried through small joints or openings in the cladding system. Upon hitting a surface the drops can shatter, with the resultant smaller drops carrying further into the cladding system. To counteract this, joints are designed with a baffle so as to provide a labyrinthine structure.

Studies by Choi[110] on the effect of wind driven rain showed that drops of small diameter (0.5mm) are capable of moving upwards, particularly in strong wind. Although the upward movement of larger drops (10mm in diameter) is unlikely, the horizontal longitudinal velocity component can be quite large and raindrops can move in a direction that is horizontal and sideways parallel to the building. This means that not only can raindrops penetrate into the depth of a cavity in a cladding system, they can also travel sideways within it.

Surface tension *(see figure 87)*

Surface tension is the property of a liquid to behave as if it were covered with a weak elastic skin and is caused by the cohesive forces of molecules at the surface. Water has a particularly strong surface tension, and this enables it to run vertically down a surface and then back horizontally over an overhanging edge. By this mechanism, water can be drawn into a joint that is not exposed to direct rainfall.

To counteract the effects of surface tension, it is usual to provide a small drip on the underside edge of a projection.

Capillarity *(see figure 88)*

Like surface tension, capillarity is due to forces at molecular level and an unbalanced attraction between the molecules of the liquid and those of a narrow tube or space. If the molecules of the liquid are more attracted to the tube than other liquid molecules, the liquid will rise. For this to happen, the walls of the tube or enclosure must be sufficiently close together. But capillarity by itself can never result in leakage, simply because water can never rise beyond the faces of the enclosure.

However, wind assisted capillarity can result in water penetration, as the force of the wind can push water further up the walls of the enclosure and beyond. To counteract capillarity, joints can incorporate an air space to break the surface tension and prevent the water rising at the interfaces.

Capillaries of less than about 0.01 millimetre draw and hold a small volume of water with such high suction that they seldom contribute to rain penetration. A greater volume of water, however, is held by the lower suction in large capillaries such as unbonded interfaces. Large capillaries are important contributors to water penetration when an additional force of even low magnitude is added.[111]

Gravity *(see figure 89)*

The force of gravity plays an important role in water penetration as it can pull water down and in through an opening that leads inwards and slightly downwards. Whilst wider joints can prevent capillarity occurring, air within the space coupled with the effects of gravity can result in water travelling vertically until it reaches a horizontal surface.

Air pressure differential *(see figure 90)*

Of the five mechanisms of water entry described, air pressure differentials are the most important and are often the most significant source of water ingress. Air will move from an area of high pressure to an area of low pressure. This movement of air can carry water with it, so that water can be drawn into a joint, glazing rebate or even to the interior of a building.

Wind blowing against or around a building creates large pressure differentials. The leakage of air through a wall system prevents the outside pressure from equalising across the cladding – rather like trying to inflate a balloon with a hole at both ends, the air leaks away. The key to dealing with air pressure differentials is to aim to equalise pressures as quickly as possible (see the material on pressure equalised systems on page 135). This is accomplished by providing a rigid air barrier on the inside and suitable gaps on the outside to allow air movement.

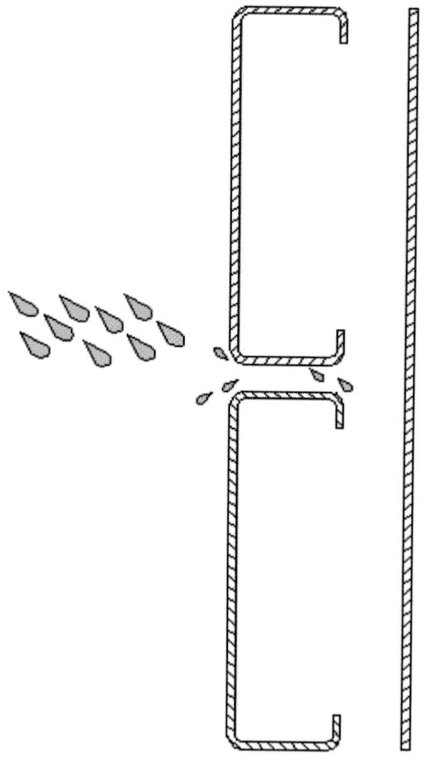

Figure 86 Kinetic energy shatters raindrops and forces water into the joint.

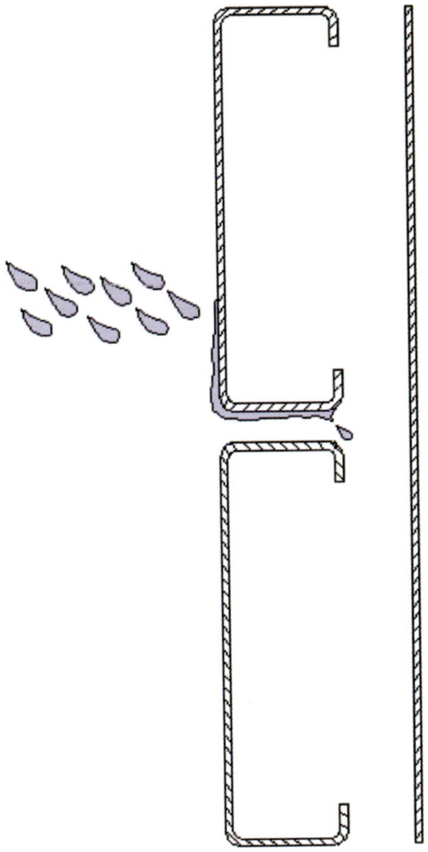

Figure 87 Surface tension permits a film of water to track into the joint.

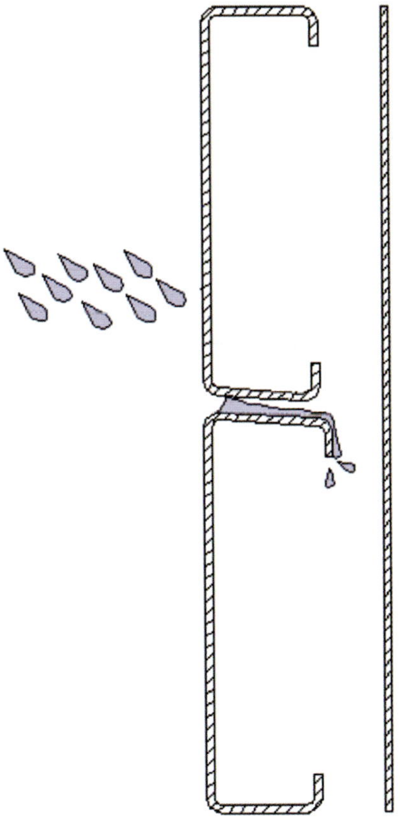

Figure 89 Water drawn in and down through a joint as a result of gravity.

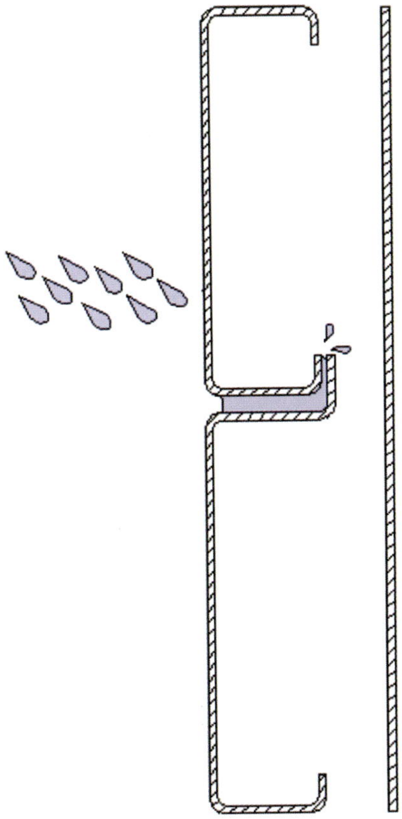

Figure 88 Water driven into a joint as a result of capillarity in a narrow joint coupled with wind pressure externally.

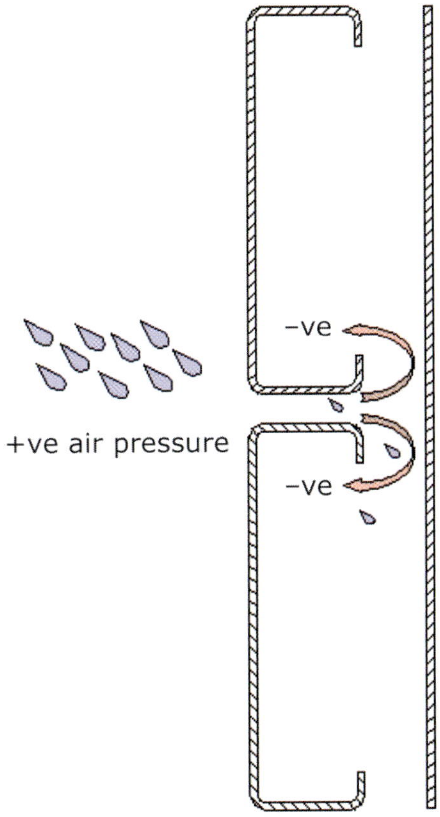

Figure 90 The movement of air from areas of high pressure to areas of low pressure enables water droplets to be carried into a joint.

15.4 METHODS OF PREVENTING WATER INGRESS

Essentially there are three options in curtain wall design:

Face sealed systems

Since the predominant materials of curtain walling, steel and glass, are impervious, careful joint design ought to be able to prevent water ingress. Early curtain walling systems tended to be face sealed and relied upon a weatherproof outer seal to prevent water penetration.

The seal must remain completely free of defects to prevent leakage paths occurring; this means that workmanship must be of a high standard. But achieving watertightness during construction is one thing, maintaining it throughout the life of the cladding is something else when the individual components are subject to temperature and structural movements, ageing of sealants, etc.

In a truly face sealed system there is no provision for drainage; any water that bypasses the outer seal could result in internal water ingress. With single glazing systems it had always been the case that glass would be fully bedded, initially with glazing putty and latterly with various synthetic glazing compounds. Early installations of insulating glass units followed the same route, with the edges of the units fully encapsulated with 'wet' applied glazing compounds. The systems are now largely obsolete simply because it was very difficult to install the sealant without the creation of voids or air pockets, which, as the building aged could become filled with water. More often than not the voids would be around the edge of the IGU, thus holding moisture in contact with the edge seal and leading to premature breakdown of the unit allowing it to mist up. Furthermore, wind induced deflections or movements in the glass could pump the water contained in the voids, forcing it towards the interior of the building.

A variation on the wet sealed system was the zipper gasket design. In this form of installation the IGUs (or single glazed panels) were fitted into neoprene rubber gaskets which completely encapsulated the edge of the glass. The gasket would be designed to hold two adjacent panels and contained a groove into which a rigid 'zipper' strip was pushed. The zipper would compress the gasket against the sides of the glass and, theoretically at least, provide a watertight joint. A small number of proprietary systems employed the zipper gasket principle but they were notorious for leakage problems. Because of their flush, or nearly flush appearance, the design found favour with architects wishing to achieve a flat plane appearance in the design of a facade.

With a zipper system, the glass is restrained by the gasket; there are no additional mechanical supports. Thus, the gasket material must be rigid enough to carry the weight of the glass as well as wind loads without deformation. In practice it is easier to accomplish the structural function than the waterproofing function primarily because it is impossible to install gaskets without joints and secondly because of the difficulty of maintaining adequate compression on the gasket throughout its contact face and throughout its life.

For a gasket to be effective at excluding water, the gasket must be held at pressure in contact with the glass or panel. The pressure of the gasket at its contact face is called lip pressure. Over time, lip pressure decreases as a result of normal hardening or weathering, compression set and or shrinkage. When installed, the gaskets are stretched; they tend to revert over time and also shrink as a result of the gassing off of volatile plasticisers.[112] The designer of the zipper gasket system is faced with a compromise; the gasket material must be rigid enough to fulfil its structural role, yet remain flexible enough to maintain adequate lip pressure.

The zipper (sometimes termed a 'lock-strip') is usually around ten points harder in durometer (the 'shore A' hardness reading – a measure of the resistance of plastics toward indentation) than the gasket itself and exerts a force on the centre of the gasket which is designed to be transferred to the tips of the gasket in contact with the glass. At a corner or cruciform joint the geometry of the joint reduces the effectiveness of the zipper and hence lip pressure can be compromised. Similar problems occur at but joints. One contemporary North American manufacturer claims that it solves the joint problem by forming factory made units to fit specified openings. Corner pieces and tee joints are made by injection moulding; this injection moulding process forces uncured rubber compound into a mould that joins the gasket ends and simultaneously vulcanises the rubber splice.

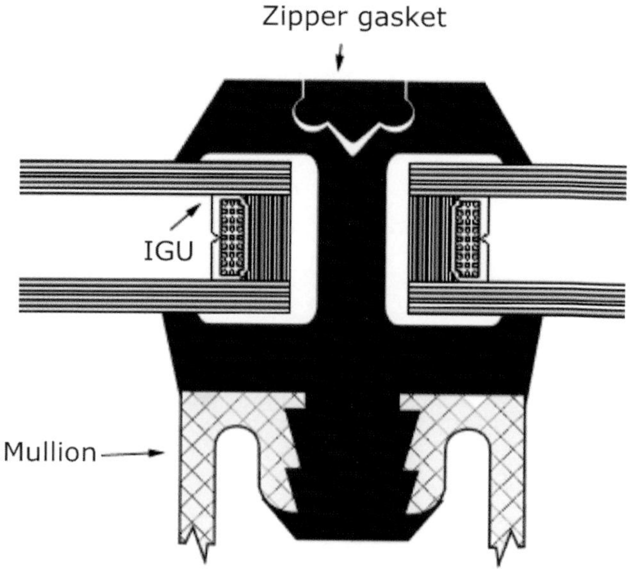

Figure 91 Typical zipper gasket system.

A face sealed, single stage design has no back up should water bypass the primary seal. Air pressure differentials will ensure that any small leakage paths are exploited. Confronting the weather head-on requires a leap of faith – faith in the engineering ability of the designer, and faith in the skills and knowledge of the installer; far better to accept the fact that 100 per cent watertightness cannot be achieved and to make allowances for small amounts of water to both leak in and drain out again.

Drained systems

An improvement on the face sealed system is the drained system which contains an air cavity behind the primary outer face. Whilst still vulnerable to leakage resulting from air pressure differentials, the cavity can at least be drained. For this arrangement to work a secondary back seal is needed.

Most cladding designers now accept that it is difficult to exclude water and therefore provision for a small amount of leakage can be made within a drained system. Typically these dry glazed systems comprise an outer decorative cover plate, and an aluminium pressure plate with two narrow rubber oyster gaskets either side clamped against the glass or insulated infill panel. The pressure plates are screw fixed through a thermal break into the mullion or transom member. A further inner gasket between the glass and the mullion or transom provides the back seal.

In drained and ventilated systems the front gaskets provide an initial barrier. The rebates and cavities are drained and ventilated to the exterior to prevent the accumulation of any water that bypasses the outer seals. Drainage is usually via small holes or slots in the underside of transoms that drain water down through the mullions.

Some systems also incorporate a foil faced butyl adhesive tape applied over the transom and mullion nosings directly beneath the pressure plate to serve as a secondary line of defence. While these systems will accommodate a small quantity of water within the glazing rebates, it is important that the pressure plate is fixed to the correct torque so that the outer gasket seal forms a good seal against the glass.

Another key issue is the adequacy of the joint design. In stick systems (see section 15.2) transoms are usually joined to mullions via metal stubs screwed to the mullion and sliding into the hollow transom profile. Since the glazing rebate is functioning as a drainage path, it is important to ensure that water cannot work its way back to the interior of the cladding system at the mullion/transom joint. Water at this point could track back as a result of pressure differentials and wind assisted capillarity as discussed in section 15.3. For this reason, manufacturers will often specify a layer of sealant at the joint.

Another variation of joint design involves the use of rubber 'infiltration blocks' which are usually placed at cruciform joints. Being single mouldings, they can help keep water away from the corner joints, but again they invariably need to be sealed properly. A failure to seal properly can often result in water leakage internally.

Water that enters the system can be drained either from the transoms via slots, or into the mullion, to be discharged at strategic points usually with a plastic drainage shoe insert. In those systems that rely upon an additional butyl foil layer beneath the outer gasket, it is important to check that the tape is cut around the drainage shoe, otherwise water can back up in the system and penetrate the building via a convenient joint.

Back ventilation is a concept mainly applied to rain-screen curtain walling systems. To deal with kinetic energy, joints are usually unsealed, but designed as a labyrinth so that drops of rain shattering on impact will not travel further into the joint. Even so, it is accepted that at times water can penetrate as a result of air pressure differentials, surface tension and gravity, so a drained and ventilated void is provided behind the outer skin. Behind the void may be located a 'technical wall' – providing insulation and a vapour barrier. To provide airtightness, joints in the technical wall would be sealed. The basic function of the rainscreen principle is to deal with all of the mechanisms of water entry.

Drained and back ventilated systems are not simply used for rainscreen cladding. Many high profile office buildings now employ a similar concept – an outer single layer of glass as the rainscreen, a void of perhaps 500–600mm and then an inner glazing system of IGUs. With these systems, positive ventilation is achieved by means of fans and exhaust grills and the air space used either to warm or cool the interior of the building according to season. This method, although expensive, has the advantage that the weather is kept away from the IGUs and so the risk of leakage into the interior of the building is much reduced.

Despite the reservations expressed earlier, one face sealing system that does appear to work is the use of structural silicone jointing – a system which has had a track record of more than 40 years.

Silicone glazing systems (sometimes called SSG) enable a flush elevational treatment with no exposed metal components such as pressure plates or caps. The system also has several advantages over conventional stick systems, for example:

- it provides a good thermal break;
- it reduces the amount of dirt and water retention in the glazing system;
- it reduces thermally induced stress; and
- it is easy to clean.

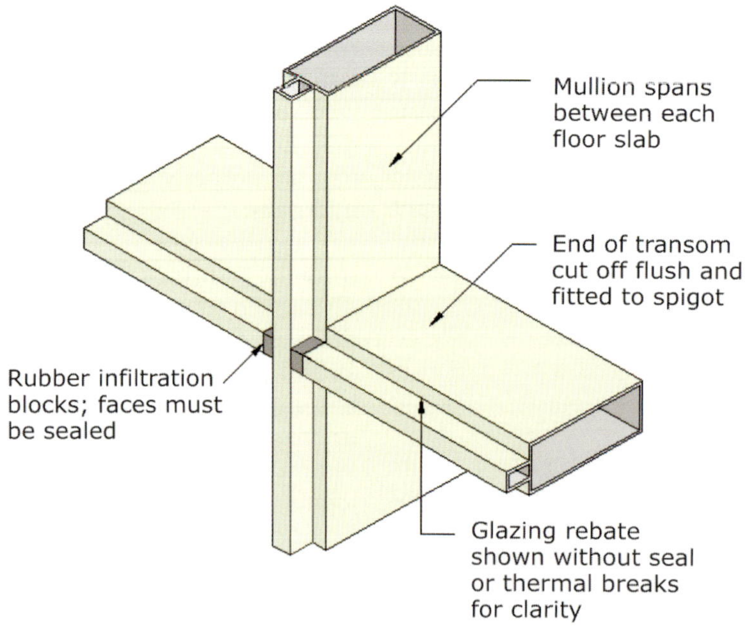

Figure 92 Rubber infiltration blocks at junction of mullion and transom.

Figure 93 shows an example of a proprietary silicone glazed system in which the IGU is usually factory bonded to a carrier frame with silicone adhesive, in turn fixed to the mullions and transoms with aluminium fixing lugs. In this case, any water that bypasses the outer weather seal between the units is able to drain away via the air chamber behind.

There are two basic types of silicone glazing: two sided and four sided. Two sided glazing provides support to the vertical edges only, whilst gravitational loads are supported by mechanical means – either a bolted system or a conventional fin support. Two sided systems should not be confused with butt-joint glazing which does not provide a structural bond to the framing system. Butt-joint glazing provides a weather seal only on two edges of the glass.

In four sided construction, the glass is bonded on all four sides, with the dead weight of the unit being carried by a structural fin and wind loads carried by the silicone joints. Sealant manufacturer Dow Corning suggests that four sided glazing systems are less likely to leak due to the adhesive/sealant which blocks air and water ingress.[113]

Not only does the silicone joint carry wind loads it must also accommodate thermally induced movements between the IGU and the structural frame. The weather seal between adjacent units is not designed to be

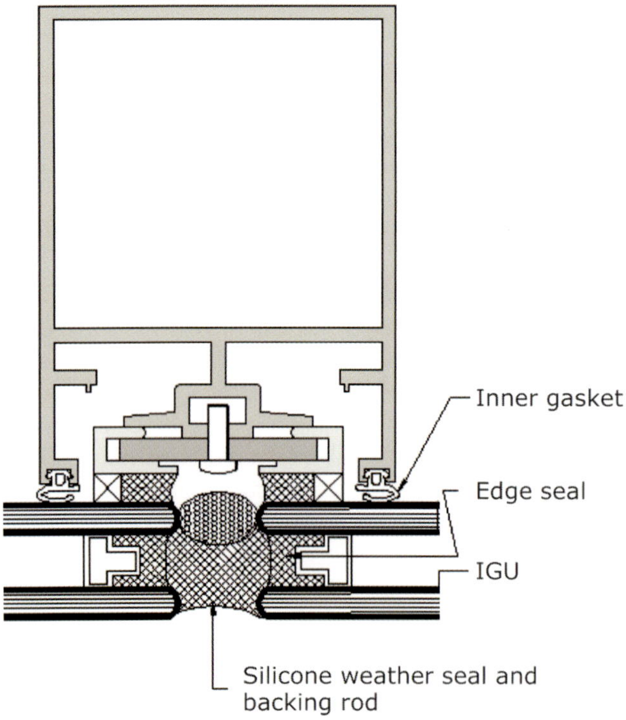

Figure 93 Example of a silicone glazed system.

Figure 94 Two sided silicone glazing – silicone to vertical joints, conventional pressure plate to transoms.

loadbearing – its function is to exclude air and water, but it must be bonded to both the edges of the glass as well as the edge seal – for this reason the compatibility of sealants must be established, and if necessary demonstrated by test.

Some forms of two sided structural silicone glazing are designed to be erected in the field. In this case, the glass is usually held in place with temporary mechanical fixings whilst curing of the adhesive takes place – this can be as long as three weeks. Once the adhesive is cured, the temporary fixings are removed and the weather seal installed. Field installation is fraught with quality control problems and over recent years, certainly within the UK, there has been a move to factory bonding.

Cleanliness and adequate cleaning with appropriate solvents is essential to avoid loss of adhesion, both in the adhesives and the weather seals. The consequences of adhesion failure of the structural bond could be severe as there are no additional mechanical fixings to hold the glass in place. Catastrophic failure of silicone glazed systems are rare, although worries over structural performance have lead to the design of structural spacers – these contain a groove which permits a mechanical clip fixing to the inner leaf.

Whilst the adhesive properties of silicone are well known and documented, problems with compatibility can still occur. (See also warm edge spacer problems in IGUs in section 16.3). Over the last ten years or so more coated glass products have been marketed in an attempt to improve thermal performance. These coatings are not always compatible with silicone sealants and careful checking is needed to ensure that a suitable bond will be achieved if the coating is to be exposed to the silicone. Similarly, some polyester powder coatings, particularly those with a high gloss, contain a wax residue which may prevent a proper bond.

Adhesion testing is usually by means of a peel test method – although this does not produce actual measurable results. More rigorous pull-off tests can be scientifically controlled and will give more reliable results. Pull-off tests usually involve bonding the samples together and then pulling them apart. The sample will fail either in adhesion (one face pulls off) or preferably by cohesion (where the silicone actually remains bonded to the two surfaces but splits through its thickness).

Pressure equalised systems

From section 15.3 it will be remembered that the prime mechanism of water entry into a cladding system is a differential between high and low pressure. If the pressure in the glazing rebate can be equalised with that outside, the movement of air (and hence water) can be eliminated. In practice, there will always be a very slight lag in the equalisation of the pressures in the glazing rebate and the exterior, but the timing can be reduced so that to all intents and purposes the pressures are classed as equalised.[114]

To achieve pressure equalisation it is therefore necessary to:

- Provide suitable slots in the pressure plates to enable air pressures to reach equilibrium very quickly.
- Divide the facade into zones that are sealed from one another (a zone would normally exist around a single IGU).
- Ensure that the rear face of the cladding system is properly sealed to both air and water.
- Design joints so that they have a high degree of resistance to the mechanisms of water entry – kinetic energy, surface tension, gravity, wind-assisted capillarity.
- Have an effective means of draining water from the system – this will usually be via transoms rather than mullions.

Fundamentally therefore, with pressure equalised systems the outer face is sealed as tight as possible against rainwater, while the inner face is sealed as tight as possible against air inflow. Nearly all modern curtain walling systems are designed utilising the principles of pressure equalisation.

The performance of the air barrier is crucial in the success or failure of the pressure equalisation system; the barrier needs to be rigid in order to provide an air cavity of constant volume – pulses of air pressure during gusting wind would take longer to dispel if the air volume changed, and if the seal were flexing under wind pressure it could also serve as a pump.[115]

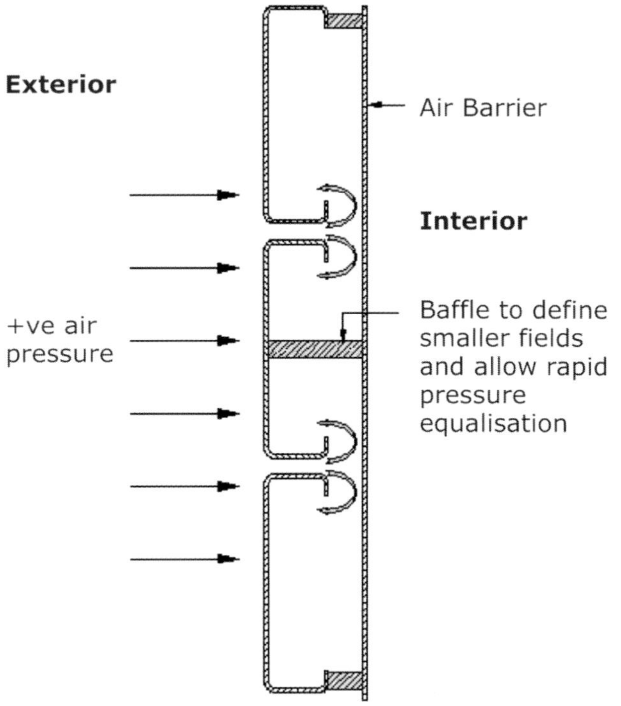

Figure 95 Method of pressure equalisation.

The effects of wind pressure on buildings are complex, with corners and the highest parts of the building suffering sharp pressure gradients. On the side facing the wind, positive pressures are experienced, whilst high negative pressures are experienced on the leeward side. Such pressure variations can cause air currents behind the outer face, with air moving from areas of high pressure to areas of low pressure. Such movements run contrary to the principles of pressure equalisation and so it is important that separate zones are created and sealed from one another; this maximises the chances of attaining pressure equalisation much faster.

Venting of the compartment permits the movement of air; vents must be large enough to prevent water filming over them. The more effective the air barriers around the compartment, the smaller the area of venting – if the compartments are poorly sealed, larger amounts of air will be needed and so more air must be drawn in from the outside.

15.5 FAILURE MODES

When a component or system fails, it ceases to perform in accordance with the design intent. The failure mode is the way in which that failure occurs. Research suggests that water penetration is the most frequent failure mode, with buildability and durability being additional concerns.[116] The underlying factors that influence water penetration failures are believed to be a lack of knowledge of cladding installers, fabrication errors and contractural pressures such as late running on site.

A study by Day in 1993 identified the different causes of failure; some 35 per cent of his sample reflected failings in workmanship, whilst 22 per cent were failures in design.[117]

Figure 96 Defective pressure plate system to a glazed rooflight. The clamping of the outer gaskets is insufficient and water penetrates the outer gasket into the drainage channel. Water in the drainage channel cannot escape owing to the lip of the roof upstand obstructing the channel at the bottom of the photograph. Note silicone top capping to try and alleviate leakage on the right of the photo.

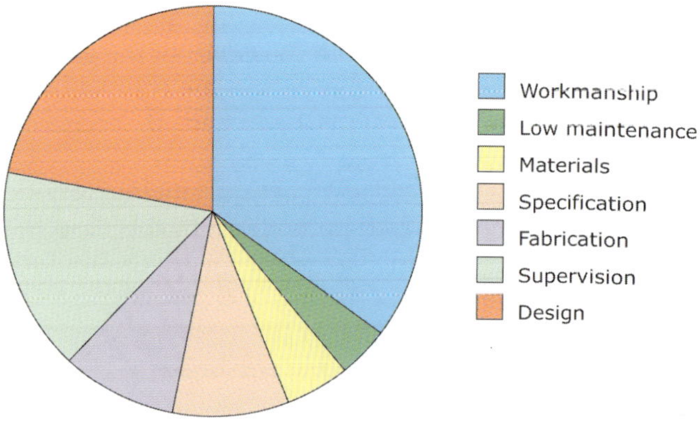

Figure 97 Causes of curtain wall failures.

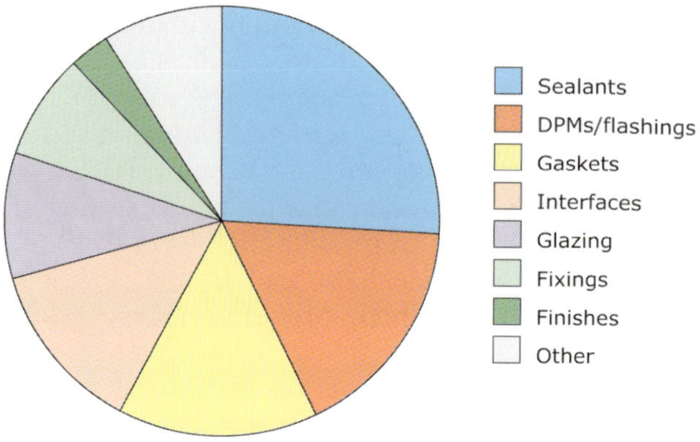

Figure 98 Elements likely to cause problems.

In 1997, CWCT studied cladding component failures and concluded that there were five main causes of watertightness failure[118]:

- defective sealant joints;
- incorrect selection and sealing of corner joints in gaskets;
- mitre joints in windows;
- poorly sealed frame connections; and
- missing or sealed drainage holes and redundant fixing holes.

The same study identified sealants, gaskets and metal finishes as the components of cladding that were most likely to fail – often as a result of inappropriate or misguided selection and use.

15.6 IDENTIFYING DEFECTS

The preceding sections have identified the main mechanisms of water ingress and the main systems of cladding. Knowledge of these principles can be put to use in diagnosing leakage problems in cladding systems that may be unfamiliar.

However, analysis of leakage patterns is seldom straightforward, as water can travel some distance within framing members before it becomes manifest internally. A methodical approach is needed, together with the courage to open up.

Stage 1 – initial reconnaissance

According to Field Marshal Montgomery, 'time spent in reconnaissance is seldom wasted'. Indeed, to identify the cause of cladding defects an initial inspection is essential. But the first inspection should not be a simple viewing of the problem, it should be tackled with some measure of science and methodology. For example, it would be little use inspecting a building only to agree with the occupant that the cladding is leaking – what is needed is an analysis of the locations and an understanding of the weather conditions that cause leaks, their duration and magnitude and whether there are any common patterns.

Whilst it may seem obvious, it is surprising how many surveyors fail to establish patterns and locations of leaks – these can often help to pinpoint potential problems. For example, the initial review may reveal that the leaks all occur at a certain level or always in line with a particular feature, a change in level or the interface between two different forms of cladding. Does orientation or level have a bearing? Are there any obvious defects? A quick check of the inner gaskets can reveal whether or not there is insufficient pressure – they should resist finger pressure, if they can be pushed inwards slightly, the external pressure plates may not be fixed to the correct torque.

Another common failing is to think in two dimensions rather than three – an art that is often easier said than done particularly as most graphical representations are drawn in two dimensions – usually because it is easier to do it that way.

Stage 2 – investigate the history and background information

Having gathered the initial information, attention now needs to focus on identifying the type and design of the system. Here an examination of the operating and maintenance manuals should help. Unfortunately, in older buildings a lack of as-built information is common, but if it is possible to identify the manufacturer, it might be possible to obtain construction details, perhaps not of the actual building but at least of the system. Needless to say, many manufacturers have ceased trading and so if all else fails, archival research may yield further information. Journals such as the *AJ* and *Building* often have building case files from back issues and these can provide some information as to the systems used, although this may be limited in terms of its reliability.

The purpose of this stage is to try and understand as much about the assembly process and design features as possible – how were joints to be formed?; was it mullion or transom drained?; are there rubber infiltration blocks?; etc.

Stage 3 – site investigation

A visual inspection of the cladding installation will reveal useful information but for many systems it will be difficult to tell too much without removal of at least the cover caps and probably some of the pressure plates as well. The object of this inspection is to check to see if the assembly of the frame, the fitting of setting blocks and the use of sealants is in accordance with the design and manufacturer's instructions. Check that pressure plates are installed the correct way up, i.e. that the slots for ventilation are not confused with the drainage slots. Check that the drainage paths are clear and not silted up and make sure that ad hoc amendments have not been made to drainage paths, such as the well intentioned but misinformed application of mastic sealant.

Water leakage can travel some distance such that the point of entry is not necessarily close to the point of egress. There are several methods of field testing that are generally recognised:

- Pressure spray in accordance with AAMA standard 501–94, *Methods of test for exterior walls*, American Architectural Manufacturers Association, USA.
- *Test methods for curtain walling*, CWCT, 1996.
- *Curtain walling. Watertightness site test.*, BS EN 13051:2001.
- *Methods of testing windows. Watertightness test under static pressure*, BS 5368-2:1980.
- ASTM E 1105–96, *Standard Test Method for Field Determination of Water Penetration of Installed Exterior Windows, Curtain Walls and Doors by Uniform and Cyclic Static Air Pressure Difference.*

The CWCT and AAMA test regime is essentially similar.

Field tests, if executed methodically, can be a very useful means of identifying leakage paths, but they require care and patience. They also need an observer on the inside, preferably equipped with a two-way radio for communication purposes. If the performance of the cladding installation is in dispute, it is best to stay within the parameters set in the standards; if the matter is non-contentious then it is possible to relax some of the requirements as long as the basic methodology is followed.

However, the two pressure spray methods, AAMA 501–94 and CWCT are rigorous in their application; they are designed for testing site installations during construction and unless used with care and judgment will create leaks in joints and gaskets that would perform adequately in normal service conditions. Similarly, if leaks are produced during the test, one needs to define whether or not the degree of leakage is acceptable (the result of extreme conditions) or whether remedial work is necessary.

Applying a pressure test to a feature that is not designed to deal with it (such as a rainscreen cladding) will inundate the system and drainage paths leading to a false result. It is usually possible to make something leak, but whether or not this produces the right answer depends upon a careful and methodical approach. It is very easy for example to form a pre-conceived view as to why a failure has occurred and then inadvertently make the evidence fit the theory.

The water spray test

Water is supplied via a compressor pump and break tank at a pressure of 220±20kPa to a suitable nozzle (CWCT specify a Monarch Type B-25, #6.030 nozzle) which produces a 30° spray cone. The cone is held perpendicular to the surface being tested at a distance of 300mm. A five minute test period is suggested for a length of component of 1.5m, i.e. passing the nozzle back and forth over 1.5m for a five minute period. This test uses around 110 litres of water over five minutes, so precautions may be needed if, for example, work is carried out above a pedestrian area or building entrance.

If no leakage is determined, the test can move on to the next 1.5m section and so on, starting at the bottom of the sample area or facade and gradually working up.

The process may identify a suspect area that can then be investigated, but if the exact position is unclear, masking tape can be applied over the joints and then progressively exposed to narrow down the area of defect.

If water leakage is determined it may be necessary to remove pressure plates and gaskets so that water paths can be examined.

As noted above, the water spray test is not suitable for all types of cladding and for rainscreens or unsealed patent glazing systems a spray bar test would be better as this reduces the pressure of water hitting the surface and produces a curtain of droplets flowing down the wall in a manner which replicates actual conditions. BS EN 13051 sets out the procedure to be followed.

The spray bar test

Using this system, a spray bar is fitted with suitable nozzles at a maximum of 400mm centres and suspended so that it is 250mm from the face of the cladding system. Water under mains pressure (2–3bar) is then applied and left to run for 30 minutes at a nozzle flow rate of five litres per minute. Unlike a water spray test, water should not be aimed directly at joints, but left to flow over them.

The cabinet test

This is usually restricted to windows and doors and involves constructing a sealed cabinet on one side of the component, extracting air from the cabinet to create negative pressure and then applying a water test (spray bar) to the outside. If the seals are defective, water can be drawn in as a result of the air pressure differential described earlier.

Cabinet tests are described in BS 5368-2; they are not particularly easy to carry out and rely upon the elimination of air leakage paths around the perimeter of the testing cabinet, or indeed around internal joints in the construction. The cabinet is usually a timber frame sealed with a thick polythene liner, although an effective system can be constructed on site using timber battens to create a gap and a layer of polythene. The polythene has the advantage of allowing some measure of visibility, so that leakage paths can be tracked. The polythene must be held proud of the window or frame in question. Once assembled, air can be evacuated using a suction pump or industrial vacuum cleaner. A manometer is needed so that the pressure within the cabinet can be measured and monitored; the test is usually carried out in 50Pa increments up to 200Pa, then 300Pa and finally in 150Pa increments up to the maximum pressure (which will need to be judged according to the specification of the glazing system). (See also Technical Update 2 – CWCT.)[119]

Checklist of critical points in rainscreen cladding[120]

- Can water be discharged from the outer cladding face; are there drips under external projections to throw water clear of the face?
- Is there a cavity behind the rainscreen of sufficient width to prevent capillary movements?
- Is there a continuous and effective air barrier? Are joints in the air barrier sealed and is it properly compressed?
- Can water drain away from the system? Is there a proper water path and is it clear from obstruction (moss, silt, etc.)?
- Is there adequate venting of the cavity? Does the area of venting take into account potential air leakages through the air barrier or changes of volume of the cavity (through flexible inner linings)?
- Is the facade compartmentalised to prevent lateral movement of air behind the rainscreen?
- Is the air barrier rigid or will it flex (and so vary the volume of the airspace)?
- Is the cladding sufficiently rigid to resist wind loads without excessive deflection?

15.7 CONDENSATION ON METAL FRAMING

From a design point of view, the metal framing members of the curtain wall (or windows) should be maintained at a temperature slightly higher than the interior surface of the insulating glass unit (IGU), so that condensation will form on the glass first rather than its frame. Although temperatures will differ according to the construction of the IGU, for a 6–12–6 unit the temperature of the frame will need to be greater than 10–12°C.[121]

The trend toward highly insulating units, using Low-e coatings, warm-edge technology and gas filled voids, tends to raise the inner surface temperature of the glass. According to CWCT, including a Low-e coating can raise the centre pane temperature to 15.2°C meaning that condensation could occur on the frame first unless a highly insulated frame design is chosen. Good practice dictates the use of better insulated framing for high efficiency insulating glass units.

Slight condensation in the form of misting towards the lower parts of the glass or window may be acceptable if it only occurs on a few occasions each year; if the condensation is severe enough to occur as running water, this is unsatisfactory and remedial measures should be contemplated. Such conditions may be the product of user abuse, for example removal of or failure to maintain ventilation and/or high internal humidity due to processes or equipment failure – something that is difficult to control at design stage.

Case study 1 – office building in Scotland

The building comprised a multi-tenanted office of circa 10,000 sq m arranged on a ground and twelve upper floors. The building's footprint reduced over three floors with a sloping glazed roof. The walls and roof were clad with a conventional drained and ventilated stick system which was designed to be mullion drained. Special shoes were built into the sloping roof mullions (rafters) to allow drainage into an aluminium gutter positioned at the transition of the sloping and vertical sections of cladding.

Water penetration was encountered at all floor levels to varying degrees, with most locations being confined to the areas below the sloping roof. Over the years, various unsuccessful investigations and trial repairs had lead to accumulations of 'wet' mastic seals over pressure plates and gaskets. The window cleaning cradles had not been maintained and could not be used for inspection purposes; the only access being via a specially designed scaffold extending vertically, raking across the roof and then vertically upwards to the full height of the building.

To begin with, selected cover strips and pressure plates were removed. Invariably, removal of the pressure plates released reservoirs of water that had been held in place by accumulations of silt and moss as well as the misguided applications of mastic, which in some cases sealed drainage paths.

The aluminium gutter was tested and found to be watertight; however the top lip of the gutter was formed of aluminium sheet held in place by the top pressure plate and gasket system to the vertical wall. When the pressure plate was properly compressed against the top glazing gasket, a good seal was formed between the glass and the gasket, but not between the gasket and the gutter. This was leading to water ingress into the system.

Tests using the AAMA procedure commenced at the bottom of the building, working upwards. Variable and random leakage patterns ensued suggesting that the basic concept of the system was sound, but that errors in construction might be responsible for localised faults.

To investigate the matter fully, various IGUs were then removed and the glazing rebate inspected. According to the manufacturer's assembly details, a bead of sealant should have been applied to the mullion transom connections – this was missing in many areas. In addition, the inner seals were assembled using pre-made corner pieces and straight gasket lengths. The contact faces between the corner pieces and the straight gaskets should have been sealed; they were not, thus allowing water to be drawn into the building as a result of pressure differentials.

Attention was then turned to the drainage shoes at the base of the rafters. These shoes were designed to sit behind the face of the gutter and to drain into it via small pieces of hose cut in the side of the gutter. If the shoes leaked, water could only travel inside the building. This was solved by amending the gutter design and extending the rafters into the gutter itself to provide a fail-safe design.

Given the age of the glazing, and the need to remove all panels to achieve a satisfactory seal to the back gaskets and frame connections, all IGUs were replaced during the remedial works contract. The extent of water ingress was such that edge seals had been in contact with moisture for some period of time and could have been compromised as a result. All back gaskets were then sealed and fillet seals inserted at the transom and mullion connections. Following reassembly and reglazing, the entire areas were again field tested.

In summary, the water leakage was due to a combination of the following:

- Poor initial assembly – a failure to seal the mullion and transom joints and a failure to make the back seals airtight.
- Poor design in the detailing of the top gutter and the creation of a bespoke drainage shoe arrangement.
- Poor maintenance – lack of the cradle meant that no cleaning had taken place; the drainage paths had become compromised.
- Well intentioned but inadequate repairs – failing to understand the mechanisms of water ingress and widespread use of sealant.

Case study 2 – office building in coastal location

The building comprised a 1970s office block of rectangular plan constructed above an enclosed shopping centre near the coast. The cladding was steel framed, but single glazed. Coloured spandrel panels and fixed vision glass was held in place by external gaskets and pressure plates.

The cladding system leaked heavily, it relied upon a face seal with no provision for accidental leakage and no drainage paths. Over the years the system had fallen into disrepair and corrosion of the frame had commenced. The client needed to rectify the problem, but low rental income and poor cost recovery prospects had forced a series of compromises in repair strategies – usually involving copious amounts of sealant. Numerous pressure plates appeared loose and some of the cover caps were missing.

Water testing was unnecessary in this instance. Being single glazed, the provison of drainage channels around the glass was unnecessary for the purposes of durability, although if drainage could have been contrived it would have been beneficial. In view of budgetary constraints, replacement of the cladding was impracticable and so a two stage external seal was provided. First, the old pressure plates were removed and discarded and the glass cleaned to remove all traces of the old mastic. Glass to metal joints were cleaned and sealed and a layer of butyl tape applied. New eagle gaskets and pressure plates could then be refitted, taking care to ensure that new fixing holes were drilled so that sufficient torque could be applied. The plates were then covered with drained cover caps. Internally, glass to metal joints around the inner seals were 'top capped' with silicone to form an internal air seal.

15.8 PROBLEMS WITH PRESSURE PLATES

A typical aluminium pressure plate system is shown in figure 99. In order to exert effective pressure to the gasket, the pressure plate must be tightened to the correct torque (specified by the designer). Pressure plates are usually secured into the mullion screw fixing channel, but repeated insertion and renewal can lead to a weakening of the fixing and it is sometimes necessary to re-drill and re-secure a plate to achieve sufficient pressure.

In most cases a snap-on cover bead is provided; this will usually be colour coated, although stainless steel varieties are available. The range of profiles is very wide but they all rely upon a friction grip. The level of friction is governed by the closeness of fit (or interference) between the cover and the pressure plate. Both fittings are (usually) extrusions that will have been created using different die, the manufacture of which is subject to manufacturing tolerances. Problems can sometimes occur in batches supplied where the tolerance in dimension errs towards the low side, leading to a poor interference fit and therefore a loose pressure plate.

Figure 100 shows an aerofoil profile cover cap with a projection of around 100mm. The manufacturer states that this type of cap must only be used in conjunction with a pressure plate of a particular profile. However, in this case, the aluminium profile has been omitted and instead a rectangular glass fibre pressure plate (resembling a strip of hardwood) has been used instead (an example of the installer mixing different

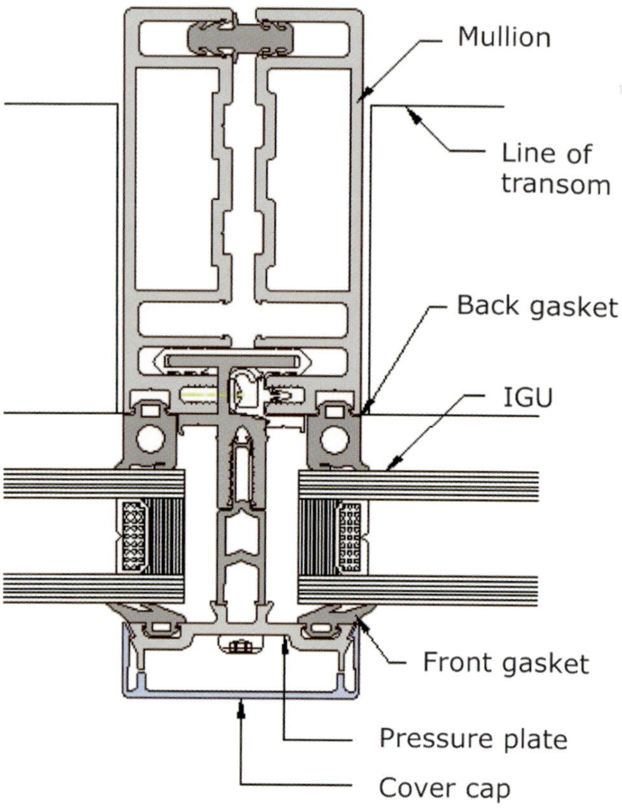

Figure 99 Section showing standard stick system with pressure plate.

Figure 100 Example of an incorrectly fitted cover plate. In this case the aerofoil shaped cover has been wedged against a glass fibre pressure plate of inferior quality.

components to save costs). The cover cap is a tight fit against the pressure plate, but there is little strength in the glass fibre to prevent the cap from being torn out.

The caps suffer from what might be loosely termed a 'Friday and Saturday Night' syndrome, as gangs of late night revellers have found that the covers at low level can be dislodged easily with a sharp kick. The more they are put back, the looser they become. Of more concern is the fixity of the covers at higher level where lateral pressure, for example from a window cradle, could pull the cover away; the pressure plates will need to be replaced with those of the correct specification.

15.9 GASKETS

Aside from silicone glazed systems, most cladding and window installations rely upon the proper performance of gaskets to maintain water and airtightness. In order to achieve a seal, the lip of the gasket must exert a pressure on the glass (lip pressure), usually in the order of 0.720N/mm to 1.760N/mm.[122]

Gaskets are usually made from elastomeric materials such as butyl rubber which, although durable, do degrade over time as a result of exposure to UV radiation, direct and reflected sunlight and ozone. Volatile compounds in the gasket are driven off (out-gassing) by these mechanisms with the result that they eventually suffer shrinkage in volume and or brittleness. Often gasket materials are stretched as they are installed with the result that they gradually shrink back into shape, opening corner or butt joints in the process.

As noted earlier, it is better to anticipate some water leakage through the outer gasket system and to take steps to drain it away harmlessly to the outside. To an extent therefore, loss of lip pressure can, in part, be accommodated. However, if the inner or air seal gasket is also defective there is a risk that water leakage can take place.

Replacement of the outer gaskets may be possible, but replacement of the air seals will probably involve deglazing (unless they are harder wedge gaskets – used when the glass in windows is bead glazed). An alternative to gasket replacement in these circumstances is the provision of a wet seal or top cap over the lip of the gasket.

Given accurate and careful workmanship a wet seal can be a useful expedient, but its use should only be contemplated if the glazing system and its weaknesses have been properly understood. Unfortunately, the application of mastic is often seen as a quick fix, carried out by inexpert applicators. For success, Nicastro recommends the following[123]:

- Use only silicone sealant that can be shown to be compatible with the original gasket.
- Use black sealant and ensure that it is applied by trained applicators.
- Prior to application, clean all contact surfaces with solvent cleaner.
- Maintain a minimum cross section of fillet of 6mm and a contact area on each face of 6mm.

Wet sealing can make a marginal difference to the shading of the perimeter of the glass and carries a slight, possibly theoretical, risk of increasing thermal stress between the centre of a pane and its perimeter.

15.10 SEALANT FAILURE AND REVERSION

Sealants are an essential part of modern construction and have a crucial role to play in the design of weather-tight joints. ASTIM defines a sealant in building terms as 'a material which has the adhesive and cohesive properties needed to form a seal'. Joints can fail as a result of poor design, workmanship or the incorrect choice of sealant for the job.

Broadly, there are several basic types of sealant failure:

- adhesion failure;
- cohesion failure;
- delamination failure; and
- reversion.

Adhesion failure is a loss of bond between the sealant and the surface to which it is applied. The failure mode is characterised by a visible separation along the line of the interface – this is termed interfacial adhesion failure. Most sealants require a primer or at least clean surfaces free of contamination by water, frost oil or dust. Omitting the primer and or failing to clean the contact surfaces prior to application may cause adhesion failure.

Cohesion failure is characterised by tears or splits within the body of the sealant, whilst the bonding faces remain intact. In these cases it is probable that the sealant has been overstressed either as a result of poor joint design or a loss of flexibility. Joint profile is important – it is generally accepted that for a sealed joint to work it should be bonded on two faces only. However, the thickness of the joint is also relevant. A thinner joint will possess more flexibility than a thicker one, but equally will be more vulnerable to the effects of exposure to UV light and the ensuing degradation.

The profile of the sealant joint will influence the performance of the sealant. For example, the triangular fillet profiles that are often found around windows perform badly in comparison to rectangular joints, particularly if they are concave.[124]

Sealants are either designated by categories of ±12.5 per cent, ±50 per cent or +100 per cent/-50 per cent of initial movement capability. Joint movement on a percentage basis is also affected by the size of the joint. For example, a 6mm movement on a 13mm wide joint will be twice as much on a percentage basis as a 6mm movement on a 25mm wide joint. Therefore, narrower joints are more prone to cohesive failure of the sealant.

Many polyurethane sealants will stiffen and show a loss in movement capability over time, making them vulnerable to cohesion failure.

Delamination failure occurs when the sealant overstresses a coating or material to which it is applied. For example, the bond between a mastic seal and a paint film may be perfect, but joint movement produces a stress that is too great for the paint/substrate bond to accommodate. Polyurethane sealants also have the inherent property of increasing two- to three-fold in sealant modulus in cold temperatures, making them stiffer and less flexible, thereby increasing joint stress.[125]

Similar effects can occur when the substrate becomes softened as a result of entrapped moisture. Polyurethane backing rods can absorb moisture and hold it in contact with certain vulnerable materials such as acrylic coatings applied to external rendering systems. For this reason, external acrylic coatings should be stopped short of a joint before the application of sealant. Failures such as these are termed interphasal failures.

Reversion failures and deterioration

Reversion is a term used to describe deterioration in the elastomeric properties of a sealant. (Some authorities term this as a loss of rheological properties.) Reversion is a somewhat difficult subject as there are mixed scientific opinions as to its cause. Essentially however, reversion is the conversion of a sealant from a cured to an uncured state, usually initiated by high temperatures, UV exposure and water vapour. Experimentation has revealed conflicting evidence as to the process; the colour of the sealant may also play a part as a result of the energy absorbing capabilities of the different pigments.

Sealant reversion can be tested fairly easily by taking a sample of the material and rolling it into a ball. If the sealant holds its new shape, it has lost its rheological properties and can be termed reverted.

Another form of reversion, again easily identified, is the reversion of silicone sealant back to a powder form. Typically, this can be seen to affect the thin joints in stone cladding and is probably the result of high temperature and moisture.

Sealants such as multi-component polyurethanes are organic materials that degrade in UV light. The degradation results in loss of flexibility and movement capability, ultimately ending in reversion.

Hardening of the sealants over time can cause a greater stress to be exerted at the bonding face with each extension and compression cycle. If the stress exceeds the allowable stress in the background material, localised adhesion failure can occur.

15.11 CLADDING AND BUILDING MOVEMENTS

Cladding systems must be able to tolerate expansion and contraction in the glass and framing members together with normal movements, deflections and long term axial shrinkages in the structure to which it is fixed. The effects of wind sway, differential settlements and elastic deformations also need to be factored into the design of both framing and fixing systems.

Thermal movements in frames must be accommodated; if there is no provision for movement in the frame design, stresses will build up in the profile; the less the provision for movement, the greater the stress and the greater the propensity for deflection or buckling. For northern climes, it is usual to apply a rule of thumb of 1mm movement provision for every 1m of length of component and to calculate stress with this in mind. The typical range of temperatures in the UK (for design purposes) could range from −25°C to +80°C. If the stress in the element in question is allowed to exceed its permissible stress, failure can occur.

Curtain wall systems are usually fixed by way of brackets back to the floor slab. Deflections at the edge of the floor slab are usually calculated at design stage – possibly span ÷ 360 or span ÷ 500, giving 18–25mm movement for a span of 9,000mm. The system must accommodate this movement, but to achieve this it will probably be necessary to incorporate split transoms as well as sliding mullion joints. A split transom will usually incorporate twin overlapping pressings and an internal concertina gasket.

Concrete structures suffer from long term creep – axial shrinkage such that movement of as much as 0.1 per cent can be expected; for example 30mm for every 30m in height[126]. Movement will become progressively greater towards the top of the building and must be accommodated in the cladding and glazing systems to prevent stress formation (see also figure 53 on page 96).

15.12 SPANDREL PANELS

A typical fully glazed curtain wall system will comprise a mixture of vision glass and opaque 'look-a-like' spandrel panels used to conceal the floor edge construction, ceiling void, heating plant, etc.

Spandrel panels pose specific difficulties since the dark colours used create high temperatures in the glass or the void behind it: 90°C is not unknown. Broadly, there are two methods of creating look-a-like effects:

- adding an opacifier to the rear face of the glass; and
- creating a dark painted void behind the glass – a shadow box.

Opacifiers are usually added to the fourth face of an IGU, being either a polyester or (mylar) film or alternatively a silicone spray treatment. Ceramic frits are also used. High temperatures can cause applied opacifier films to blister, whilst oil or latex paints are short lived as UV rays break down the molecular bond between the paint and the glass. Most opacifier failures are associated with shrinkage, weathering, oxidisation or adhesion failures in films that are sensitive to moisture, heat or light.

Annealed glass can be used for spandrel panels, but only if certain criteria are met. A ventilated void would be essential to remove warm air and glass edges would probably need to be polished to reduce the risk of thermal stress fractures. If annealed glass were to fail, sharp shards of glass could fall out and so some form of scrim would need to be applied to the back face. Given the uncertainties of performance, heat strengthened or toughened glass is a more likely choice for spandrel panels. Heat strengthened glass is able to deal with higher working temperatures than annealed glass although it can still shatter into shards; it does however display less distortion due to roller wave.

Shadow box construction is difficult to achieve in practice and creates further problems of condensation control and heat build up. High temperatures in the void behind the glass can permit volatile compounds to gas off and cause staining to the back of the glass when they condense.[127]

15.13 CLEANING EQUIPMENT

A detailed study of the various types of window cleaning installations is beyond the scope of this book. However, there are various matters which need to be considered in conjunction with the cladding design.

For buildings over 30m in height, a fixed installation is generally essential, but cradles will need additional attachment points such as slotted mullions, button guides, tie bolts or pins.

Typical maintenance standards for curtain walling demand cleaning of the metalwork up to five times a year. However, most installations are suitable in conditions where windspeed does not exceed 12m/sec. While this condition may be fine at ground level, conditions may be very different at upper stories in a tall building. If it takes five to ten days to access all parts it is unlikely that weather conditions will be favourable on this number of days five times throughout the year and hence the maintenance and inspection provisions may be compromised.

Figure 101 Roller wave distortions in toughened glass.

Figure 102 Cladding systems must be cleaned regularly to prevent deterioration. Specialist cleaning may help to improve the finish on this colour coating but it is probably too far gone to be restored to its original condition.

16 Insulating glass units

At a glance

- Insulating glass units are an integral part of any glazed cladding system, but poor detailing and assembly of the units can lead to rapid deterioration, with the formation of condensation and scumming taking place between the two glass layers.
- The edge seal detailing of an insulating glass unit (IGU) comprises two seals, a primary seal between the glass and the spacer and a secondary seal around the outside edge of the glass. Seals are vulnerable to moisture damage if held in contact with damp surfaces or water that cannot drain out of the glazing system. Failure of the seal within 10 years is not unknown in glazing systems that rely upon the full bedding of an IGU in glazing compound.
- To help prevent moisture entering the void between the glass panes, spacer bars are used. Spacers contain a dessicant to absorb moisture. With increasing emphasis upon thermal performance, spacers are now being designed as 'warm edge' units which have a higher thermal resistance. Some types are manufactured from thermoplastic butyl cord; these materials contain the dessicant within the body of the butyl cord and are thought to perform more effectively. However, fairly recently problems of the spacer 'walking' into the field of view have become known. For commercial reasons, manufacturers publish little information on this problem, but some large, high profile buildings are known to suffer from it. Remedial work, which involves replacing the units, is expensive.
- Aside from moisture problems, glass units can fail as a result of temperature differentials between the perimeter and centre of the glass panes. Warm edge spacers can increase the risk, although failures often occur in conjunction with another problem – edge damage to the glass for example.
- Heat strengthened and toughened glass is able to withstand a greater range of temperature than ordinary annealed glass. However it can suffer from visual defects such as roller wave (this can be significant on mirror glass surface) or a problem of spontaneous fracture due to glass impurities, often nickel sulphides (NiS).
- NiS are small particles that become metastable during the toughening process. As the glass cools the particles, which shrink in high temperatures, try to return to their original sizes. Depending upon the size of the particle and its location, spontaneous failure of the glass can then occur, often several years after installation.

16.1 INTRODUCTION

Insulating glass units began to be used in the UK during the 1960s, although at that time, and well into the 1970s, single glazing held prominence. Whilst single glazing can be expected to last for many years, the useful life of an insulating glass unit (IGU) is governed to a large extent by the quality and durability of the edge seal. Once this breaks down, condensation, misting and scumming of the inner glass surfaces occurs. This process is irreversible and leads to the need to replace the glazing.

With the trend towards greater thermal efficiency, manufacturers have in recent years introduced noble gases into the void of the IGU in an attempt to improve thermal performance. Whilst significantly improved performances can be obtained over conventional systems, they will lose efficiency if the gasses leak out – this can be difficult to check during a normal site inspection.

While the normal life of an IGU is expected to be around 25–30 years, longer periods are not unknown. However, the major concern arises when the units begin to fail after 10–15 years.

A number of variables affect the life of an insulating glass unit:

- design and physical properties of the materials used;

- the degree of care employed in manufacturing the unit; and
- service conditions.

Specifically, the variables relate to:

- the moisture resistance of the seals;
- the amount of desiccant used and its moisture absorbent capability;
- the length of the vapour transmission path through each material; and
- the width of the path between the moisture barrier and the edge of the glass.[128]

16.2 COMPOSITION

Whilst there are a large number of different combinations of seal and spacer design, the essential components of an IGU are shown in the diagram below.

The purpose of the edge seal is to bond the two layers of glass together and to ensure that air leakage into the inter-pane area does not occur. Within Europe, and particularly in commercial buildings, the edge seals are usually two-stage seals, whilst in North America these are less common.

Spacers are usually of aluminium, although plastic materials are becoming more common in the drive to reduce heat losses around the perimeter of the panes. In older units, steel or even lead spacers were used.

The purpose of the outer or secondary seal is to perform the structural function of bonding the two leaves together and restraining them throughout the service life of the IGU, including during transportation and installation. In a majority of panels the edge seals are of silicone – this is universal in the case of structural silicone cladding systems.

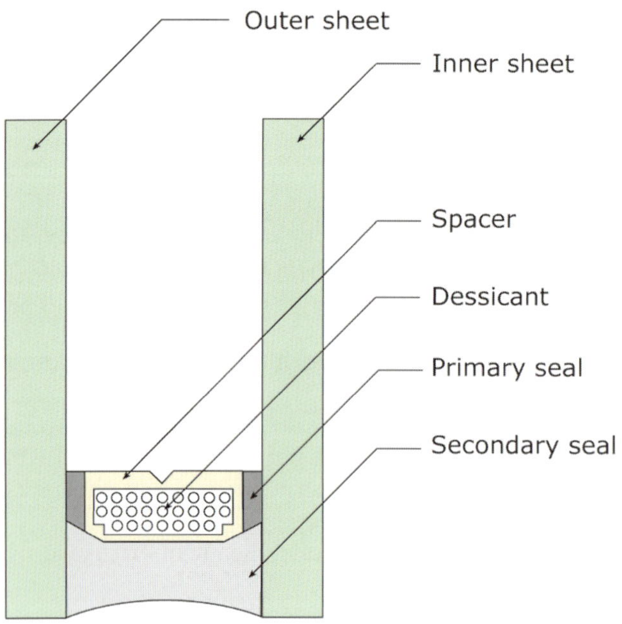

Figure 103 Basic components of an insulating glass unit.

Some IGUs are also supplied with a foil edge tape. The tape is designed to facilitate the curing of the secondary seal and to help protect it during handling and from the effects of UV radiation. However, edge tapes have been found to cause problems as they can form channels which hold water in contact with the secondary seal.[129] Additionally, some PVC based edge tapes shrink in service, placing additional stress on the edge seals.

All insulating glass units contain desiccant to help absorb any residual moisture vapour arising during manufacture or through vapour leakage into the void. Silica gel or molecular sieves are often used individually or in combination with one another. The desiccant is able to absorb a small measure of moisture but only up to about 20 per cent of its dry weight; beyond this, the dew point of the air within the unit will rise. If the dew point reaches 0°C, condensation will form on the inside of the outer pane if the weather is cold enough. Initially, such condensation may go unnoticed, but over a period of time, repeated condensation causes scumming.

The primary seal is designed to provide a barrier to the transmission of gas and the diffusion of water vapour; polyisobutelyne is often used for this function. Most premature failures in IGUs occur as a result of the loss of adhesion of the primary seals.

It would be easy to conclude that once the insulating glass unit has been lifted into position and fixed, the job expected of the edge or secondary seal would be much reduced. In fact, the opposite is probably true. The seal is located within its own microclimate in which conditions can be far removed from the ambient climate. Throughout the service life of the unit it will be subjected to changes in temperature, air and wind pressure, exposure to water and, through a process of internal reflections, UV light. Pressure fluctuations will cause the individual panes to bow. (The stresses so caused are generally greater for small units or those with a high aspect ratio.) Stresses are therefore concentrated at the interface between the glass and the secondary seal. The stress concentrations are not simply the cumulative effects of the agencies described above; they are disproportionally higher due to a process known as synergism.[130]

Various studies have recorded the highest stresses on the edge seals as occurring during a combination of exposure to moisture, sunlight and high temperature over a period of several years. The edge seal is usually contained within some form of frame – in Europe, temperatures at the edge seldom exceed 40°C, but in hot climates, temperatures of 80°C can occur for extended periods. If the edge seal fails, the unit may not lose its structural integrity (unless the failure is profound) but water vapour may penetrate the desiccant, leading to saturation and the failure of the unit.

Because of their polymeric nature, common edge seal materials, silicone, polysulphide and polyurethane, are all able to store water, although silicone usually performs much better whilst polysulphide is generally able to hold much more. This means that the sealants can expand – swelling of 38–55% has been recorded in polyurethanes held in water at 60°C whilst silicones displayed 6–12%. Swelling of the secondary seal will place the primary seal under stress; adhesion failure can then occur with the result that air and moisture enters the void and reduces its dew point.

The performance of the IGU can therefore be seriously compromised if the edges are allowed to be in prolonged contact with moisture. In contemporary glazing systems it is usual to provide a gasket system with ventilated drainage paths to allow moisture to escape or evaporate. However, over a period of time, drainage can be reduced as a result of silting, organic growth, corrosion product, etc. For this reason periodic cleaning is essential. Since the drainage paths are not normally accessible for cleaning it is necessary to clean the external surfaces regularly to prevent the accumulation of material that may find its way into the system.

Some sealants can be affected by chemical attack from atmospheric conditions, cleaning fluids, paints or stains, so it is important to ensure compatibility of sealants with the expected environmental and physical conditions that will be experienced in use.

Earlier curtain wall and glazing systems were less advanced; in the 1970s it was common to find fully bedded systems, whereby the IGU was surrounded by glazing compound with no provision for drainage. Whilst in theory, full bedding was likely to exclude water, it was very common to find voids in the sealant as a result of air entrapment during glazing. These voids could contain water; in many cases this would be 'pumped' in as a result of pressure fluctuations or deflection of the glazing. Furthermore, the sealants would gradually breakdown as a result of exposure to UV light, giving further opportunity for water ingress.

Figure 104 Failed IGU. Note condensation trapped between panes and salt crystallisation on top of spacer.

Some early curtain wall system utilised insulating glass units incorporating lead spacers. The units were fully bedded in an oleo resinous glazing compound. The compound broke down rapidly as a result of a combination of UV light and high temperatures compounded by thermal conduction in the lead bars. These glazing systems were inherently defective and contained numerous voids which could contain water. The average life of this type of unit before failure is around 15 years. Extensive IGU replacement is usually necessary, but without changing the entire glazing system it is impossible to incorporate drained and ventilated glazing rebates. By adopting a vapour permeable sealant to the outside of the rebate, coupled with an impervious seal to the inside, some measure of improvement can be gained.

16.3 WARM-EDGE TECHNOLOGY

The desire to reduce heat losses and improve the thermal efficiency of insulating glass units has lead to developments in various technologies: Low-e coatings, gas filling and 'warm-edge' spacers are some examples.

The edge spacers described in the preceding section are usually aluminium types, but these conduct heat and lead to heat losses around the perimeter of a panel. The use of a non-metal system will result in the lowest thermal conductance at the edge, and spacer systems using designs that are comprised of metals with lower conductance (steel as opposed to aluminium) will help, although to a lesser extent.

Over recent years, investigations by various research organisations has demonstrated that insulated spacers can reduce the overall heat losses through an IGU. It used to be the case that U values could be measured through the centre of a glass panel, but this method gives an unreliable result and is now discontinued in favour of an assessment of the panel as a whole. By using insulated spacers, heat losses can be reduced over the pane to give a more uniform performance, the dominant design factor then becoming the framing material. The term warm-edge technology has been coined to reflect the research into different methods of reducing heat loss through spacer bars.

Warm-edge spacers have found a niche in the residential market, but to begin with they were slow to be adopted in commercial buildings. Commercial markets are now expanding and the technology is becoming more popular, with a growing number of high profile buildings being glazed with IGUs with warm-edge bars.

Spacer/sealant combinations with low thermal conductivity values will keep the edge of the inside pane of glass warm when external temperatures are cold. Of course, at the middle of the glass, glass surface temperatures will be the same as would be found with conventional units (assuming the air gap and type of glass is the same). But the lower the conductivity of the edge spacer used, the less of a difference will be seen as measurements are taken from the middle of the unit moving toward the edge of the window.

Achieving uniformity of glass surface temperatures has significant benefits as it reduces stress in the glass. In cold weather, conventional spacers result in less thermal expansion of the glass around the perimeter than at its centre. In very warm weather, these same differences exist only in reverse. The differences in the amount of thermal expansion or contraction lead to the build up of stress within the glass and can contribute to the formation of stress cracks. By using insulated spacers the difference in expansion or contraction between the centre and edge can be reduced, thus reducing the stress.

Whilst warm-edge technology can reduce stress in the glass, the same cannot be said for the spacer itself. In a conventional unit, the thermal properties of the spacer mean that there is a smaller difference between internal and external temperatures. An insulating spacer alters this condition so that whilst the outer pane of glass is at the same temperature, the inner pane will be at a more uniform internal temperature. The different rate of expansion of the two panes creates a stress on the edge seal. Flexible warm-edge spacers tolerate this condition well, whereas rigid spacers are at a disadvantage.[131]

There are various types of warm-edge spacer, some containing stainless steel and various butyl formulations. Thermoplastic spacers (TPS) were introduced in 1995 and comprise a butyl-based thermoplastic material formulation that is UV-stable with integrated desiccant applied directly onto the glass by a computer-controlled extrusion nozzle. It is non-metallic and designed to provide enhanced heat insulation. It is also an effective gas and moisture barrier. Thus, whereas a conventional IGU contains a spacer, dessicant, and primary and secondary seals, a TPS assembly contains only two: the butyl spacer and the outer, secondary seal.

Unlike the dessicant systems in a conventional IGU, the dessicant within the thermoplastic edge seal is incorporated as a fine powder within the seal itself, usually polyisobutylene. Moisture that is able to bypass the secondary seal is readily absorbed by the dessicant at the bottom of the seal, but the rate of diffusion of moisture gradually reduces as the diffusion distance becomes longer. Because the dessicant at the lower edge is at 100 per cent humidity, fresh moisture is not drawn into the material – in other words the entry barrier becomes increasingly difficult to pass. Thus, thermoplastic edge seals are believed to offer greater resistance to condensation problems than conventional units and on this basis their life expectancy can be considerably longer.

Similarly, the manufacturers' will claim that units constructed with thermoplastic edge seals offer significant benefits in terms of gas tightness. This is essentially because the permeability of the seal is so low. Given a thickness of 8mm the butyl's low gas permeability becomes so dominant that the permeability of the outer sealant is no longer relevant.[132]

Despite the obvious advantages of warm-edge spacers, there is evidence of an increasing number of failures arising from the displacement of the edge seal or 'spacer walking', in which the relative movements of the two glass panes causes the spacer to work its way inward and into the line of vision. The problem takes the form of an initial bulge in the spacer, followed by progressive movement and leading, in some cases to the migration of the spacer away from the edge of the glass. Whilst being visually unattractive, such movement could affect the gas tightness of the IGU or its structural and thermal performance. IGUs that are affected by this problem will probably have to be replaced, and in a building where large panes could cost upwards of £3,000 to replace, the likely costs would be significant.

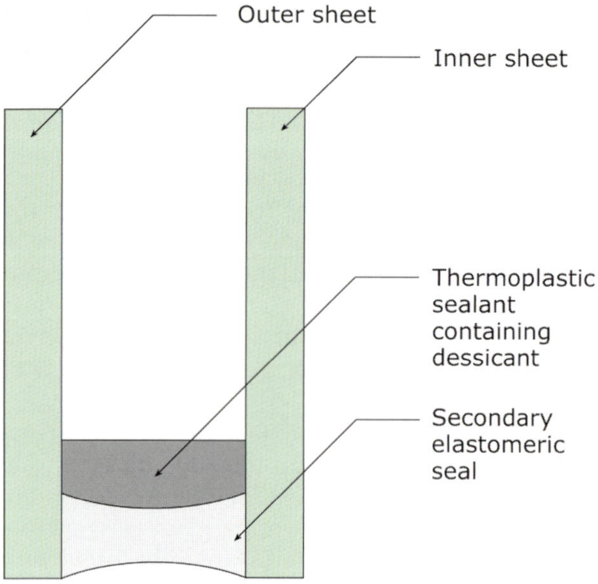

Figure 105 Warm-edge technology.

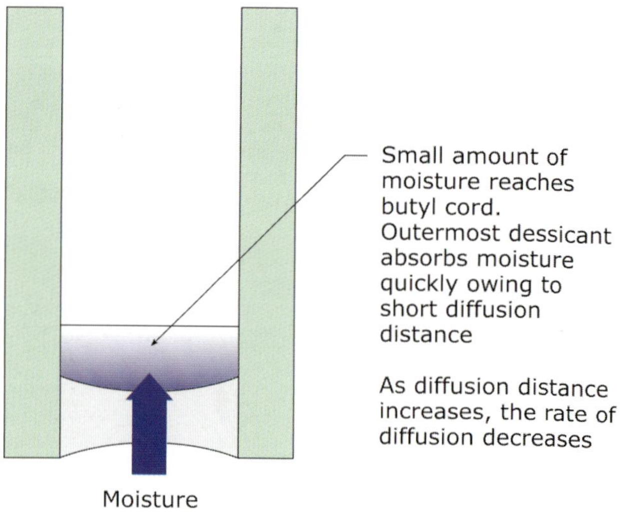

Figure 106 Moisture diffusion into a thermoplastic warm-edge spacer.

Problems such as this tend to result in litigation or warranty claims against the manufacturer; information as to exact causes can therefore be hard to come by – manufacturers and building owners will wish to keep matters such as this confidential. Commentators have speculated that the problem may be due to one or a combination of the following:

- thermal stresses;
- poor adhesion of the seal during manufacture – possibly due to contamination of the glass from the use of incorrect cleaning rags, moisture or solvents on the edge of the glass during handling;
- incompatibility between the spacer and any coatings applied to the inside faces of the glass – most soft-coatings have to be stripped from the margins of a pane before it is assembled in to an IGU as edge seals do not bond adequately to soft coatings.

According to CWCT, although warm-edge spacers enhance the overall insulating properties of the unit and reduce the risk of condensation, they will increase the risk of thermal breakage.

Given the numbers of buildings now employing warm-edge spacers, it is possible that instances of failure will become more widely publicised.

16.4 VISUAL DEFECTS IN GLASS

Glass colour

As a result of its iron content, glass naturally has a greenish tinge which can vary according to batch and manufacturer. By reducing the iron content, the 'greenness' of the glass can be adjusted, although this is usually more expensive. Similarly, glass can be body tinted by adjusting its constituent materials during manufacture. Body tinted glasses are produced by adding metal oxides to the molten glass at the time of manufacture. Different colours can be achieved including grey, bronze, green and blue. However, the thickness of the glass affects the strength of colour, with thicker varieties tending to have a stronger colour. Care is therefore needed to ensure that replacement glass, even if it is the same body tint, is the same thickness as the original in order to avoid noticeable colour changes.

Figure 107 Early stages in the displacement of a warm-edge spacer. Note the bulge in the spacer in the bottom corner of the right hand pane. (Photo courtesy of Richard Alison.)

Figure 108 More extensive deterioration – the spacer has moved well away from the corner and into the line of sight. (Photo courtesy of Richard Alison.)

Low-e and other coatings also affect the appearance of glass as well as the colour of reflections and objects viewed through the glass. Variations of colour and thickness between batches are possible.[133] Variations in the thickness of coatings will also affect the way in which light passes through the glass. Low-e coatings are designed to reflect radiation in the infrared frequency and to allow visible light through. If the coating is too thick, or its thickness varies, there will be a tendency for light towards the red end of the spectrum to be reflected, which could result in apparent streaking given appropriate light conditions.

Another problem that can affect Low-e glass occurs when the coating extends over the face of the glass to the perimeter of the pane. In most cases the coatings should stop short of the edge of the glass by 8–10mm to avoid the spacer although manufacturing tolerances can make this difficult to achieve. In cases where there is no external pressure plate (for example in structural silicone glazed applications) a coating that extends behind the black primary seal of the spacer bar may, in strong sunlight, set up a metallic interference reflection, commonly reddish-pink or blue. If the edge strip is too wide, i.e. there is a gap between the black primary seal and the edge of the coating, bright strips of light can be seen around the perimeter at night.

Appearance of heat strengthened or toughened glass

Toughened glass is ordinary annealed glass that has been heated in a furnace and then cooled rapidly by jets of cold air passed over its surface. The process ensures that the surface of the glass cools faster than the centre glass thereby creating differential stresses within the pane. Heat strengthening is a similar process but to lower temperatures. The effect of either process is to produce a stronger product, able to resist both impact loads and thermal stresses to a higher level than annealed glass.

However, in polarised light (as can sometimes occur at certain times of the day depending upon weather and the angle of the sun's rays) coloured 'leopard spots' can be visible. These marks are simply a visual effect caused by double refraction – birefringence – resulting from the different stress patterns in the glass. This effect results in light being transmitted unequally in different directions and is unavoidable. Leopard spots are not a defect and no remedial works or mitigating specifications are possible.

Roller wave distortion is another common issue with toughened or heat strengthened glass (the latter tending to be less of a problem), but unlike some visual characteristics, roller wave distortion is controllable to a certain extent and common building standards give guidance on acceptable limits. During manufacture, the annealed glass is passed into an oven on a series of rollers – these can give a regular pattern of distortion at centres of about 300mm with variations from plane of 0.5mm over that length. High quality producers can improve the tolerance considerably and in the best cases a deviation of 0.1mm can be achieved in glass of 8mm thickness and over (see figure 101).

Newton's rings

IGUs contain air or another gas that is sealed into the space between the glass panes during manufacture. The pressure of the gas within the cavity will be dictated by the external air pressure at the time of assembly – somewhere between 960mbar and 1,040mbar. Once transported to site, the external air pressure could be greater or lower than the original pressure and this can lead to deflections in the panes. Higher pressures will produce a concave effect in the panes whilst lower pressures will cause the panes to adopt a convex profile. Similarly, changes in temperature within the unit will affect the pressure of the gas and hence produce expansion or contraction. Small changes in temperature and pressure are sufficient to cause significant changes in the images reflected from the windows due to glass deflection.

In a large IGU, the two glass leaves may be so displaced by air pressure as to touch in the middle. If this happens, then Newton's rings may be visible in this area. Newton's rings are an interference pattern caused by the reflection of light between two surfaces; they are roughly circular, coloured bands like oil films on water, but normally less intense in colour. The coloured bands appear because of the different wavelengths of light interfere at different thicknesses of the air between the surfaces. They occur only near the centre of a unit and cannot appear if the cavity, internal pressure and/or glass thickness is sufficient.

Brewster's fringes

The appearance of an oily effect on an IGU is not necessarily a fault but more likely due to an optical phenomenon known as Brewster's fringes. Brewster's fringes only occur with glass of very high optical quality. For the effect to occur in IGUs, both leaves must be of high quality, flat and parallel.

Brewster's fringes appear when wavelengths of light meeting up with each other are exactly 180 degrees out of phase (a condition known as light interference). The effect is similar to, although usually much smaller than, the interference fringes that can sometimes be seen on toughened glass windscreens.

Multiple reflections of light within one glass pane can combine with those similarly reflected within the other, with such small path differences as to cause interference. The effect is of faint or oily coloured bands or irregular shapes, which can be located anywhere over the surface. It is rarely noticeable in normal lighting conditions.

Iridescence

Iridescence is associated with thin films and is most commonly seen in an oil film on water or a soap bubble; it is an optical phenomenon caused by multiple reflections from multi-layered, semi-transparent surfaces which modulate the incident light by amplifying or attenuating some frequencies more than others. The appearance of any iridescence on glass may not be visible in all lighting conditions.

Breakage of glass

There are four principal types of glass in use in buildings:

- annealed;
- toughened;
- heat treated; and
- laminated.

Of these four, annealed is perhaps the most important as it from this that all the remaining variants are derived. Both heat treated and toughened glass are manufactured from annealed glass, whilst laminated glass can be either annealed or toughened or indeed a mixture of the two – both layers being separated by a polyvinyl butyl (pvb) interlayer.

An analysis of the crack patterns in glass can give a good indication as to the mode of failure. Ordinary annealed glass tends to break into sharp-edged shards, whilst toughened glass breaks into a multitude of small fragments or dice. However, it does not necessarily follow that all of the fragments will separate on failure, and large numbers of dice can clump together. Clumping is especially prevalent in situations where the glass has been fritted, or coated on one surface, perhaps for reasons of solar control or privacy. Thus, whilst a small fragment falling from height may cause little or no lasting injury, a larger clump weighing much more could be far more serious. A clump of fragments 15mm thick and around 120mm across will weigh somewhere in the region of 0.33kg.

Upon shattering, a toughened glass panel expands slightly and this can help to lock the small dice in place. However, in frameless or overhead applications, the glass will simply fall away. Glass in overhead situations, such as a glazed canopy, could have very serious effects if it failed, simply because gravity will take over, and even expansion of the plate following cracking will not be enough to restrain the glass.

Another problem area is the use of toughened glass in freestanding balustrades, for example in prestige office buildings. These panels tend to be unrestrained, and so it is almost inevitable that they will collapse in the event of a spontaneous failure. Given that the panels are protecting a vertical drop, say to an atrium or staircase, the failure could be very serious indeed.

In laminated glass, the pvb interlayer is designed to hold the fragments of glass in place when the glass is broken or shattered.

CWCT identify the potential causes of breakage of annealed and toughened glass as follows[134]:

- thermal stress;
- site damage;
- poor glazing;
- edge damage;
- impact;
- uniform load; and
- nickel sulphide inclusions (toughened glass only).

Some glass panels are point fixed; they are held in place by cylindrical fixings ('spiders') or may be hung rather like a curtain from point fixings at the top. If the epicentre of the cracking is around one of these fixing points, failure could be the result of:

- glass to metal contact;
- misaligned fixing;
- overtightened fixing;
- movement of one of the fixings; or
- lack of provision for expansion.

Thermal stress

For thermal stress failure to occur, different parts of the glass pane must be held at different temperatures. Such conditions exist quite easily between the centre of a pane and the edges because the edges are usually contained within a frame and are shaded from solar radiation. As the centre of the pane heats up it expands and forces the cooler, shaded edge to expand as well; this induces a tensile stress. A tensile stress of around $0.7\mathrm{Nmm}^{-2}$ would be expected for every degree rise in temperature. The critical stress for annealed glass is around $20\mathrm{Nmm}^{-2}$ – equivalent to a temperature difference of around 30°C. For toughened glass the critical stress is much higher, and a centre to edge difference of 90–150°C is needed before failure occurs. However, small imperfections around the perimeter of the glass can trigger failure at lower temperature differentials in both toughened and annealed glass.

Thermal stress fractures will often start at a small imperfection in the edge of the glass because the induced stress will be concentrated at this location. Typically, single thermal stress cracks will radiate out from the edge damage, commencing in a direction perpendicular to the edge. In cases where failure occurs in the absence of an imperfection, multiple cracks can be initiated.

Solar control glass is particularly at risk of thermal damage as this material absorbs more solar radiation and heats up more as a consequence. Insulating glass units tend to permit a higher heat build up still, particularly if one of the leaves is a solar control glass.

The spacer bar can help to act as a thermal bridge – in cold conditions it will conduct heat from the interior to the outer face and thus reduce the temperature difference between the centre and the edge of the glass. Conversely, insulated warm-edge spacers do not do this, and so the risk of thermal damage can be increased.

The window frame material plays a significant part in the reduction of risk of thermal cracking. A frame fitted to a good heat sink such as an external leaf of a cavity wall will allow heat to leak away from the frame thus maintaining or increasing the temperature differential between the edge and the centre of the glass. Similarly, a light metal frame will not perform as well as a dark coloured one as the latter will absorb more solar radiation, thus reducing the temperature differential.

Differential thermal movements in glass can be induced by shadows and reflections. Shading from balconies, brise soleil or other elements of structures can permit quite large temperature differentials and must be considered at design stage by using appropriate stress analysis software. Similarly, reflections may increase the temperature of the glass and need to be considered.

Site damage

Careless cutting or handling of the glass during installation can lead to several types of damage, some of which can permit localised stress concentrations to occur, in turn affecting the performance of the glass in bending. A shelled or vented edge carries a high risk of subsequent failure as does a severely feathered edge produced during cutting.

Avoiding glass to metal contact is important as this can cause failure. The correct location of glass setting and spacer blocks is therefore important, as is the correct selection of setting block material. If the block is too soft, the glass can sink into it and may make accidental metal contact. Similarly, if the blocks are too hard they may damage the edge of the glass.

Accidental weld splatter on glass can cause difficulties as well being unsightly; the particles of metal that have become fused to the surface of the glass may corrode and thus induce a localised area of stress high enough to fracture the glass.

Impact damage

There are two types of impact damage:

- hard body; and
- soft body.

Both could be accidental or deliberate. A soft body impact (such as a person falling against the glass) carries with it the risk of laceration should the glass fall out of place. For this reason BS 6262-4 sets guidelines for the appropriate type of glass according to location and risk. Hard body impact usually implies some form of projectile or rigid implement, for example an air pellet, stone or hammer.

The crack patterns arising from hard and soft body impact can give clues as to the cause. Typical patterns are illustrated in figure 109.

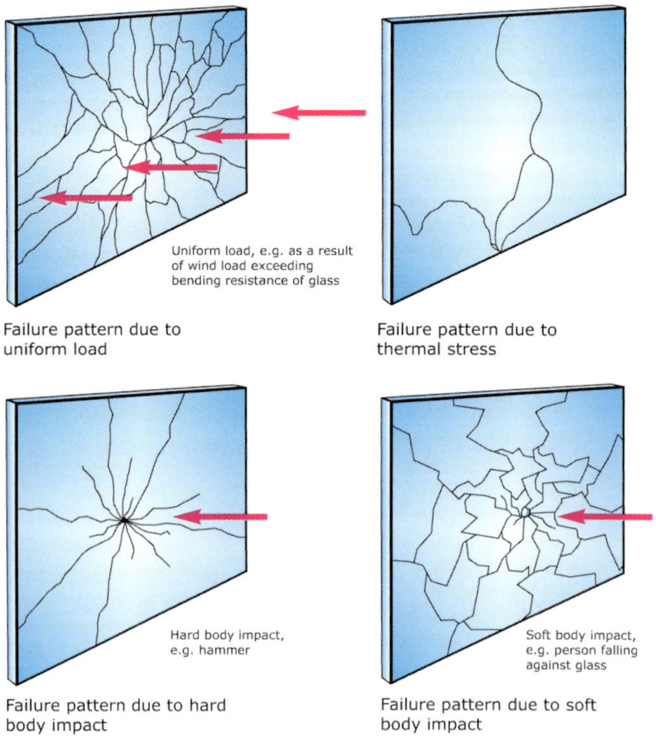

Figure 109 Crack patterns on annealed glass.

Uniform loading

As its name suggests, uniform loading is load that is applied uniformly over the surface of the glass; it could result from wind loads or occasionally snow loads depending upon the location and orientation. In much the same way as a beam will deflect under uniform load, glass will do the same. If the thickness to span ratio of the glass is incorrect, or if the uniform load is far greater than expected, excess deflection can cause the glass to break.

Up to a certain limit, the deflections can be regarded as bending, but when the deflections exceed 75 per cent of the thickness of the glass, the stresses change from bending stresses to in-plane stresses (sometimes called membrane stresses). The ratio of height to width, or aspect ratio of the glass panel, is critical for determining these in-plane stresses. In annealed glass, the crack pattern arising from uniform mechanical stress can resemble the cracking of toughened glass, with crack patterns radiating outwards, joined by transverse cracks rather like a spider's web (see figure 109).

Nickel sulphide inclusions

Over the past few years, the glass and glazing industry has invested considerable research time into the problem of spontaneous failure of toughened glass arising from the problem of nickel sulphide inclusions. Nickel sulphide impurities exist in all forms of glass – they are small discrete particles that are insoluble in molten glass. The chemical symbol NiS is commonly used as shorthand for nickel sulphide inclusions, although this may not be strictly correct in all cases.

If the glass fractures as a result of expansion of the nickel sulphide inclusion, those fragments immediately adjacent are more hexagonal and at the epicentre of the breakage the two larger particles form a distinctive butterfly shape linked by a central straight-line crack. If large enough, the inclusion may be seen in the form of a black speck, or its presence may be confirmed by optical microscopy.

Whilst other impurities may be present in glass, nickel sulphides have the capability to be far more disruptive. When heated in the furnace to around 650°C, the particles become metastable, i.e. they are seemingly stable, but as the glass cools they gradually change phase back to a thermally stable form. The transition is accompanied by an increase in volume of between two to four per cent.

The manufacturing process of toughened glass involves taking annealed glass and heating it in a furnace followed by rapid cooling by cold air being blown over the glass. The metastable particle is therefore trapped before it has time to reach its stable form; it wants to expand but cannot do so because it is restrained by the glass – it is locked in place.

Naturally, the increase in volume places the glass that surrounds it in tension. The pressure caused by the volumetric change in particle size creates small semicircular flaws around the inclusion. The cracks are stable until a certain critical size is reached, and this will depend upon the stress situation around the inclusion within the glass.[135] The glass core is already in tension as a result of the toughening process and if the maximum tensile stress exceeds the strength of the glass, rapid, almost instantaneous cracking propagation occurs, with the glass shattering into small dice.

At room temperature, the transformation from metastable back to stable form will take months to years. Because of the differences in thermal coefficients of the nickel sulphide inclusion and the glass, a small spherical hollow space is formed around the particle immediately after cooling. The volume change in the inclusion as it changes back is initially taken up by the spherical void – hence a delay of a couple of years or more before failure occurs. Once the particle has filled the void it can begin to exert a force on the glass and this initiates a series of small semicircular flaws around the inclusion. The cracks are stable until a certain critical radius has been reached – the mirror radius.

The expansion of particles will not affect ordinary annealed glass or heat strengthened glass which has been strengthened to a surface compressive strength of up to $60N/mm^{-2}$. However, the properties that set toughened glass apart from other forms of glass are its safety characteristics and high internal tensile stresses. Further, it does not follow that all nickel sulphide inclusions will lead to failure of the glass as much will depend upon the size of the particle and its position within the pane. Generally, the impurity will become a problem if it is situated very close to the glass surface and the pane is subjected to external tensile forces such as bending stresses or in-plane stresses under wind load or as a result of dimensional changes following from temperature changes.

If in-plane stresses act in the vicinity of the inclusion, there is a risk that the glass will fail. If the particle is located close to the surface of the glass, there is a risk that cracks will propagate as a result of the in-plane stresses.

Regardless of the effect of wind-induced stresses, the stress created by the expansion of the particle must be added to the stress already present within the body of the glass, so that it is possible for the glass to fail if the stresses are sufficient. As a result of the toughening process, the stress across the thickness of a piece of toughened glass is not uniform – the maximum tensile stress in the middle of the glass section is half that at the surface. For example, a pane with a surface compressive stress of $100Nmm^{-2}$ (which is the industry norm) will experience a minimum tensile stress of $45Nmm^{-2}$ in the body of the glass. The stress caused by expansion of the particle could be much more damaging if it occurs towards the outer faces of the glass where stresses are already high.

Aside from the position of the inclusion within the thickness of the glass, its position relative to the edges or centre of the glass pane is also relevant. For example, in a storey height glass panel, the stresses at the centre of the panel will be greatest under wind load, extending towards the corners.

As noted above the 'conversion' of metastable particles back to their stable form will take many years at room temperature. The process can be accelerated by heating the glass to a temperature high enough to permit the conversion to occur within an hour or two, yet not so high as to relax the stresses induced by the toughening process. This process is called heat soak testing and is the method adopted by many major glass manufacturers to try and limit the number of failures that might otherwise occur in service.

Heat soak testing helps to weed out those panels that are likely to fail at some point and to this extent it is a destructive test – it cannot be applied to glass in situ so has to be regarded as a part of the manufacturing process.

The process is usually capable of achieving a conversion rate of 90 per cent, i.e., the particles do not fully convert, but the incidence of nickel sulphide inclusions of the critical size is reduced to about 10 per cent or less of the original content. Theoretically, the heat soaking process should reduce the risk of breakage through nickel sulphide inclusions to no more than one breakage per 400 tonnes of thermally toughened glass.

Detection and cure

Although photographic methods have been used to diagnose the risk of nickel sulphide inclusions, these are very expensive and time consuming and as yet have not been developed commercially. Currently therefore there are no reliable commercial tests to enable the risk of breakages to be assessed. Interrogation of the operating and maintenance manuals may reveal the glass specification and confirm whether or not it has been heat soak tested. Questions put to the occupier of a building may reveal useful information as to the previous occurrence of failure.

However, knowledge of the possible existence of nickel sulphides does not escape the fact that serious injury can result from spontaneous failure, particularly if the glass is located in an overhead situation or is unrestrained in some way. There is very little that can be done to protect an unframed member (such as a glass balustrade), but horizontal, inclined and vertical glass can be retro-fitted with film provided that the edges of the film can be suitably anchored. If anchoring is not possible, the provision of film could actually increase the risk of injury, as shear forces around the perimeter of the film could cause the glass to fall out as a large and potentially very heavy panel.

17 Upper floor and roof decks

At a glance

- The advent of steel framing introduced a need to reduce the dead weight of floors whilst at the same time maintaining their fire resistant capabilities.
- Early systems utilised a system of steel joists surrounded with coke breeze concrete. There were various forms of this, but primarily they were similar in concept. The loadbearing function was provided by the steel joists, typically at 600–900 centres and spans of around 2.7m.
- Coke breeze is corrosive in damp conditions and it is possible for severe corrosion to damage the web of the embedded steel such that its strength is compromised. Tell tale parallel cracks on a plaster soffit will usually indicate filler joist construction and a need for further investigation.
- Later flooring systems involved hollow clay pots as void formers and woodwool slabs as permanent shuttering. Both of these systems could give rise to honeycombing of the concrete ribs and subsequent loss of fire protection or durability. The main difficulty is one of grout loss and the risk that the aggregate used in the concrete mix is too large to pass around the rib reinforcement. Since the ribs are concealed by the woodwool or clay spacer tiles, diagnosis and repair during construction is unlikely.
- Lightweight reinforced autoclaved aerated concrete planks were a popular choice for roof decks in some buildings, particularly schools and shopping centres. In section, the planks resembled ordinary hollow cored planks save that they were made from aerated concrete. Some records refer to the planks as 'Sipporex' after the manufacturer. The planks are no longer specified in UK construction, but many examples still remain. Some early examples suffered from deflection and associated cracking due to inadequate bond between the reinforcement and the aerated concrete.

17.1 INTRODUCTION

Whilst timber had for centuries been the standard means of constructing upper floors in all forms of building, by the 19th century it became desirable to provide floors of improved fire resistance particularly within mill and industrial buildings. Initially, cast iron and brick vaulting was used, but this later gave way to wrought iron and concrete followed latterly by mild steel and then various forms of reinforced concrete. As noted elsewhere, the widespread use of mild steel had, by the early 1900s, fuelled a demand for floors of lighter weight and fast erection.

In-situ concrete floors provide good sound and fire protection but they are heavyweight structures that need some form of temporary support during construction. To reduce the overall dead load of the structure, void formers can be used. Whilst there are a variety of void formers available today, those that gained most popularity were hollow clay pots, woodwool formers and waffle slabs.

Hollow clay pot floors were common in the post-war years but their use gradually declined and in the mid-1980s the London Brick Co ceased manufacturing them. Made from extruded clay forced around a moulded die, they formed a durable and robust method of construction when used between reinforced concrete ribs and with a structural concrete topping. Floor structures using hollow pots remain common in continental Europe.

Some hollow pot floors date back to the early 20th century; in 1906 Keine floors were first used. Later variants introduced concrete ribs between pots as tensile reinforcement in much the same way as contemporary floors. The early floors incorporated hoop iron or steel in the mortar joints between the hollow clay blocks.

During the 1950s–1970s, woodwool was often used as permanent formwork as a convenient means of providing thermal insulation properties to exposed floor slabs such as in podium car parks, ground floor slabs, etc.

Both hollow clay pots and woodwool formers can suffer from similar disadvantages; both can be difficult to detect in existing construction and both can lead to a need for expensive repairs.

The floor comprises an arrangement of narrow, closely spaced (about 750mm typically) joists, with very thin concrete slabs between them to complete the floor surface. The construction of the floor system is started by building a form work platform on which void formers are placed in the desired pattern; reinforcement for the joists and slab are then placed. Concrete is cast around and above the void formers to create a ribbed slab with a smooth upper surface. Tee beam construction comprises a series of ribs and flat sections, with the shuttering formed of either pre-made woodwool slabs or specially formed units. This form of construction suffered the same drawbacks as flat slabs, except that additional care was needed in selecting the size of the steel reinforcement bars in relation to the dimensions of the rib, such that there was adequate space for say 20mm of aggregate to pass around the bars and encase them. Furthermore, loss of cement grout through gaps in the formwork or by soaking into the woodwool could again create honeycombing.

Because the permanent shuttering conceals the steel reinforcement, it is difficult to know whether full compaction and coverage of the steel has been achieved. Until it was withdrawn in March 2003, BS 3809:1993 recommended the cutting of small access points in the woodwool shutter to enable quality checks to be undertaken.

Broadly speaking concrete floors constructed with hollow clay pots may fall into one of two categories:

- ribbed slabs where the topping is considered to contribute to the structural strength of the floor; and
- where the topping is not considered to contribute to structural strength, as a series of concrete ribs cast in-situ between blocks which remain part of the completed structure – the tops of the ribs may be connected by a topping of concrete (not necessarily of the same strength as that used in the ribs).

Whilst floor specifications varied, it was quite common to find clay spacer tiles between the larger pot components. These spacer tiles were designed to set the width of the concrete ribs that would be cast later. Incorrect placing of the steel reinforcement, the use of large aggregate or poor compaction could lead to honeycombing of the concrete rib, with a potential loss of fire resistance, durability or, in extreme circumstances, strength. The defect is very similar to that experienced with woodwool but the problem is not often reported and hollow pot floors have not suffered the same lack of confidence that befell woodwool.

Two applications for woodwool are typical:

- flat slab construction; and
- tee beam construction.

The former type is self-explanatory and will often be found as additional insulation to car parks. In these locations, woodwool cement board (WWCB) was quite often left exposed or painted and so should be readily identifiable. However, when used internally, detection may not be as easy as a common form of construction was to plaster direct to the woodwool soffite or to apply plasterboard linings thus concealing the shuttering from view. In such circumstances a surveyor should not treat the painted, plastered soffite at face value and simply assume that it is of reinforced concrete. Careful probing with a sharp awl pushed through the plaster may reveal the true nature of the substrate.

The problem with WWCB when used in this way is that it is impossible to know whether the steel reinforcement to the slab has been placed properly, or that it is properly embedded in concrete. The steel reinforcement ought to have been laid on spacer blocks, but these may have been displaced during the placing of the concrete, or missed out altogether. The effect of this would be that the steel would rest directly

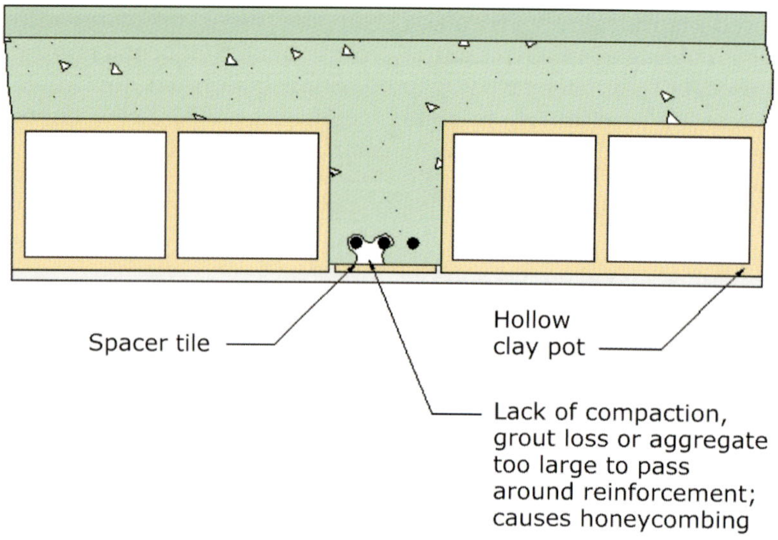

Figure 110 Possible voiding to ribs when spacer tiles are used with hollow pots.

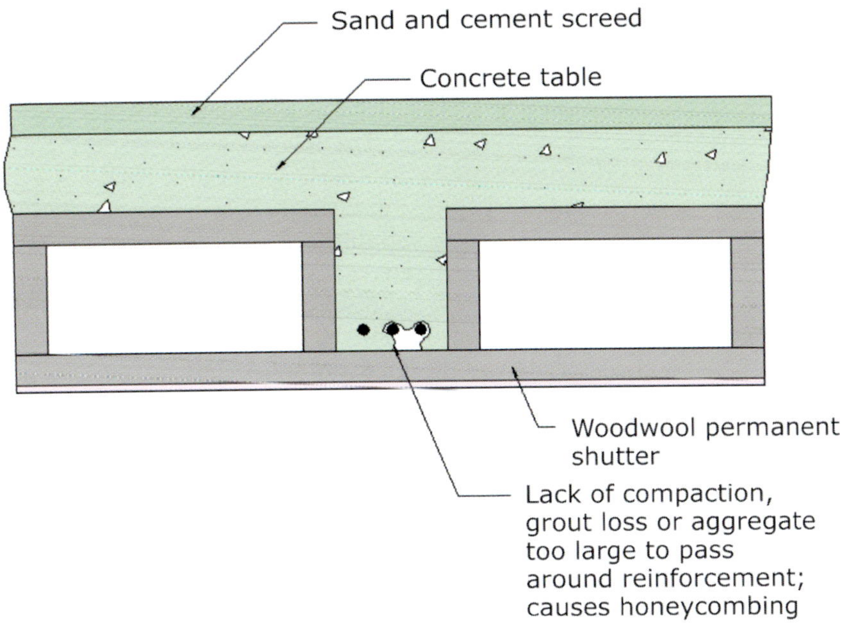

Figure 111 Possible voiding of concrete to rib (woodwool slab construction).

Figure 112 Woodwool slabs as permanent formwork. In this case the ribs are visible.

on the top surface of the slab such that concrete would not be able to be placed around it properly. As a result, strength and fire resistance will be compromised.

Non-destructive testing techniques can be employed to identify whether any problems are likely to exist, although physical sampling may be more reliable. Removal of the shuttering and repairs using hand placed mortars or sprayed concrete may be necessary if defects are found.

As far as evaluation is concerned, the significance is that what appears to be solid concrete – and may sound like solid concrete when tapped lightly with a hammer – may actually be weak, brittle clay tile or woodwool slab. Care needs to be taken that during refurbishment or alteration, new loads, particularly point loads placed on a floor slab, can be accommodated without punching through a hollow area.

17.2 COKE BREEZE AND FILLER JOIST CONSTRUCTION

During the second half of the 19th century and up until about 1925, fire resisting floors of filler joist construction were popular. A system known as the 'Fox and Barrett' floor was patented in 1844 and was used widely for the construction of commercial and public buildings up to about 1880.

Older proprietary systems are sometimes encountered. The Frazzi system of flooring was introduced in the 1890s and used up until the First World War. The system comprised a series of steel filler joists and hollow pots, often with clay cover pieces to protect the bottom flanges of the floor joists. The manufacturer's details illustrate a floor made up primarily of coke breeze concrete, similar to that used in filler joist construction.

Whilst there are a few varieties of filler joist floor, the basic form of construction of each is similar, comprising small (up to 150mm deep steel or iron joists) arranged in parallel lines at 0.6 to 1m centres. The space between the joists is filled with concrete which, to reduce the overall weight, was often constructed from coke or coal breeze – the fines left over from the production of coke for blast furnaces. The joists were usually arranged such that they had a flat soffite, usually plastered with a wooden wearing layer on the top surface. The top flange could either protrude slightly or be totally encased in concrete. Sometimes, and particularly in public buildings, the floors were finished with terrazzo or granolithic surfaces.

The difficulty with the form of floor is not so much the concept (filler joist floors at the Royal Albert Hall were found to provide a floor capacity of as much as $5kNm^2$) but the lack of corrosion protection that was afforded to the steel joists. Coke breeze concrete is not a particularly strong material but this did not matter because the strength of the floor was provided by the steel joists.

Industrial furnaces of the period were not particularly efficient, meaning that substantial amounts of unburnt coal or coke could sometimes be left following combustion. The resultant materials could have a high sulphur content, which upon oxidisation turns to sulphur dioxide. If moisture is present, sulphur dioxide converts to sulphurous acid, which in turn combines with more oxygen to create sulphuric acid, a highly aggressive chemical, likely to attack steel.

Normally steel is protected against corrosion in concrete because it is highly alkaline. Coke breeze is different insofar as it can be acidic and that it is very porous – a factor that led to its choice in the first place. Furthermore, the material has low resistivity, which means that high corrosion currents can occur.

Investigations have demonstrated that in some cases, the unburnt portion of carbon in the concrete could be as much as 30 per cent – this throws in to question whether the so-called fire-proof floor would actually be partly combustible!

Site inspection

Given the lack of corrosion protection it is not surprising that steel embedded in coke breeze can decay very significantly in the presence of moisture. Whilst the exposed flanges may appear to be sound or only lightly corroded, it is sometimes possible for the web to have corroded away completely. For this reason it is important to make an accurate diagnosis of the condition of the steel – usually by physical opening up and non-destructive testing methods.

The form of construction can usually be identified by a series of parallel cracks along the lines of the filler joists where they have corroded. Evidence such as this should certainly be investigated in detail, particularly in cases where change of use or refurbishment is considered. To reduce the risk of corrosion it is vital that the coke breeze is kept dry – easier in internal environments but a possible vulnerability where terrazzo or tiled floors are regularly swabbed down.

Methods of repair

For obvious reasons it would be difficult to break out the entire coke breeze surrounding a corroded joist or joists. Resin injection treatments are available to be drawn in under vacuum to encapsulate the joist and to protect it against further corrosion. However, if the joist has been seriously weakened by section loss, replacement may be the only suitable expedient. Otherwise, a strategy of regular inspection and great care to keep the floors dry would appear to be a sensible precaution.

17.3 RAAC PLANKS

In February 1999 the Standing Committee on Structural Safety advised owners of both school and non-school buildings which have pre-1980 RAAC plank roofs to arrange for these roofs to be inspected if this has not been done since 1994, although generally the deterioration of RAAC planks may not jeopardise structural safety.[136] Tests by BRE have

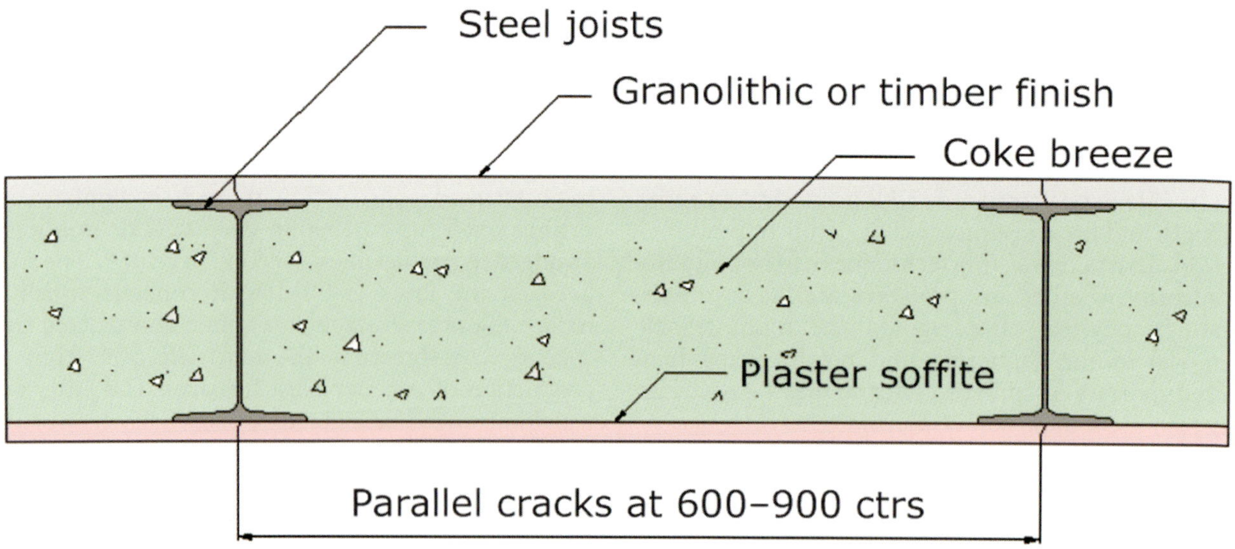

Figure 113 Typical filler joist construction.

supported this view, but firm recommendations have been given for the regular inspection of roof planks of this type – particularly those erected prior to 1980.

The stability of more modern imported panels has been called into question and because of this RAAC planks are not considered to be durable materials; references to them in BS 8110 and prEN 12602 have been dropped for fear of lending 'respectability' to what is viewed to be a relatively short-life material.

Possible defects to look out for include:

- Slippage of the reinforcement under load. (Because the material is soft and easily crushed, it does not secure the reinforcement as solidly as with conventional concrete.) Slippage could lead to deflection.
- Crushing of bearing ends.
- Corrosion of light steel reinforcement.
- Transverse cracking on the soffite.

Tests undertaken by Matsushita et al. have shown that the microstructure of AAC changes during carbonation, resulting in degradation such as cracking, with the degree of carbonation increasing over time.[137]

Following the identification of RAAC planks, the following initial checks should be undertaken:

- Check for evidence of cracking on the soffite.
- Check for evidence of deflection in the planks (it may be necessary to measure this accurately as part of a more detailed investigation).
- Determine the age of the structure – planks used before 1980 are more susceptible to deflection and cracking.
- Check the condition of the roof coverings – prolonged leakage could encourage corrosion of the steel reinforcement.
- Check the bearing widths. If less than 40mm seek further advice.

Should the above pointers suggest that a more detailed investigation be in order, the procedures contained within BRE Information Paper IP10/96 can be followed.

Possible recommendations may include the following:

- Inspect every 5 years, or if located in a harsh environment check each year.
- Reduce dead loads by removing stone chippings, felt over-layers, plant, etc.
- Check for transverse cracking and if appropriate measure deflection of the slabs.

18 Floor screeds and finishes

At a glance

- The main purpose of a floor screed is to provide a finishing layer on to an existing structural base. Screeds are not intended to be wearing courses and will usually require some form of protective finish.
- There are essentially four types of screed: bonded, partially bonded, unbonded and floating. Bonded screeds are usually up to around 40mm thick, whilst unbonded screeds are usually in excess of 50mm. All types can curl as a result of shrinkage and all require careful mixing, placing and curing to reduce the risk of cracking and curling.
- Failure to control curing properly could, depending upon atmospheric conditions, lead to premature drying. However, the surface of the screed is likely to become drier than the main body of the screed with the result that the top surface will shrink more than the lower parts, thus creating curling forces.
- Rapid drying will also result in cracking. Differential curling between adjacent sections of screed will affect rigid or flexible flooring systems. If the screed has curled away from the base sufficiently for it to deflect under load, there is a very high likelihood that the screed will crack – again with the risk of damage to the floor finish.
- Resistance to applied loadings can be improved by increasing the thickness of the screed, using fine concrete or selecting a proprietary polymer modified screed.
- Applied floor finishes are sensitive to movements or defects within the screed upon which they are laid. Typical defects include: rippling of flexible floor finishes over crack lines; damage to flexible floor finishes over movement joints; blistering of flexible, impervious floor finishes; osmosis in applied seamless floor systems and disruption and delamination of rigid flooring materials due to substrate shrinkage.
- Magnesite floor finishes were once popular in housing and can often be found in industrial units dating back to the 1930s. The material resembles asphalt in appearance; it is made of caustic magnesia, additions (fillers), an aqueous salt solution of bivalent metals – generally magnesium chloride – and additives (like colouring materials). Magnesium chloride is an acid, hygroscopic and corrosive salt and can corrode ferrous metal that is in contact with it. The material is also very sensitive to moisture and will deteriorate in damp conditions, for example where impervious floor finishes entrap moisture.

18.1 INTRODUCTION

The primary purpose of a screed is to provide a finishing layer onto an existing structural base. Screeds are not intended to be wearing courses, they must be protected with another material such as carpet, vinyl or other traditional finishes. Given the range of problems experienced with screed as well as the time needed to lay, cure and dry out, it is not surprising that there has been a tendency to move away from screeds altogether in favour of power floated finishes, where the surface of the concrete or the topping is finished to a high degree. However, power floated floors can be difficult to achieve in all cases, and the areas around columns or other obstructions can be difficult to finish adequately. The marks left by the power floating paddles can often be seen showing through vinyl floor finishes unless a self-levelling compound is added. Furthermore, it may be difficult to correct a lack of tolerance in the floor structure or to accommodate services within the floor thickness.

Materials constructed using ordinary Portland cement have long been known to shrink as they dry out. Sand and cement screeds are no exception to shrinkage problems, but whilst shrinkage can often cause cracking in larger concrete elements, floor screeds, because of their relative thinness, can be prone to a problem of curling up at the edges. To the designer, the spectre of a curling floor screed can come as a nasty shock as more often than not it becomes apparent at a critical time, such as the fitting out of a new building.

There are four types of screed that need to be considered: bonded, partially bonded, unbonded and floating screeds. (Within the UK a screed is defined as a layer of well-compacted material applied in-situ to a structural base or other substrate as opposed to a topping which provides a wearing layer)[138]. Bonded screeds are usually up to around 40mm thick, whilst unbonded screeds are usually in excess of 50mm. Floating screeds are applied to a non-structural base such as a layer of insulating material or a damp-proof membrane. All types depend upon careful laying and curing in order to avoid shrinkage and curling problems, it is not a case of simply specifying one in preference to the other to avoid curling. Generally, a bonded screed will have more resistance to curling than an unbonded screed, although when laid to a greater thickness, the screed will have a greater resistance to the curling force than a thinner unbonded screed. The table below compares the four types of screed.

Over the years a variety of different screed mixes have been designed and manufactured – each with its own properties, benefits and limitations, for example:

- cement/sand, fine concrete;
- synthetic anhydrite;
- lightweight, flow-applied cementitious or anhydrite;
- no-fines.

18.2 WHY DO SCREEDS CURL?

It is important to distinguish between the curing of a screed and the drying of a screed as the two processes are very different. Curing is the chemical reaction process (hydration) that takes place and is essential for the proper service life of the completed product. The process of hydration uses about one third of the mixing water used in the mix and should take place over a minimum of four days but usually seven. It follows therefore that if the water is removed too early (by drying out), full hydration may not take place and the durability of the screed may be compromised. To prevent premature drying it is usual to cover a newly laid screed with polythene for at least the minimum four days but preferably seven. (Spray-applied curing membranes may adversely affect adhesion of overlaid flooring and should not be used.)

In general terms, the rate of drying is influenced by the ambient weather or temperature conditions such as relative humidity, exposure to air currents, direct sunlight, etc., the type and thickness of the floor slab itself if there is no separating layer and the finish applied to the screed.

Failure to control curing properly could, depending upon atmospheric conditions, lead to premature drying. However, the surface of the screed is likely to become drier than the main body of the screed with the result that the top surface will shrink more than the lower parts, thus creating curling forces.

Rapid drying will also result in cracking – some cracking is almost inevitable in a large area, but excess cracking, particularly map pattern cracking, is an indicator of poor curing practice. Shrinkage can also become manifest at daywork joints, or, if the screed has been laid in defined bays, at bay joints (this practice was popular during the 1960s but is generally out of favour today unless underfloor heating systems are used).

The optimum rate of drying as set out in BS 8203 is one day per millimetre of thickness up to about 50mm. Screed thicknesses greater than 50mm will take appreciably longer to dry out, more so if they are laid on a relatively new floor slab which in itself has not fully dried. However, optimum conditions may not be achievable on site, with the attendant pressures of programming and the needs of following trades. Applying floor finishes at too early a stage brings the risk of damage and deterioration to the floor finish. A number of proprietary quick drying screeds have therefore been developed to enable floor laying within a much reduced time period. The advantages of fast

Comparison of screed types

Type	Characteristics	Comments
Bonded screed	25–40mm thickness applied to a mechanically prepared concrete substrate and bonded with a water/cement slurry grout.	Screeds thicker than 45mm have an increasing risk of becoming debonded. Fully bonded screeds up to 75mm thick may be possible with the use of fine concrete as opposed to sand and cement.
Partially bonded	Minimum 50mm thickness applied to a water/cement slurry grout that has in turn been applied to a concrete surface that has been brushed before hardening to provide a key.	Domestic housing only. No official guidance as to thickness, but risk of debonding is high, best to maintain a minimum thickness of 50mm and use in conjunction with flexible rather than rigid floor coverings. Partially bonded screeds offer a high risk of cracking, debonding and curling.
Unbonded screed	Minimum thickness 50mm either fine concrete (preferred) or sand and cement. Laid upon a smooth base with separating membrane. May be reinforced.	Curling and hollowness is likely, particularly at joints, around the perimeter and at crack positions. The risk of curling can be eliminated by incorporating fabric mesh (not chicken wire) at mid-depth of the screed.
Floating screed	Minimum 75mm thick reinforced with D49 or D98 mesh at mid-depth laid on a separation layer and then an insulation layer.	65mm thickness is usually satisfactory for domestic purposes. Chicken wire reinforcement offers little benefit to serviceability, better to provide steel mesh continuously over the floor – this must not be interrupted at daywork joints.

drying systems must not however be seen as a substitute for proper preparation and laying practice.

If the cracks are wide enough to prevent the aggregate particles bridging and interlocking, there is a risk that the curling forces could break the bond of the screed to the sub-base and permit the lifting of the screed around the perimeter of the bay or crack. Curling is less likely to occur if the screed has been allowed to develop full strength properly and allowed to dry out naturally. The provision of steel mesh within a screed (see table on page 162) will prevent differential curling of screeds at crack positions or at daywork joints although it will not contribute significantly to preventing the subsequent deflection of screed that has curled, for example at the perimeter of a bay. The mesh will however help to limit cracking. (At one time it was common practice to include chicken wire within the body of a screed. This material has very little benefit in controlling cracking or curling and is best avoided in favour of fabric mesh such as D49 or D98.)

Differential curling between adjacent sections of screed will affect rigid or flexible flooring systems according to the severity of the problem. Rigid surfaces such as ceramic tiles will be less tolerant of movement, although even flexible sheet materials can be damaged in the longer term by relatively small movements. Importantly, if the screed has curled away from the base sufficiently for it to deflect under load, there is a very high likelihood that the screed will crack – again with the risk of damage to the floor finish.

Resistance to applied loadings can be improved by increasing the thickness of the screed, using fine concrete or selecting a proprietary polymer modified screed.

Whilst the section of screed that has curled could in principle be ground down to remove the lipped edge, the fact that it has curled away from the sub-base means that it will no longer be bonded and may well sound hollow underfoot. Service life could therefore be compromised. In these circumstances, the provision of a low viscosity epoxy grout may be successful in re-attaching the screed, although this should only be considered where there is reasonable certainty that the screed was of an appropriate quality to begin with, i.e. it was adequately mixed and compacted.

According to CIRIA, the application of an impervious floor covering can sometimes trap moisture within a screed causing it to uncurl[139] – although to specify such treatment would seem to be a fairly high risk policy in case it did not work. The future serviceability of the screed would need to be considered carefully.

Another factor that may influence curling is the lack of proper preparation of the concrete base prior to screeding. Large amounts of dust on the surface, traces of laitance or contaminants or excessive suction may all reduce the bond or prevent it occurring in the first place. Surface preparation is as important as proper curing and must be carried out carefully. Measures include:

- washing the surface with water to control suction;
- grit blasting to remove laitance and loose material;
- mechanical scabbling to provide a key and clean and expose coarse aggregate in the surface;
- provision of a bonding cement/water slurry (although this must not be allowed to dry out before the screed is laid); and
- use of a bonding agent such as epoxy, PVA or SBR (PVA is not recommended in areas that may be subject to wetting).

18.3 SCREED COLLAPSE

From the discussion on screed curling given earlier it can be seen that proper curing, drying and preparation are important factors in the successful laying of a screed. However, appropriate mix selection and water cement ratio are also vitally important as is the method of mixing the materials. Some proprietary polymer modified pre-mixed screed are availiable for uses where curing and drying time must be constrained (for example where fitting out operations do not permit the normal curing and drying time that is needed).

The water to cement ratio is important as whilst sufficient water is needed to ensure hydration, too little water will affect strength whilst too much will make the screed difficult to place and to level, and will increase the risk of drying shrinkage and therefore cracking and curling. Usually, the aggregates should be batched by weight rather than volume and when mixed be capable of being moulded in the hand but not so as to release water when squeezed.

The method of mixing the screed can play a large part in its future serviceability. If the mortar is too dry it is possible that, aside from hydration problems, only the surface layer will become properly compacted, such that the top 5–10mm appears to be properly formed, but beneath this compaction and density is poor. Depending upon the future use of the building, the poorly compacted material may eventually fail under compression giving rise to undulations in the floor or shallow depressions – elephant's footprints[140]. Such problems will often occur in areas of point loading, such as beneath chair legs or equipment stands. Once collapse occurs, further traffic causes the perimeter of the collapsed area to break down and thus extend the area of damage.

Given the variability of site mixed screeds (both in terms of mix and compaction) collapse of a screed can be very localised. Whilst early breakdown can occur, it is equally possible that a screed that has seemingly performed well for many years may suddenly deteriorate given perhaps a change in layout of the floor, different degrees of usage or change in use.

The problem is more likely to become manifest in floors that are covered with flexible sheet materials. Hard surfaces such as tiles are not immune (if the deficiencies are over a large enough area) but generally they will help to spread point loads somewhat and may permit a screed to perform slightly better in terms of loading than would otherwise be the case.

A second feature of poor mixing is the existence of balls of cement powder – particles of cement that have clumped together in the mixer. These result from there being either insufficient water to disperse them properly throughout the mix, or simply the mixer being incapable of permitting even distribution. A conventional mixer will mix wet materials satisfactorily, but the configuration of the blades in the mixing drum is such that they will not break up the clods of material that form with granular materials in damp conditions. For this reason, special forced action trough and paddle or pan mixers are needed.

18.4 REPAIR OF COLLAPSED SCREED

Whilst a variety of possible floor repair strategies can be considered, the choice of method will very much depend upon the extent of the problem as well as site constraints. Work within an occupied building will inevitably mean a need to control noise dust and vibration, whilst time constraints will impose other restrictions upon the range of suitable materials.

CIRIA Report 184 Chapter 9 sets out some useful guidance upon the selection of appropriate repair strategies. For small pockets of damage, the affected areas can be cut out and filled with a suitable rapid drying screed of which there are a number of proprietary types. However, if larger areas of unsound material exist, more significant areas of replacement may require a different approach. Any investigation into localised collapse must involve consideration of remaining, apparently sound, areas to determine the likelihood of future defect.

In some cases it may be possible to restore serviceability to an unsound screed by the application of a low viscosity resin to soak into the body of the screed and to consolidate it. Epoxy (or polyester) resins are expensive materials and although capable of fairly rapid application, noxious fumes do need to be considered, making this form of treatment applicable only in certain conditions. If a screed is seriously compromised, and if service conditions are demanding, impregnation treatments may not be appropriate anyway as they may still fail to deliver satisfactory performance.

Assuming total replacement with a new screed is not practicable or desirable, another alternative might be to apply a flowing restoration screed – these proprietary products can often be laid down to about 3mm thickness (although it is difficult to see what meaningful benefit could be achieved at this thickness). A nominal thickness of 10mm would be more usual. Again, these products are relatively expensive, although the savings in time and disruption can often mean that when considered in the round, they can provide a more cost-effective alternative than full screed removal and replacement. Flowing screeds are specialist applications and in most cases specialist advice is needed in selection and application. Of course, the provision of an additional thickness of screed will have related effects – adjustments to door heights, skirting boards, floor boxes and the like and so may not be suitable in all cases.

If full replacement of the screed is the only option, cementitious screeds should be selected and specified with great care to avoid some of the problems of cracking, delamination and failure discussed earlier. Drying time is likely to be a problem in refurbishment works and so proprietary quick drying screeds may be preferable to conventional mixes.

18.5 FAILURE OF COMPOSITE NO-FINES SCREEDS

In contemporary construction the use of raised access floors in office buildings is almost universal. These floors are often set at 100–150mm above structural slab level, meaning that at core positions the designer is faced either with the option of extending the raised floor into the core or raising the floor up to finished level ready for the application of hard surfaces such as terrazzo (not so popular now) or hard tiled surfaces such as natural stone. There are various ways in which the core floor level can be raised – the provision of lightweight blocks with a cementitious screed being one method or alternatively a no-fines screed (for example lytag) laid to an appropriate thickness and with the top 13mm finished (whilst the lytag is still green) with sand and cement. The use of lytag screeds is popular since in many cases the same material is used for the main floor plates in conjunction with permanent steel shuttering.

However, the provision of a 13mm surface layer can cause problems, with delamination occurring between the sand and cement and the base material. Lytag screeds do not have a particularly strong surface and are easily damaged. If the finishing layer exceeds 13mm, shrinkage forces can disrupt the screed base material with the result that not only does the dense material become detached, but a portion of the screed beneath can be destroyed.

Since this problem is most likely to occur in a highly trafficked area, most probably with expensive natural finishes, delamination of these forms of composite screed can be problematic.

Delamination of the screed will result in the loosening of the joints in the floor finish, edge breakdown of the finished flooring units (stone or tile) as well as the propensity for cracking being reflected through to the finished surface.

As with many screed failures, investigations should concentrate not only on the failed areas, but also adjoining sound parts in order to gauge the propensity of the floor to develop further problems. Remedial work could take one of two forms:

- localised lifting of finishes, removal of defective layers and their replacement; or
- replacement of the entire area to the correct thickness and specification.

If localised repair is considered to be appropriate, it is important to cut out the affected area and to ensure that the remaining screed surface is firm and without friable material. Once all loose material is removed, the surface should be coated with a polymer modified bonding slurry and a suitable polymer mortar - the main concern being to limit the potential of the screed repair material to shrink and permit cracking or risk further delamination.

Repairs in excess of 25–30mm thickness could cause shrinkage problems and in this case, particularly if the damaged area is extensive, it would be better to take up the entire screed to its full thickness (in some cases partial replacement of material to a minimum 50mm over the whole area may be possible) and to replace it with a new no-fines screed, paying particular attention to the need to restrict the surface layer to 13mm.

Replacing the screed with conventional screeds could cause problems in terms of loading and should be avoided unless the additional floor loadings can be justified in structural terms.

18.6 PROBLEMS WITH FLOOR COVERINGS

As noted earlier, a floor screed is not intended to act as a wearing course; for this reason, a suitable floor finish, offering both decorative and wearing capabilities, is an essential part of a finished floor. Whilst failures can of course occur in the finishes themselves (due to incorrect or inappropriate use, workmanship or product failure) more often than not the root cause of failure is with the substrate itself. Whilst flexible finishes may have some measure of tolerance to cracking or even curling, rigid finishes such as tiles or in-situ finishes such as terrazzo are very vulnerable to deterioration due to defects in the substrate.

Possible defects include the following:

- rippling of flexible floor finishes over crack lines;
- damage to flexible floor finishes over movement joints;
- blistering of flexible, impervious floor finishes;
- osmosis in applied seamless floor systems; and
- disruption and delamination of rigid flooring materials due to substrate shrinkage.

Rippling of flexible floor finishes

From earlier comments it will be remembered that curling of unbonded floor screeds is a prevalent problem. Such curling can occur some time after initial laying and so it is quite possible that the flexible floor covering could have been laid over an area that was either prone to curling, had started to curl or was fully curled. In the latter case, the problem may have been masked by the application of a latex screed or levelling compound prior to floor laying.

Given the time that is needed for a screed to dry out fully, the provision of an impervious floor covering could seal residual moisture into the slab and screed, causing a redistribution of the moisture content throughout its thickness. It will be remembered that curling occurs because of a differential moisture content between different surfaces and so if moisture is redistributed, the curling effect may be reversed. If this happens, the floor covering will be slightly wider than is needed; it cannot be compressed back to make it narrower and so it will remain elevated as a ridge. Repeated trafficking will cause the ridge to wear and perforate, leading to more extensive deterioration.

Similarly, if the floor was flat prior to floor laying, the curling of the screed would stretch the floor covering (assuming it was bonded). Subsequent reversal may be accommodated depending upon the qualities of the covering, but it may not return to its previous width, leading again to the formation of a ridge.

One might expect that thermal movements in a substrate could create similar rippling problems, although there seems to be slightly more uncertainty as to this. Early work by BRE tended to suggest that thermal movements did not replicate the problem, although in more recent years there appears to be more of an acceptance that thermal or normal diurnal movements could have an effect.

Lifting of rigid floor finishes

Rigid floor surfaces such as ceramic or stone tiles and terrazzo are often bedded in adhesive with grouted joints that contain very little, if any, flexibility, with the result that if the substrate expands, cracking may occur in the grouted joints. However, rather more disruption can occur in cases where the substrate shrinks significantly after the tiled finish has been laid - perhaps where for reasons of programming the tiles were laid before full drying shrinkage had taken place.

Depending upon the measure of adhesion between the tiled surface and the screed, shrinkage of the latter will exert compression forces on the edges of the tiles. If the compression forces are able to overcome the adhesion force, the pressure will be relieved by the tiles suddenly popping up - often along the line of a daywork joint or formed joint in the screed. Such effects can happen suddenly, several years after completion, although the most usual period is about a year from completion. Long term expansion of the tiling can produce similar effects.

Blistering of thin sheet flooring

The appearance of bubbles or blisters in sheet flooring materials is often a sign of moisture related problems – either the lack of an effective damp-proof membrane (DPM), water entrapped in the floor screed or possibly leaks from service pipes buried in the screed. Occasionally, blistering can be due to lack of sufficient or patchy adhesive application during laying or failure of the adhesive due to incompatibility with the substrate or flooring.

A loss of adhesion in bonded floor coverings means that in trafficked areas the sheet gradually stretches and eventually works up into a raised area or blister which then wears rapidly and perforates. The presence of moisture beneath the sheet can affect certain types of adhesive, gradually causing them to emulsify and lose their effectiveness.

One frequent cause of this type of problem is a failure to allow sufficient time for a screed and floor slab to dry out before the floor covering is laid. It is usually the case that coverings should not be laid until the relative humidity of the slab is at or below 80 per cent. In the case of a ground bearing floor slab, the existence of a DPM beneath the floor could restrict the drying out of the slab with the result that an impervious floor covering effectively seals the moisture in the floor construction.

Case study 1

The premises comprised a small supermarket about two years old. Within a year of opening the floor coverings in the vicinity of the refrigerated display cabinets began to lose adhesion and display excessive wear. The resultant perforated coverings were a risk to health and safety and needed to be replaced. The client was concerned that the blistering was the result of a failed damp-proof membrane or leaking service pipe.

Following removal of the damaged floor coverings, the screed surface was exposed and a series of hygrometer probes were inserted into holes drilled into the screed to determine its relative humidity. In addition, the floor was cored to determine its overall composition and the existence of a DPM. The floor was found to be a ground bearing slab 300mm thick with a 75mm cementitious screed. Initial relative humidity (RH) readings were in excess of 95 per cent.

Over a period of three months the RH of the floor screed gradually fell as the water was free to evaporate from the surface. After four months the RH had dropped below 80 per cent and the floor was successfully relaid.

In this case, the considerable thickness of floor slab plus screed compounded the problem of moisture retention after construction. The provision of the impervious floor covering meant that evaporation could not occur and therefore the floor coverings had been laid too soon. However, the luxury of four months drying time was unusual; in most cases operational demands will mean that remedial measures must be put in place far earlier.

In the above example, alternative approaches might have been appropriate to restore the damaged floor in a much quicker time. A water resistant adhesive might have been simple, but a more effective approach would have been to have applied a surface damp-proof membrane to the entire floor, covering it with a smoothing compound and then re-laying the impervious floor covering.

Case study 2

The premises comprised a retail unit with a ground floor staff room at the rear. Staff complained of musty smells and when the floor coverings were taken up they were found to be very wet. The staff room was housed in an extension constructed about ten years previously. A conventional cementitious screed had been laid over a concrete ground bearing slab and polythene DPM.

A series of hygrometer probes were inserted and left to stabilise for 72 hours before RH readings were taken. These revealed an RH in excess of 95 per cent. Repeat measurements were taken every two weeks but these revealed little improvement and in two cases a worsening of the situation.

Given the lack of improvement it became apparent that the problem was not simply due to entrapped moisture dating from construction. Taking into account the thickness of the slab it was felt unlikely that the DPM was at fault. Comparison of the RH readings in the floor with RH readings of the ambient air failed to establish a correlation suggesting that a further source of moisture was at fault.

As an experiment, copper flow and return pipes serving the radiator in the staff room were isolated. The pipes passed through the thickness of the screed and could possibly have been pinholed. Over the following weeks the floor RH readings progressively fell indicating that a failure of the pipe had indeed occurred. Once the defective pipe had been replaced and the RH had dropped to below 80 per cent the floor covering was replaced.

18.7 SITE INVESTIGATIONS

Assessing the performance of a screed that may be showing signs of defect is to some extent a subjective task based upon experience and data gathered from site as well as consideration of the likely uses of the finished floor. However, one of the difficulties faced by the surveyor is the possible variability of mixes and workmanship issues. In addition, the lack of information as to the quality of preparation, curing and drying, as well as knowledge as to the thickness of the screed, which could vary considerably, makes assessment problematic.

The first component of a screed investigation will therefore involve data gathering in the form of area, use, specification of mix, curing records, an examination of the construction programme (if available) as well as a preliminary site survey. A sketch plan or general arrangement plan will provide a useful base upon which to mark crack patterns and details (width and direction) as well as areas of hollowness (in the case of a bonded screed) which can be determined by tapping with a suitable implement.

For a new screed, tests to determine whether or not it has lost adhesion should be carried out at least four weeks after laying and preferably as late as practicable in the programme to ensure that most drying shrinkage has taken place. Loss of adhesion is not necessarily a problem unless it is over a wide area and is accompanied by unacceptable lifting at joints.

BS 8204–1:2003 provides acceptance limits as to line and level according to future use. This standard may be inappropriate for older buildings constructed prior to 2003 but it nevertheless provides a good guide. Remember that flatness and levelness are not the same things – a floor can be flat (i.e. a constant plane) but the plane is not necessarily perfectly horizontal – it could be inclined. For this reason the level of the floor can be checked with levels – preferably to millimetre accuracy, but a 2m straightedge should also be used to determine tolerances over a smaller area.

If instructing a level survey, be aware that a close grid will yield a very large amount of information often in a form that is somewhat indigestible. A set of spot levels can be difficult to read and will involve careful analysis to establish a pattern. Specialist floor surveys are possible to submillimetre accuracy, with survey data produced on a coloured contour plan to make visualisation much more rapid and effective. Departure from stated tolerances does not necessarily mean that the floor screed should be condemned – this is more of a judgment call. If the departures are frequent and significant the decision to call for replacement or remedial work will be easier.

Aside from dimensional issues, the performance of the screed in terms of crushing strength and propensity to curl or crack must be assessed. Commonly, cores can be taken to examine the degree of compaction, layering, existence of cement balls, mix proportions, aggregate size, etc. Taking cores is preferable to breaking out with a hammer and bolster as it involves less damage to surrounding surfaces. The alternative is to grind out a rectangular test piece, although this is a much more dusty operation (beware of smoke detection systems!). The difficulty with the core approach is that each sample is very local – a large number of test sites would be necessary to ensure that representative results had been obtained. Cores can be useful in assisting the identification of failed screeds, BS 4551–2[141] provides a methodology for the determination of constituent mix proportions. Note that under that code, the minimum mass of collected sample per 50m^2 of floor area is 100g, made up if necessary by samples of not less than 50g each.

The alternative is to undertake in-situ compression tests in the manner set out in BS 8204–1:2003[142]. Essentially, the method is based upon the use of a device known as the BRE Screed Tester and the methods first described by BRE in their Information Paper 11/84[143].

The screed tester comprises an annular weight of 4kg or 2kg (depending upon the type of screed) which is free to slide up and down a 1m long steel rod. The base of the rod has a surface area of 500mm^2 and is placed in contact with the screed. The weight (which is held by a catch) is then dropped down the rod to impose a load on the screed. This test is repeated a further three times in the same location. The depression that is formed in the screed as a result of these repeated shocks can then be measured and compared with published tables to give an indication of the soundness of the screed. Whilst the apparatus does not directly give any information as to mix proportions or any of the other factors described above, experience has shown that as all of these factors are related to performance, measurements to determine the soundness of the screed are highly relevant.

Whilst random readings may give a quick indication of the performance of the screed, it makes sense to establish a regular grid pattern of readings to achieve a more scientific and meaningful result. The grids, or corridors to use the BS parlance, should be at 3–5m intervals (i.e. to give an area of 9–25m^2). BS 8204–1:2003[144] adopts a methodology that is slightly different to that contained within IP11/84 and contains procedures for applying the test to floating screeds, which the earlier guidance did not. Under the standard, random tests are to be carried out in bays (defined by site circumstances or by the division of the floor into a grid). Not less than three sets of readings are taken of areas not exceeding 25m^2. If faults are found within a particular grid area, the frequency of tests can be increased as appropriate. It is preferable not to test areas within 300mm of any cracks or joints as cracking can result. However, it is important to test areas that will be subjected to high levels of traffic such as doorways and corridors, or where point loads are expected.

BS 8204 provides tables of acceptance criteria – for screeds that are subjected to heavy loads or where surface breakdown is unacceptable, the limit for the depth of the depression caused by the test is 3mm, whereas for lighter duties 5mm may be acceptable. Sometimes, particularly if the floor is worn and/or the anvil strikes a particle of exposed aggregate, the depression depths can be more than would otherwise be the case. A single failure should not be regarded as evidence that the floor has failed the test – the standard allows for an additional 1mm of depression to no more than five per cent of the total sample.

Whilst the tables provide a good measure for comparison purposes, it is not simply a matter of selecting an appropriate use category and comparing performance against this. For example, if the screed is to be covered with thick materials such as 25mm woodblocks, natural stone or terrazzo tiles up to 28mm thick, it will be less likely to suffer from damage in use and therefore some loss of strength and integrity in the screed is less likely to be of significance than where, say, the floor is to be covered with sheet vinyl. Some measure of subjectivity is therefore required, and all relevant factors must be considered before forming a final conclusion.

Floating floors need to be considered as a special case because of the risk of punching shear during the test. If the screed is less than 65mm thick, the results of a compression test may be less reliable – the standard method of dropping a 4kg weight may create a cone shape fracture line beneath the anvil, accompanied by a significant change in tone and additional and sudden indentation. For this reason, particularly with screeds where light uses are intended, a 2kg weight should be used.

The table below gives the maximum permissible indentation limits for in-situ crushing resistance tests, derived from BS 8204–1:2003.

Serviceability is not just confined to measures of crushing resistance; the severity of curling and cracking should also be assessed. Since curling is almost inevitable with some types of screed, a pragmatic approach is needed as to whether or not the movement is acceptable. Just because curling has occurred it does not follow that total replacement is required, unless there is a risk of fracturing under superimposed loading or where lipping has occurred to an unacceptable degree. Given that all floating or unbonded screeds will sound hollow when tapped, undertaking a survey with a hammer or rod may give misleading results. A two metre straight-edge across joints or cracks may give a better indication of curling in these circumstances.

Aside from cracking arising from physical influences (movements in the structure or substrate) the propensity of a screed to cracking is generally influenced by restraint to early thermal contraction and drying shrinkage. Because it is isolated by a slip membrane, a floating or unbonded screed may be less prone to this type of problem.

Other factors which influence cracking are the water content of the mix. Too much water leads to greater drying shrinkage whilst certain aggregate types have a high propensity for shrinkage.

Cracks in a screed do not necessarily mean failure and again need to be interpreted carefully to determine whether they are the result of excess shrinkages or other faults. Fine cracks should not affect serviceability and rarely need filling. Wider cracks need to be assessed in terms of likelihood of continued movement or other effects such as structural movement.

18.8 OSMOSIS

Osmosis can be defined as the flow of water from a more dilute solution to a more concentrated one

Maximum permissible indentation limits for in situ crushing resistance test

Description (from BS 8204–1:2003)	Category	Type of floor covering	Maximum indentation for bonded screed	Maximum indentation for floating screed
Areas expected to take very heavy foot traffic and/or heavy trolleys, or where any breakdown of the screed would be unacceptable: for example, hospital operating theatres, X-ray rooms, main hospital corridors and rooms where radioactive material is handled.	Heavy	Thermoplastic sheet/tiles	3mm	3mm
		Adhesive bedded 9mm ceramic tiles	3mm	3mm
		20–25 mm timber block flooring, 16mm ceramic tiles, 20mm natural stone or 28mm terrazzo tiles	4mm	4mm
				4mm
Areas expected to take heavy foot traffic and/or medium weight trolleys: for example, public areas, corridors, main lift and lobby areas, canteens and restaurants, public rooms in residential accommodation, classrooms, hospital wards and offices.	Medium	Thermoplastic sheet/tiles	3mm	3mm
		Adhesive bedded 9mm ceramic tiles	4mm	4mm
		20–25 mm timber block flooring, 16mm ceramic tiles, 20mm natural stone or 28mm terrazzo tiles	5mm	2.5mm (using 2kg weight)
Other areas subjected to foot traffic and light trolleys: for example, light office use, consulting rooms and domestic housing.	Light	Adhesive bedded 9mm ceramic tiles	4mm	4mm
		20–25 mm timber block flooring, 16mm ceramic tiles, 20mm natural stone or 28mm terrazzo tiles	5mm	2.5mm (using 2kg weight)
		Thermoplastic sheet/tiles	4mm	4mm

when separated by a semi-permeable membrane. A semi-permeable membrane is one which allows water molecules to pass through, but which prevents larger molecules of contaminants to pass. In the case of flooring, the surface zone of concrete immediately below the floor finish acts as the semi-permeable membrane.

For osmosis to occur, three conditions are required:

- a concentration of water soluble material – this could come from resins used in the flooring, contamination or from the concrete base;
- a semi-permeable membrane; and
- a source of water – such as from the concrete base (even apparently dry concrete will have 3–5% free water by weight; sufficient for osmosis to initiate).

The formation of blisters ranging from a few millimetres to 50mm or so in resin flooring is usually a symptom of osmosis. The blisters can be up to 12mm high and may contain aqueous fluid under pressure, although older cases may be dry.

Osmosis can occur in floors coated with thin (up to 6mm) resin finishes, perhaps three months to two years after laying. Osmosis in bonded rubber or PVC floors is not unknown, but rarer.

Remedial work involves the cutting out of the affected area and its replacement. The surface will need to be properly prepared prior to treatment – often with the use of 'blast track' equipment – a form of grit blasting using an enclosed machine. Double priming the surface or using trowel applied, thicker coatings are alternative methods of treatment.

18.9 MAGNESITE FLOORING

Although no longer popular within the UK, magnesite floor finishes and screeds are sometimes found within industrial buildings, workshops and the like, particularly within Europe. It is made of caustic magnesia, additions (fillers), an aqueous salt solution of bivalent metals – generally magnesium chloride – and additives (like colouring materials). Magnesium chloride is an acid, hygroscopic and corrosive salt; the material was popular in housing (particularly local authority housing) between 1945 and 1960 where it was sometimes mixed with red and yellow pigments to give a mottled appearance.[145] In some cases, sawdust, wood flour or even asbestos were used as a filler. The thickness of magnesite ranges from 10–25mm (single coat work) and 50mm in two coat work.

Magnesite screeds are never used externally; they offer a dust free, high abrasion and impact resistant surface of low heat conductivity and good resistance to mineral oils, solvents and fuels.

After hardening, the pores in the magnesite screed are filled with magnesium chloride solution. However, above a relative humidity of 32 per cent, magnesium chloride remains damp and can cause corrosion in ferrous building materials that may be in contact with it. The hygroscopic nature of magnesite makes it unsuitable for use in damp conditions. If the material is covered with an impervious floor covering, entrapped moisture can cause the magnesite to degenerate. Sometimes, swelling and expansion of the finish can occur.

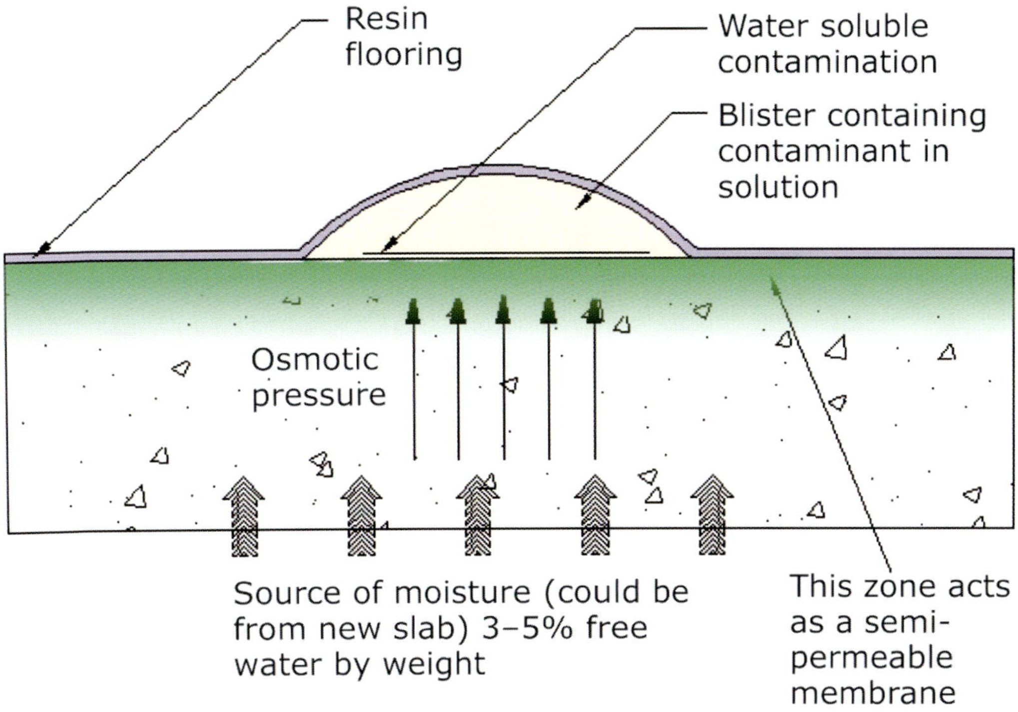

Figure 114 Osmosis in resin floor finishes.

Occasionally, timber floors within light industrial buildings were improved by laying magnesite screeds reinforced with wire mesh.

To prevent the migration of chlorides from magnesite into a reinforced concrete sub-base it is very important to ensure that a suitable impervious barrier is provided between the two materials. Failure to do this can cause corrosion problems, possibly evidenced by disruption in the flooring caused by expansion of a buried component or reinforcement bar.

In the article *Corrosion Damages Caused By Cast Magnesite Floor Screed*, Beul and Menzel describe the effects of corrosion in a modern industrial floor slab finished with magnesite but without a separating layer.[146] The building had been constructed in 1997 and by 2000 had begun to show evidence of small bulges in the floor finish close to columns. The bulges were found to coincide with the location of steel studs within the concrete base, the studs having relatively low cover of less than 20mm. The uniformity of the apparent damage on all the areas covered with magnesite screed and evidence from chloride profiles in different locations established a correlation between progressive chloride contamination, depassivation of the buried studs and subsequent corrosion. In areas where the studs were located at greater depth, corrosion had not occurred, although repeated tests in 2003 demonstrated that chloride penetration had increased.

19 Raised access floors

At a glance

- There are numerous different types of raised access floors ranging from types that utilise bricks held in a plastic tray to conventional metal pedestals with metal encased chipboard or gypsum cored tiles.
- Within the UK, the main standard for raised access floors was at one time the PSA MOB standard which set out four levels of duty. For most speculative office buildings, a medium grade floor was usually specified. More recently, the specifer also has the choice of tiles to BS EN 12825, which theoretically at least expands the range of performance to some 72 possible combinations.
- For most fully accessible floor systems, the easy removal of individual tiles is critical. The tiles should be capable of lifting, rotation through 90° and reinsertion without difficulty.
- The floor should not rely upon the perimeter walls or partitions to provide lateral stability; it should be capable of supporting itself without additional lateral restraint (other than within the floor system). For this reason pedestals are normally fixed down, either mechanically or with adhesive (sometimes both).
- Adhesive fixed pedestals should be checked to ensure that they are properly bonded or that the bond strength is not greater than the strength of the floor screed. (Lightweight floors, i.e. lytag concrete, have a low strength and will often fail in preference to the adhesive.)
- Cut tiles need proper support and therefore perimeter details need additional scrutiny to ensure that proper support (additional pedestals) have been fitted and that the tiles are tight and secure. Cut tiles should not be less than 150mm wide to avoid fixing difficulties, and should generally be mechanically fixed rather than simply retained under gravity.
- Loose tiles at door threshold positions could be particularly dangerous and so perimeter stringers should also be placed in these positions.
- The serviceability of raised access floors can be reduced by factors such as deflection in the structure, pedestal movement (during fit-out or maintenance operations), damage to tiles during handling and/or excess tackifier adhesive working its way between tile joints.
- In data centres, the formation of zinc whiskers (microscopic filaments of zinc) on galvanised steel floor components can cause serious problems to sensitive electronic equipment if they are disturbed and enter air circulation.

19.1 INTRODUCTION

Originally introduced in the 1960s, since the late 1970s and early 1980s raised access floors have been commonplace in office buildings and are now virtually the norm for new commercial developments. Their use has been extended to laboratories, cleanrooms, studios, data centres and many similar environments where there is a desire to allow cabling, pipes and services to be run beneath a floor and where a degree of flexibility is required to accommodate future changes.

There is a wide variety of different types and specifications of raised flooring systems ranging from the fully accessible to the semi-accessible (or in some cases limited accessible) but in generic terms most current systems comprise a walking surface of 600 × 600mm tiles supported on pedestals of some kind. Access tiles may be of timber products, for example plywood or chipboard, and sometimes encased with metal or cementitious or calcium sulphate cored metal-faced panels with a variety of finishes including carpet, vinyl or plain steel. The table overleaf details the common types of raised access floor panels.

Pedestal systems vary from the sophisticated adjustable metal jack types to systems as simple as an upended clay brick held in place with a plastic tray. Less accessible types of floor may involve the use of larger sheets of flooring material supported on foam faced timber battens – although these systems are mainly used in residential situations to provide sound insulation and to create a void for services. The key

Common types of raised access floor panel

Panel type	Features
Steel encased, chipboard core	Very popular flooring system for general office based (medium) applications. The chipboard core is bonded with epoxy or polyurethane adhesives to a galvanised steel outer casing. Different thicknesses of steel and core can provide a variety of different loadbearing characteristics. Panels are usually held in place by gravity or locked down with corner screws or toggles if additional fixity is required. Panel casing must be sealed with tape when the panel is cut; panel lips can be damaged easily by misuse or careless fitting. Gravity fixed tiles can be pre-finished with carpet or vinyl floor finishes.
Chipboard	Less common use in speculative offices or where limited accessibility is likely to be required. Although flooring grade chipboard is used, the panels are more susceptible to moisture movements than encased versions.
Steel faced cementitious flooring	Very popular in Grade A developments during the 1980s and 1990s. Constructed from a steel flat top plate, with a profiled bottom sheet (often in the form of a series of hemispherical depressions) and filled with foamed concrete. Robust floor type, with good fire resistance and acoustic properties. Tiles are often locked down – screw heads can become filled with tackifier adhesive and difficult to turn. Edges of the tiles prone to lipping and distortion if misused.
Steel faced anhydrate core	Similar to cementitious types but with a calcium silicate or calcium sulphate core. Good fire and acoustic properties.

distinguishing system between these floors and products for commercial use is that in the latter system, modular tiles can be lifted easily to enable voice, data and small power services to be reconfigured with minimum effort and disruption. In some cases, the use of a raised floor enables the provision of floor mounted cooling or air-conditioning systems, with treated air being discharged at floor level rather than overhead.

In some installations stringers are needed to add additional stability to a floor (particularly where there is a deep void of say 600mm and above) or where the underfloor area is used as a plenum – in this case an air seal is needed to prevent air escaping between the tile joints and the stringers may not be required to provide a structural function. Stringers are usually of galvanised steel, clipped or screwed to pedestal heads. Cut perimeter tiles will not fit onto a standard pedestal and so it is usual to provide alternative pedestal heads with perimeter stringers to provide adequate support to the cut tile.

Whilst being simple in concept, the installation of raised access floors is not always straightforward and can lead to difficulties in subsequent serviceability. In some circumstances the risk of collapse can pose unacceptable risks to health and safety, or risk damage to sensitive electronic equipment. For this reason, care in the selection of a suitable system and its installation is essential.

19.2 SPECIFICATIONS

For many years, the primary specification for raised access floors was a document known as the PSA MOB (Method of Building) Standard PF2[147]. First published in 1982 it was subsequently amended and re-released in conjunction with the Access Flooring Association in January 1992. This document set basic performance criteria for various types of floor and gave four different grades of floor according to specific end use. The standard was onerous and exacting in terms of

Figure 115 Proprietary raised floor system using aerated blocks held in a plastic tray. (Photograph courtesy of Allen Gilham.)

installation and tolerances, although some manufacturers claimed that it resulted in floor systems that were overengineered. Floors designed to the MOB standard required independent testing.

Latterly, the specifier has the ability to select raised floors according to a new standard, BS EN 12825[148], which was published in 2001. This document gives a greater range of options but differs from the MOB standard in a number of important respects; for example, floors designed to BS EN 12825 do not require independent testing and there are no specified tests for humidity and temperature effects on chipboard cores. Some manufacturers offer products complying with either standard.

The two documents are similar in some respects, but whereas there are only four grades under the MOB guide (PSA MOB PF2 PS/SPU, see the adjacent table) there are theoretically 72 different possible combinations of performance standard under the European standard (BS EN 12825), see the second table below). Technically therefore, the risk of making an incorrect specification choice under BS EN 12825 is much greater. Furthermore, the two documents do not interrelate.

Features of a raised access floor specified to PSA MOB PF2 PS/SPU 1992

Grade and identification colour	Use	Uniformly distributed loading
Light	General office accommodation without heavy equipment.	Not less than 6.7kN/m^2
Medium	General office accommodation where it is expected that heavy office equipment will be in use, data preparation rooms, educational a ccommodation and public areas.	Not less than 8kN/m^2
Heavy	Computer rooms, telephone exchanges, public areas and control rooms.	Not less than 2kN/m^2
Extra heavy	Computer rooms with heavy equipment and other special uses.	Not less than 2kN/m^2

19.3 CRITICAL POINTS

For most fully accessible floor systems, the easy removal of individual tiles is critical. The tiles should be capable of lifting, rotation through 90° and reinsertion without difficulty. However, for this to be possible, the tiles need to be manufactured to within fine tolerances. Although it would be reasonable to expect that reputable manufacturers are able to achieve this, it is not always the case, and examples of poorly fitting floors are not uncommon. However, assuming that the dimensions of the panels are within tolerance, the following factors might affect the fit of the tiles:

Temperature and humidity during installation

Installation of the tiles during hot weather could cause problems if the floor plates subsequently contract during colder weather, leading to tightening of the joints between panels. Similarly, if the panels are not fully encased, small dimensional changes arising from changes in humidity could give rise to either loosening or tightening of joints. Generally, for installation to proceed, the tiles should be stored and laid in dry and watertight areas at a temperature above 5°C and humidity below 75 per cent RH. It is usual to provide a perimeter expansion gap of 10mm – a failure to do this could lead to difficulties in the ability to lift tiles and replace them.

Movement or sway in the pedestals

The installation of the floor should not rely upon the perimeter walls or partitions to provide lateral stability; the floor should be capable of supporting itself without additional lateral restraint (other than within the floor system). Possible causes of sway could be due to the design of the pedestal, for example being extended close to its limit, failing to tighten off the lock nut, detachment of the pedestal from the floor or lack of stability (in the case of a deep void additional stringers are usually needed to control lateral movement). Alternatively the movement could be due to too much play in the pedestal head (as a result of missing fixing screws or poor design). Most MOB standard floors will have positive fixing lugs on the tile and or pedestal and many medium and higher grade floors have additional corner fixing screws.

72 possible combinations of panel specification under BS EN 12825:2001

Class of element	Ultimate load at the point of failure	Classes of deflection	Classes of safety factor	Classes of panel deviation in size
1	≥ 4kN	A, B or C, i.e. deflection of 2.5mm, 3mm or 4mm respectively	Class 2 or Class 3 (take ultimate load and divide by safety factor to give working load)	Class 1 and Class 2 (different allowances for thickness, dimensions, etc., with class 1 being more onerous)
2	≥ 6kN			
3	≥ 8kN			
4	≥ 9kN			
5	≥ 10kN			
6	≥ 12kN			

Figure 116 These tiles are showing signs of poor alignment; the tile in the top right of the photograph has been forced in against the restraint lug, distorting the galvanised steel sheet cladding. (Photograph courtesy of Jane Dalgliesh.)

According to the MOB specification all pedestals should be mechanically fixed as well as adhesively bonded unless it can be shown by testing that additional mechanical fixings are not needed. The usual method of fixing is to bond the pedestal plate to the floor with a resin adhesive, and it is quite often the case that an installer will omit mechanical fixings altogether. Whilst the bond so formed can be very strong, the surface of the structural floor may have been contaminated with dust prior to fixing, or, and particularly with lightweight concrete floors (lytag), the surface of the concrete is weaker than the adhesive and permits the pedestal and the adhesive to shear away.

Similarly, a very smooth power floated floor could create adhesion problems and manufacturers will often call for the surface to be roughened prior to the application of the adhesive.

A field test for pedestal fixity has been developed; it is included in the MOB but not the BS EN document. Essentially a small rig is set up above a pedestal supporting a 3kg bag of sand. The bag is supported by a 1m length wire and swung through a 90° arc to impact the side of the pedestal, resulting in a simple pass or fail result. The standard suggests a 0.5 per cent test frequency. The test is onerous, and unsurprisingly unpopular with installers, but if there is doubt over the adequacy of fixing, the swing bag test should be deployed. For pedestals in excess of 300mm additional mechanical fixing will almost certainly be required. Loads on pedestals are not always in the vertical dimension – rolling loads can induce sideways movement and the sudden overturning of a pedestal could prove very serious for users or for fixed equipment supported by the floor.

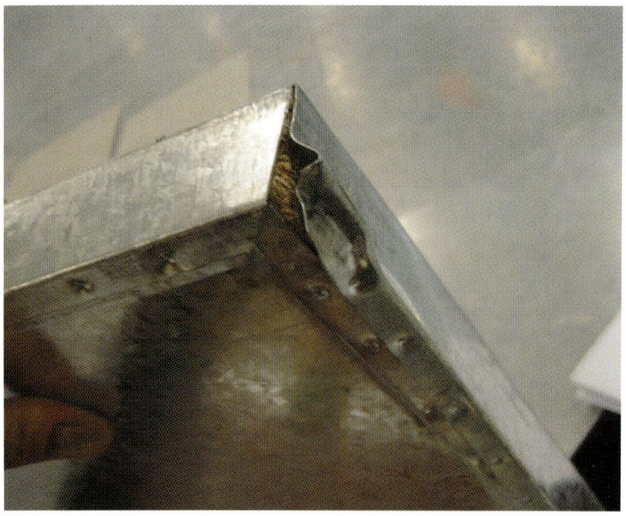

Figure 117 The incorrect location of a fixing nipple in relation to the panel edge has resulted in damage.

Deflection of the structural floor could, if severe or coupled with some of the other factors described above, serve to lock the panels in place.

If floor plates could be constructed to perfectly modular sizes it would not be necessary to cut tiles at thresholds or perimeters. Unfortunately such accuracy is seldom possible and cut tiles are often required at perimeters and thresholds. Cut tiles need proper support and therefore perimeter details need additional scrutiny to ensure that proper support (additional pedestals) have been fitted and that the tiles are tight and secure. Cut tiles should not be less than 150mm wide to avoid fixing difficulties and should generally be mechanically fixed rather than simply retained under gravity. Loose tiles at door threshold positions could be particularly dangerous and so perimeter stringers should also be placed in these positions.

Additionally, if the tile has been cut the core will have been exposed – additional site sealing is therefore required to protect the core from moisture and or abrasion damage.

Site work

Whilst the installation instructions for raised floors will caution against the creation of long rows of opened tiles or the formation of islands during fitting out and cabling operations, such activity is seen frequently. Foot traffic or rolling loads from pallet trucks, bins and the like can in these cases allow slight movement to occur in the pedestals, with resultant difficulty in the reinstallation of tiles.

Tiles should normally be lifted out and replaced vertically and not hinged back into position – such action can distort the edge of the tile and lead to a raised lip.

The sudden release of compression in the entire floor plate, if a row of tiles is removed, could lead to permanent but slight movement, making it impossible to reinsert the row of tiles without physical reduction in size. For this reason alone, continuous rows of tiles must not be removed.

Dragging electrical and data cables around pedestals should be avoided to prevent pedestal damage or

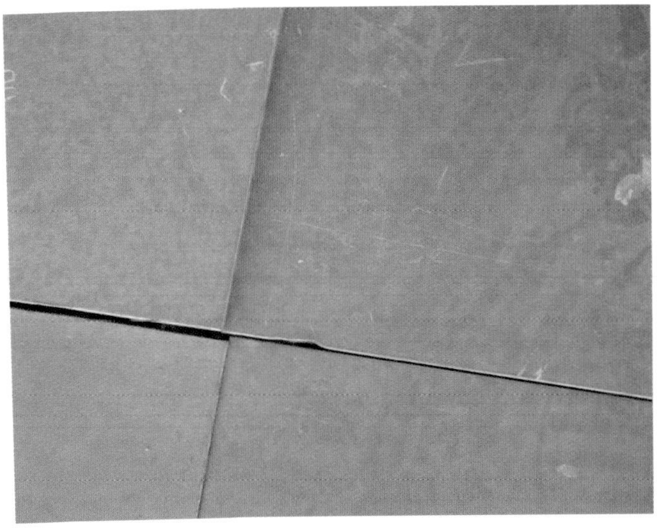

Figure 118 Severe lipping and edge damage to galvanised steel-encased chipboard floor tiles in a new building. Attempts to prize up the floor panel have damaged one edge, whilst excessive gaps between the panels are visible. Slight in-plane movement has occurred during fitting out, possibly as a result of a continuous row of tiles having been removed. Alternatively, the lipping may have been the result of incorrectly positioned tiles placed on 'nipple' based caps during the natural raising and replacing of floors during the fit-out phase.

displacement. Raised access floors usually need to be earthed in order to comply with statutory requirements (in the UK, 16th edition IEE Wiring Regulations). Pedestals will often contain a copper or other conductive contact system beneath the tiles to avoid the need for an electrical earth contact to each and every pedestal. However, contacts can become damaged in older installations or during refurbishment, so in some circumstances earth continuity testing will be advisable.

Carpets, both broadloom and in tile form, are common floor finishes in general office areas. They are usually attached to the floor with tackifier adhesive in dabs or continuous rows rather than over the surface of the tile as a whole. Needless to say the accumulation of adhesive or overzealous application can lead to contamination of the joint and the heads of corner fixing screws, making the removal of tiles very difficult. Care should be taken to ensure that old adhesive is removed properly and that new carpet is laid in accordance with manufacturer's requirements. Contact adhesives can be very difficult to remove and may result in panel delamination during attempts to remove carpets or finishes bonded with these adhesives.

19.4 AIRTIGHTNESS

In some cases, raised floor voids can be used as a plenum for air-conditioning systems; in such cases there will be an air pressure differential between the air within the plenum and that within the office itself. Whilst conditioned air is usually distributed via flexible ducts and special nozzles in selected floor tiles, excessive air leakage through panel joints can render the system inefficient and so special attention needs to be given to panel joints. At one level, neoprene tapes around the edges of tiles may be sufficient, but in areas that are likely to be accessed regularly, wear and damage to the tapes could be a problem. In the alternative, stringers fitted with gasket seals can be fixed to the pedestals so as to provide an effective seal around the entire perimeter of the tile.

Airtightness testing is usual as part of the commissioning of the building prior to completion. Whilst the manufacturer may provide air leakage data based on laboratory tests, field tests of the as-installed floor would serve to give a better indication of actual performance and also assist in the diagnosis of additional leaks through service penetrations, cut perimeter tiles and the like.

19.5 DUST

Raised access floors do not necessarily prevent dust and small pieces of debris penetrating between tiles and accumulating in the underfloor void. Whilst in many applications, a small amount of dust may not be critical, more specialised applications may require much higher standards of cleanliness. Floor surfaces should usually be sealed to prevent dusting from cementitious materials and care needs to be taken when lifting floor tiles where the floor void is used as an air plenum and conceivably small amounts of dust and grit could be blown up into the face of the operative as a result of a sudden change in air pressure.

Furthermore, a collection of dust and grit together with excess tackifier adhesive can work its way into the joints between panels, locking them in place and making removal very difficult without risk of panel damage.

For data centres and computer rooms where delicate electronic equipment is housed, the problem of dust can be significant, particularly where that dust could be conductive.

19.6 ZINC WHISKERS

For managers of data centres, the unlikely sounding phenomenon of zinc whiskers can be a serious problem, leading to regular and apparently unexplained systems failures in power and data supplies.

Zinc whiskers are essentially tiny conductive filaments that 'grow' on the underside of floor tiles and floor pedestals. Although the mechanism behind metal whisker growth is not well understood, the filaments are actually crystals of zinc that form as a result of compressive stresses imparted to a zinc passivating layer added as part of a corrosion protection process. It is thought that if the zinc layer is pure enough to have a regular crystalline structure, it will relieve these stresses by pushing filaments outwards. Growth can also be encouraged by mechanically induced stresses, thermal stress and stresses induced by the diffusion of different metals. Growth on zinc was reported by Lindborg in 1975[149]. In his paper, Lindborg suggested that to initiate zinc whisker growth, a minimum stress must be present in the zinc coating. Unlike other metallurgical problems, whisker formation does not require either dissolution of the metal or the presence of an electromagnetic field.

The filaments can be dislodged during routine maintenance work and carried via air currents (particularly where the floor void is used as a plenum) into electrical equipment, switchgear and the like to cause a short circuit.

The whiskers are typically no more than a few millimetres long (2mm or so) and a few microns (usually 2μm) in diameter – for this reason they may not be visible to the naked eye, and almost certainly not without proper lighting. There is considerable uncertainty as to why they occur, how they grow and why not all zinc surfaces are affected. A similar problem is the formation of tin, gold, cadmium and antimony whiskers, but since they are most unlikely to be used on floor tile components the features of these metals are not discussed here.

Numerous studies have been undertaken into the propagation of zinc whiskers but no reliable conclusions have been reached as to when and to what extent growth will occur.

19.7 ZINC WHISKERS AND RAISED FLOORS

In order to provide strength, floor tile encasement as well as the supporting pedestals and stringers are often constructed of steel; to provide corrosion protection the steel surfaces are usually coated with zinc – hence hot dip galvanising (HDG) (where the metal is immersed in a bath of molten zinc) or zinc electroplating (where the metal is immersed in an electrolyte). Research has shown that electroplated products are more inclined to display evidence of whisker growth and some reports suggest that HDG surfaces (which can be recognised by the lozenge pattern of zinc on the surface) are immune. However, research by Brusse and Sampson[150] and Lahtinen and Gustafsson[151] has identified that HDG can propagate whisker growth, but for this to happen the zinc must be exposed to chlorine and sulphur, small and slow temperature fluctuations (which would cause the fretting of grains in the metal) and compressive stresses in the coating.

As their name suggests, the whiskers tend to grow in length – but they do not necessarily start to grow immediately after the installation of the flooring component. Incubation times can vary from months to years, but once growth commences the filaments can continue to grow at up to 1mm a year for a period of time. However, growth is not predictable and some zinc surfaces may never develop the filaments. Raised floors can pose a particular problem because of the large surface area and the possibility that growth can continue undisturbed for a number of years. With some raised floor installations approaching 40 years, the risk of disturbance could be high.

As noted above, visual identification is not easy, but close examination of white powder-like or fuzzy deposits may reveal the existence of the whiskers, particularly on the underside of floor tiles or other components. Visual examination under bright lighting may reveal a bright-grey zinc surface that has not yet begun to grow whiskers, but the discovery of a dull-grey surface may indicate the presence of whisker growth.

Figure 119 Zinc whiskers growing on the surface of a zinc sample. (Photograph courtesy of the NASA Electronic Parts and Packaging (NEPP) Program http://nepp.nasa.gov/whisker)

Sparkling or twinkling under strong directional lighting may also reveal contamination.

X-ray elemental analysis together with scanning electron microscopy (SEM) is the most common method used to confirm the presence of zinc whiskers in dust collected from suspect areas. A less expensive but also less reliable method is to collect a wipe sample of suspect surfaces. The presence of zinc can be identified in a laboratory but it is not possible to see individual whiskers or to confirm the source.

The fibres are easily dislodged but removal by cleaning does not necessarily deal with the problem as there is a risk that they could grow again. Whilst the deposition of fibres in an underfloor void may never cause a problem, there is a risk that sensitive electrical or electronic equipment could be affected if the fibres are allowed to work their way into system casings where they could bridge electrical contacts – even more likely now considering the gradual drift towards miniaturisation and lower working voltages. The fact that electrical components are smaller and closer together means that they are more likely to suffer from an electrical short circuit even by short fibres.

Zinc whiskers will vaporise at currents of about 15–25mA, but in so doing, the arcing patterns that result can damage fine printed circuit boards.

Diagnosis of the problem may follow reports from data managers of unexpected equipment failure, particularly following the installation of new equipment or alterations to cabling – actions that may lead to the disturbance of the whiskers. Because they are so small, the filaments are easily carried by air distribution systems, not only within the floor void, but within the rooms. Given their size, the filaments are likely to be respirable and capable of entering the lower parts of the lungs. So far, studies as to the health effects of filament inhalation have not identified any serious or long term effects, but there is some doubt as to the reliability of investigations and a possible lack of suitable medical evidence. Available material regarding the inhalation or ingestion of zinc in oxide or dust forms suggests that it is generally harmless.

Remedial works

As noted above, cleaning may not provide a permanent solution to the problem, but it could be a valuable first action to reduce the effects. Vacuum cleaning with HEPA (high efficiency particulate air) filters may enable large concentrations to be removed, but there is a risk of further filament release when the floors are opened up and so specialist cleaning will be necessary to carefully controlled standards.

In severe cases, where the consequences of failure could be very severe, replacement of the raised floor may be a preferable alternative to cleaning or the application of barrier coatings. The selection of HDG components is thought to be a safe option, but alternative protective coatings such as epoxy coatings may be a better choice in high risk environments.

20 Vibration in composite floors

At a glance

- Floor plates constructed from steel permanent shuttering and lightweight or conventional concrete are very popular. Problems of vibration in office floors are most often encountered in long-span steel floors with composite decks. Precast concrete floors are also vulnerable in long span configurations.
- However, a by-product of this 'slimming down' is vibration within a floor. Vibration is generally not harmful, but it can be very disconcerting to occupants; generally, the psychological effects are worse than the actual physical problem and can affect a person's ability to work.
- Workers such as traders, who rely upon small lines of text on visual display units, can suffer difficulties due to movement of the screens; for this purpose damping devices fitted to screen arms are often required.
- Vibrations are usually measured in terms of acceleration – sometimes expressed as a percentage of the normal acceleration due to gravity – 9.8m per second per second .Vibrations as low as 50mm/s^2 can be unacceptable to persons at rest, whilst those engaged in aerobics are much more likely to be tolerant of vibrations of 1,000mm/s^2 or more.
- A person walking across a floor imposes a series of repeated or impulsive loads that build up a vibration response within the floor. A series of events no more than two seconds apart can be considered impulsive and during normal walking foot contact with the floor can be expected to occur every 0.6 seconds, i.e. a frequency of 1.6Hz (1.6 complete cycles per second).
- During each cycle of loading more energy is fed into the system and the magnitude of vibration grows, until it reaches a maximum value – this is influenced by the amount of damping in the floor. When the 'forcing' cyclic frequency approaches that of the floor, resonance occurs – this is the point at which the vibrations are most severe.
- The physical response of a floor to imposed repetitive impact loads will depend upon a number of factors – span, deflection, stiffness, mass, etc. Floor coverings make little difference although some slight improvements can be derived from partitions, heavy furniture and the like.

20.1 INTRODUCTION

Floors prone to vibration are best considered prior to fitting out works – a detailed analysis will identify areas most at risk so that the design and location of corridors or walkways can be configured to avoid these.

Over the last few years the construction of multi-storey office buildings has almost become standardised, with the use of profiled steel metal decking and concrete slabs spanning between steel beams the accepted norm. Such construction is highly engineered, designed to reduce self-weight, economical and capable of delivering large uninterrupted spans.

Coupled with the development of composite floors is the realisation that vibrations in the floor can create unfavourable psychological responses from the occupants. Vibrations are simply up and down motions of a floor caused by the application of forces – either the activities of the occupants or arising from external influences such as traffic, underground trains and the like. Vibration of this nature is not indicative of structural failure and may be within accepted building limits but, nevertheless, to a lay person regular vibration can be disturbing and can affect their ability to work.

Vibrations are usually measured in terms of acceleration – sometimes expressed as a percentage of the normal acceleration due to gravity – 9.8m per second per second .Vibrations as low as 50mm/s^2 can be unacceptable to persons at rest, whilst those engaged in aerobics are much more likely to be tolerant of vibrations of 1,000mm/s^2 or more. Generally, it is not so much the person causing the vibration as the person who may be watching or sitting near by who will notice the effects more.

Vibration that occurs for more than 16 hours during the day or an uninterrupted period of 8 hours during the night can be very disturbing at a low threshold – this is termed continuous vibration. These forms of vibration are usually due to a physical factor – perhaps a worn bearing on a heating pump or a failure in the anti-vibration mounts in a plant installation. This type of vibration is not discussed in this book.

All buildings are subject to external vibration influences, for example heavy road traffic, adjoining demolition work, underground trains, etc. Most people will be familiar with the slightly bouncy feel of a domestic timber floor but will not be troubled by it. However, the vibrations that occur within office floors of composite construction can be created by something far less dramatic – the occupants themselves. More often than not, vibrations can be traced back to the simple action of a person or group of persons walking purposefully across a floor or undertaking normal activities such as dropping a box of copy paper a short distance onto a floor.

The problem of vibration in composite floors should not be compared with the vibrations that can be created in dance floors, grandstands, etc., where large numbers of people congregate and where they may jump together in tune to a regular musical beat. If the resonant response of the floor approaches or mirrors that of the natural frequency of the floor, loads can be magnified significantly and can lead to dangerous structural conditions such as fatigue (similar to bending a wire back and forth until it breaks). In these conditions, vibrations in excess of 2,000mm/s^2 are likely. The change of use of a building to a fitness studio or gym may need to take into account the effect of a number of people jumping or running in unison, for example during an aerobics class.

Human perception to vibration is a difficult subject. Very low levels of vibration are usually identified quickly, so perception is fairly acute. Changes in the magnitude of vibration are however very much more difficult to identify and will vary from person to person or even nationality to nationality.[152]

Whilst the implications in structural terms of vibration may be insignificant, an office floor that is perceived to be 'lively' could easily bring about feelings of insecurity even at low frequencies. At higher frequencies of vibration, human reaction is generally linked to associated noise levels rather than the physical vibration itself.

20.2 HARMONICS

A person walking across a floor imposes a series of repeated forces – periodic forces that build up a vibration response within the floor. Such vibrations are termed impulsive – they display a rapid build up to a peak followed by a slow decay[153]. A series of events no more than 2 seconds apart can be considered impulsive and during normal walking foot contact with the floor can be expected to occur every 0.6 seconds, i.e. a frequency of 1.6 Hz (1.6 complete cycles per second).

During each cycle of loading, more energy is fed into the system and the magnitude of vibration grows until it reaches a maximum value – this is influenced by the amount of damping in the floor. When the 'forcing' cyclic frequency approaches that of the floor, resonance occurs – this is the point at which the vibrations are most severe. For impulsive vibrations, resonance occurs not only at the cyclic value but at multiples (harmonics) of it; for example, a person walking at 1.6Hz can produce resonance at 3.2Hz (2nd harmonic) and 4.8Hz (3rd harmonic). The higher the harmonic, the lower the magnitude of resonance. Since many floors have a natural frequency of 4–8Hz, problems are more likely to happen with the 3rd harmonic. By contrast, an aerobic class could produce resonance at 2.5Hz, 5Hz and the 3rd harmonic, 7.5Hz, making the 2nd and 3rd harmonics more critical.

So, in order to avoid vibration problems, the frequency of a floor should be greater than the highest frequency of the 'forcing' harmonic.[154]

20.3 SITE ANALYSIS

In order to analyse the performance of a floor, it is first necessary to investigate and determine its natural frequency; this can be determined using a 'heel stamp' technique. The natural frequency is sought because energy input at this frequency can cause the magnified 'worst case response' i.e. resonance.

The impulse of a sudden force such as a heel stamp will cause the floor to vibrate to a minute amount at its natural frequency; this can be detected with the use of sensitive monitoring devices known as accelerometers. A magnified response can arise when the second and third harmonics of the walking pace match the floor natural frequency.

The charts in figure 120 show the responses obtained from a heel stamp test on a typical concrete/steel shuttered floor in a speculative office building in London. In this case, the heel stamp consists of the movement of a person raising his body weight onto toes, then coming down onto heels quickly. In the example given, the body weight of the person undertaking the test was 88kg. The test was repeated three times.

The heel stamp excitation procedure can be used to extract the floor's natural frequency and the critical damping percentage. The latter is calculated by taking into consideration five cycles from the second peak response. The first impulse arises as the heel engages the floor and is omitted in the peak-to-peak analysis. The floor is then seen to respond in what is termed a damped oscillatory mode and this relates to the lower frequency motion of the floor. After the second peak, progressive damping effects occur and the peaks gradually diminish. (The positions are shown by the twin horizontal lines on the graphs).

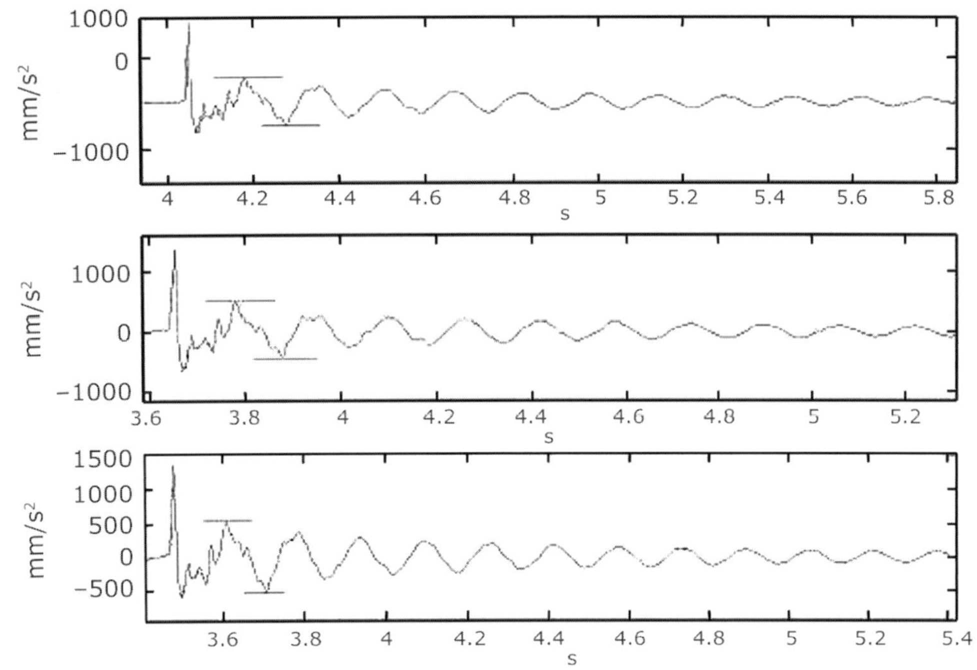

Figure 120 Heel stamp acceleration time history for three repeated events.

Heel stamp	Second impulse peak to peak	Critical damping %	Floor's natural frequency
Heel stamp 1	1070 mm/s^2	3.7 %	6.28 Hz
Heel stamp 2	908 mm/s^2	3.6 %	6.33 Hz
Heel stamp 3	1069 mm/s^2	3.7 %	6.38 Hz
Average value	1015.66 mm/s^2	3.66%	6.33 Hz

Having ascertained the natural frequency of the floor, tests can then be undertaken to determine the response on the basis of one or more persons walking. During walking, a fit and able person will raise their foot by about 50mm before dropping it back down again on the floor. Thus, their mass is lifted and dropped back again at each pace. We know that for every action there is an equal and opposite reaction, and so every time a person's foot touches the ground, the force applied is met by an equal and opposite reaction from the floor. The pace at which the person is walking may be treated as the frequency – each step and the time interval between each step dictating the frequency of the response of the floor. By matching the frequency of the walking steps with that of the floor, it is possible to simulate resonance. Such an exercise is usually carried out with the aid of a metronome, which emits an audible click at a set frequency, for example 125 paces per minute.

In the above example, the natural frequency of 6.33Hz is in the mid to upper range of floor natural frequencies that are typical of long span floors. Therefore it is not as vulnerable as some long span floors can be to footfall induced vibration. Floor natural frequency as low as 4Hz is not unknown and floor damping as low as 1.5 per cent has been recorded in some buildings. The floor damping percentage of 3.66 per cent is low, but again not untypical of a steel framed concrete decked structure. The particular floor was unoccupied; an occupied floor should exhibit a higher level of damping.

20.4 DAMPING EFFECTS

The physical response of a floor to imposed repetitive impact loads will depend upon a number of factors – span, deflection, stiffness, mass, etc. The tendency to reduce mass to obtain better structural economy (and hence cost) is at odds with the principles of avoiding vibration. Whilst it will be remembered that vibrations set up by people walking across the floor are highly unlikely to be anything more sinister than a psychological effect, the problem is still very real and needs to be addressed.

Generally, it is impracticable to provide damping to a floor that displays tendencies to vibrate. From a theoretical point of view, stiffening support beams will help, but it would be difficult to do this within an occupied building or at reasonable cost.

The provision of floor finishes will make a negligible difference to the problem (other than by a very marginal increase of mass) although the cushioning effect may change a person's response to the problem even when in reality there has been very little change. Other subtle changes might help to reduce complaints; for example, removing or changing fittings that are inclined to rattle, securing loose floor tiles and avoiding cantilever light and computer screen supports.

Increasing the mass of the floor is one method, but this is unlikely to be possible in a floor that has already been constructed. Similarly, floors rarely incorporate any particular damping measures (although they may do where external vibration sources are likely to cause problems). In practice, a measure of damping may be provided by non-loadbearing elements of structure, the occupants (to a small extent) and by the performance of the structure where this differs from that which was intended in pure design terms. To provide rigidity to a structure, engineers will often provide reinforced concrete walls (shear walls) or cores; this enables them to simplify the design process for the steel frame by treating structural connections as pin joints as opposed to moment connections. In practice, pin joints are often not pin joints in the true sense of the word and some load transfer is inevitable – this can modify the performance of the floor.

Floors prone to vibration are best considered prior to fitting out works – a detailed analysis will identify areas most at risk, so that the design and location of corridors or walkways can be configured to avoid these. Measures such as the careful location of office equipment can also serve to keep foot traffic to more satisfactory routes, avoiding, for example, long span support beams and utilising areas that are naturally stiffer.

Problems of vibration in office floors are most often encountered in long-span steel floors with composite decks; however precast concrete floors are also vulnerable in long span configurations. The Steel Construction Institute's *Design guide on the vibration of floors* sets out calculation methods for the prediction of the effects of transient loads.

Case study – 1980s office building, West End of London

The building was a purpose built office of around 6,000m^2 arranged over nine floors with a basement level. The building was multi-occupied with a variety of different business types from general offices to trading floors.

The tenant of one of the trading floors complained that periodic vibrations were affecting their ability to read the small rows of figures displayed on their computer screens. Each trader had a display of three monitors, each arranged on a complicated bracket device attached to the individual workstation. No other tenants were affected, and the perception was that the vibration was due to the operation of a particular air-conditioning cassette unit on the floor above.

The landlord's initial response was to try and replicate the vibration by operating the air-conditioning unit and monitoring the response. No perceptible vibrations could be reproduced. The next step was to set up a series of monitoring points around the office using accelerometers wired back to a data logging unit. The fortuitous reoccurrence of the vibration incident enabled a graphical record of the condition to be obtained. The next step was to try and identify the source. By this time, tenants on two other floors had miraculously discovered vibration problems and so monitoring of those floors was put in place and a detailed study was undertaken over a weekend period when extraneous background 'noise' could be eliminated. Various items of main plant were turned on at roof level, macerators in female toilets were operated and local on-floor appliances turned on. All failed to replicate the vibration pattern discovered earlier.

Analysis of the key incident showed that it was significantly below the levels that would be considered unacceptable by comparison with published guidance. The most likely cause was human activity within the office, but the problem was magnified by the computer screen arrangements, each display arm tending to amplify the very small vibrations that were occurring. It was probably the magnified response that was alerting tenants to the problem rather than the vibrations themselves. Word of mouth between tenants led to concern that the building was in some way at fault, hence the 'discovery' of vibration on the other floors.

By designing damping mechanisms to the support arm it should have been possible to eliminate the vibration or at worst to bring it down to a level that did not affect the user. The building was not defective and landlord intervention was not required.

21 Ground bearing floor slabs

At a glance

- Refinements in concreting technology have resulted in an ability to lay large areas of concrete in single pours. However, since the drying of concrete is usually accompanied by shrinkage of the cement paste, some method of movement control is needed to reduce the effects of cracking. There are a variety of placement methods but generally movement will be controlled by the use of formed or induced joints (at about 6m centres) or by the use of steel fibres in the mix.
- Cracking due to early drying shrinkage is common but can be reduced by careful mix design. The cracks that develop are not particularly serious and can often be ignored.
- Induced joints are formed by cutting the surface of the slab into bays. The saw cuts should penetrate into the surface of the slab by 25 per cent of its depth. It is usual to prevent contamination of the joints by applying a soft, flexible sealant and then to replace this with a harder variety after 12 months or so when most of the shrinkage will have taken place.
- Delamination of power floated floor slabs can create a serious serviceability issue. Sheets of thin surface concrete ranging from a few square metres to several square meters can delaminate and break up under traffic loads.
- To be successful, power floating of a floor slab needs to be carried out at the optimum time – usually when no bleed water or sheen is visible on the newly placed concrete and, as a rule of thumb, when the top surface allows an indentation underfoot of around 4–6mm. If the floating operation is carried out when the concrete is too wet, it will simply tear off the surface; too dry and it will not smooth the high spots.
- By failing to level off the concrete sufficiently before power floating, the float will tend to draw concrete from high spots into hollow areas. Since a degree of stiffening will have already occurred, the 'new' layer in the hollow area may not bond effectively thus creating a plane of weakness causing delamination later.
- In certain circumstances, the surface of the concrete may set before the process of bleeding has been completed. If this happens, the presence of trapped bleed water could again give rise to a plane of weakness. Such a problem could even occur prior to power floating. Water trapped below the surface could be disturbed by the action of the paddle blades from the floating machine dragging over the surface, extending the water below the surface and pulling up a blister. The blister may not be noticed, leaving a thin crust of concrete that subsequently breaks down.

21.1 TYPES OF FLOOR

Whilst ground bearing slabs are essentially similar, their method of construction varies according to the likely end-use and the accuracy of building that is required. On the one hand there are what may be termed conventional slabs with steel reinforcement mesh whilst on the other there are fibre (usually steel) reinforced slabs designed to reduce the need for movement control joints.

Concrete floors behave in similar ways in that they are all subject to drying shrinkage. If no provision for this is made during construction the slab would crack in a random fashion. By introducing fabric reinforcement or steel fibre, cracking can be controlled and, whilst not entirely eliminated, at least made to occur in set locations.

The following table summarises the principal types of ground bearing floor construction.

Movement control joints can be either formed or induced. A formed joint is a distinct break in a slab usually created by a pair of metal plates, whilst a sawn joint is usually cut to a depth of 25 per cent of the floor thickness and filled with soft sealant. Upon contraction the floor should crack along the line of the saw cut. For the sealant to operate correctly it

Type	Characteristic
Large area jointed floor	Fixed formers at 50m centres (typically). Usually mesh reinforced and subdivided into 6m bays by the cutting of sawn induced joints. Shrinkage at perimeters or formed joints usually in the order of 4–5mm.
Large area jointless floor	As above, but steel fibres used to control shrinkage so no sawn joints. Shrinkage at perimeters or formed joints usually in the order of 20mm.
Wide bay floor	Same as large area, but greater accuracy achieved by placing side rails at 12–15m centres.
Long strip floor	As large area jointed, but rails placed at 6m centres to obtain greater accuracy. Bays can be cast hit-and-miss or sequentially if permanent rails are used. Sawn induced joints at 6m centres.
Two layer floor	Two layers of concrete slab laid one over the other. Side rails are fixed to the lowest slab before placing the second slab. This method achieves the greatest accuracy.

should be placed against a polyethylene backing rod – this ensures that only the facing sides of the cut are bonded to the sealant.

Since most shrinkage will take place during the first 12 months or so of the life of the building, any sealant placed in the induced joint will need to be flexible enough to accommodate the expected strains and hard enough to prevent the shoulders of the cut breaking down under wheeled traffic. It is normal to use sealant of shore hardness below 30 during the defects liability period and then to specify its removal and replacement with a material of shore hardness 60 or so after about 12 months.

21.2 SURFACE REGULARITY

For certain warehousing operations surface regularity is critical to the proper operation of materials handling equipment (fork lift trucks, masts, pallet trucks, etc.). Concrete Society Technical Report 34 sets out certain performance requirements according to end use and provides useful guidance on design.

High bay warehouses are particularly demanding in terms of flatness requirement; TR34 gives an example of a 2.5mm difference in level creating a 25mm displacement at the top of a 10m fork lift. With dynamic movements in the forklift, the actual displacement could be three to four times this amount.

One frequent point of difference is whether or not flatness and levelness are one and the same thing. A flat floor is not necessarily a level floor; flatness is usually measured over 300mm, whilst levelness is measured over 3,000mm.

21.3 CRACKING

Cracking in concrete is a complex subject and often results from the effects of various faults acting in combination. Generally, and leaving aside cracks resulting from structural faults, cracking in floor slabs will be either the result of plastic shrinkage, early or long term drying shrinkage, chemical effects (such as ASR) or corrosion.

Plastic shrinkage cracking is rare in power floated floor slabs as the action of mechanical trowelling recompacts the surface and eliminates the cracking.

Cracking in floor slabs is most usually the result of restrained shrinkage of a slab. Early loading may pin the slab to the sub-base and features such as columns, inspection chambers or re-entrant corners can create conditions of restraint which permit the tensile stress to exceed the tensile strength of the concrete. This is not to say that structural cracking does not occur; overloading or ground movement could cause localised failure, although the pattern of cracking will usually suggest that something serious is amiss as opposed to straightforward shrinkage cracking.

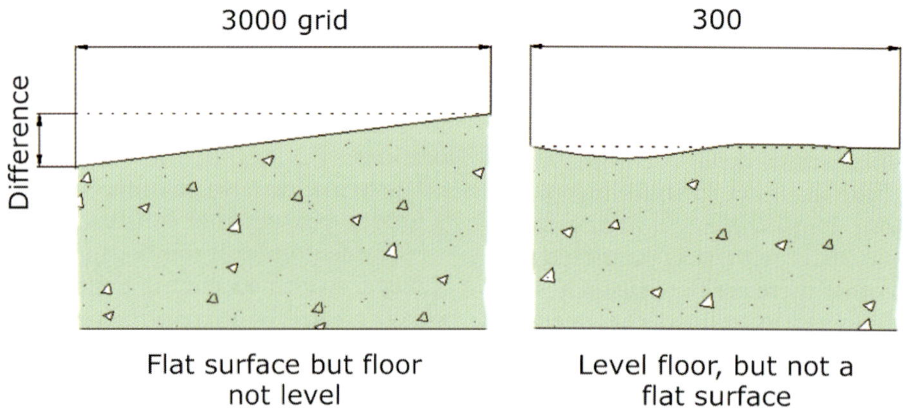

Figure 121 The difference between a level floor and a flat floor.

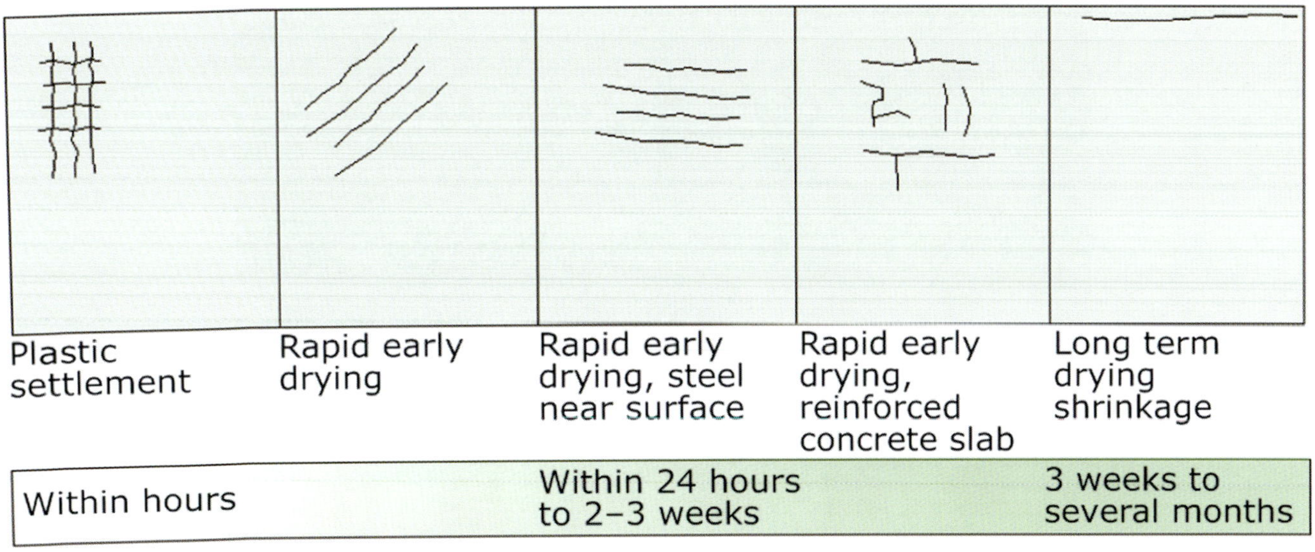

Figure 122 Drawing indicating the nature of different crack patterns.

Plastic shrinkage cracking occurs within a few hours of placing the concrete and is usually associated with high rates of evaporation. Similarly, plastic settlement often occurs when there are high levels of bleeding and some form of restraint to the settlement of particles, for example reinforcement. Plastic settlement and cracking can only occur whilst the concrete is still plastic, i.e. before it sets. If the rate of evaporation exceeds the rate at which bleed water rises to the surface, shrinkage will occur, inviting the risk of cracking.

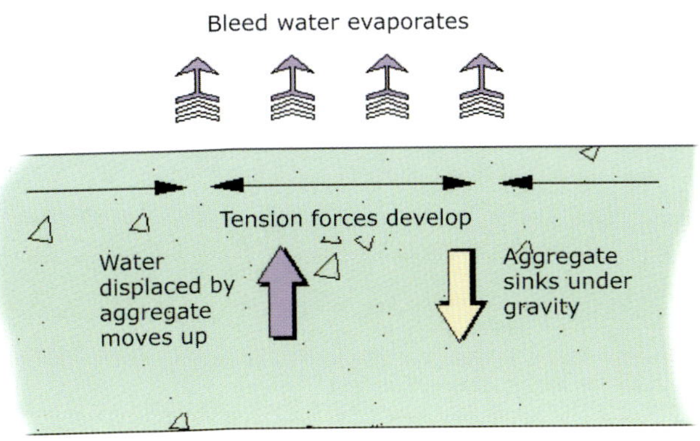

Figure 123 Mechanism of bleeding in a concrete floor slab.

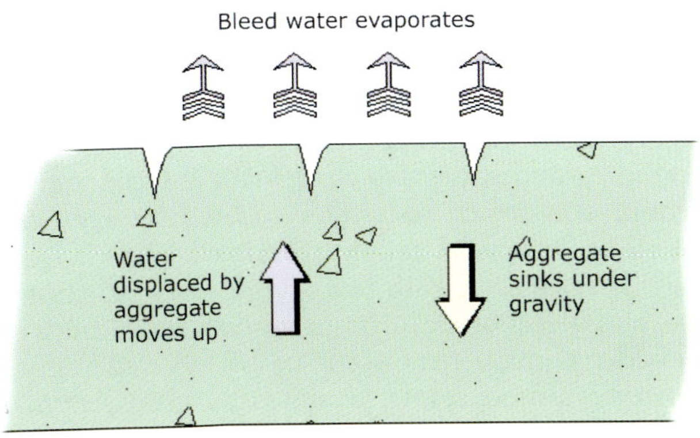

Figure 124 Second stage of bleeding, plastic shrinkage cracks develop.

To avoid problems of rapid evaporation it is usually preferable to protect a new slab from the drying effects of wind and sun, therefore placement after completion of the roof and walls is desirable. Proper curing is essential, so polythene sheeting and/or a suitable spray on membrane must be used. However, whilst proper curing is essential for the correct performance of a concrete slab, the curing process does not affect contraction; it merely improves the tensile strain capacity of the concrete. The best means of minimising shrinkage is to keep the water and cement contents as low as possible.

Cracking as a result of drying shrinkage will usually take place within several months of construction and for up to three to four years, the rate depending upon relative humidity. When concrete is mixed, more water than is needed for the chemical reactions is added to make the mix workable. This residual water is held in capillary pores and gradually evaporates from the surface. Removal of water is accompanied by a reduction in volume, hence shrinkage of the concrete. If the concrete is restrained in some way, cracking will occur; this is usually at right angles to the direction of the restraint. For ground bearing slabs, the cracks are most likely to occur when the length of the panel is much greater than its width; the crack running parallel with the short side and through the centre of the panel or sometimes diagonally at corners.

Shrinkage cracking will usually stabilise, but damage to shoulders of the crack could occur and when it is thought that cracking has become dormant, the cracks can be ground out and filled with a suitable epoxy sealant. If shoulder or arris cracking occurs before the anticipated movement has ceased a more flexible sealant may be desirable. If the matter is not attended to, deterioration can be quite rapid.

Figure 125 Deterioration of shoulders to induced joints possibly due to slight curling of the adjacent slabs and joint sealant being too soft, offering insufficient support.

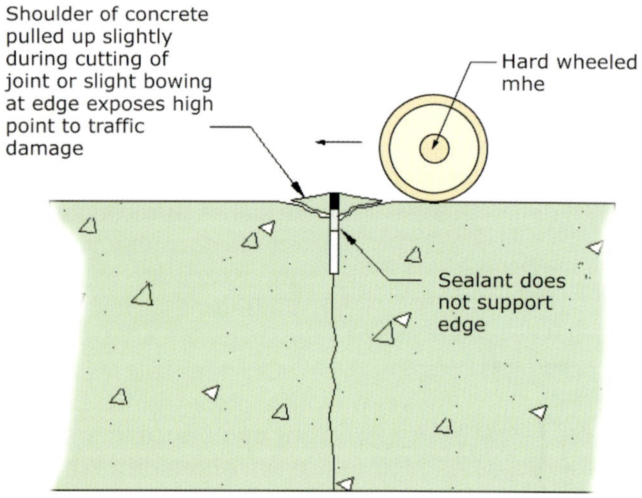

Figure 126 Damage to edge of an induced joint following from curling or raised edges pulled up during the sawing process.

Crazing

Crazing of a power floated surface is a common characteristic of this type of floor and is not normally detrimental to use and serviceability. Crazing is characterised by a map pattern of cracks around 0.1mm in width and extending only a few millimetres into the depth of the slab. The individually linked map 'polygons' are around 50–75mm across.

Crazing will occur wherever there is a change in the properties of concrete close to its surface. The effect of power floating can, if overworked, lead to a thin layer of laitance (a dilute mix of cement paste and fine aggregate particles). The surface tends to dry out quicker than the main body of the slab thus imparting tensile forces in the surface; these forces are relieved by cracking.

Curling

Curling of floor slabs is fairly common and usually becomes manifest at joint positions. The problem is due to different rates of shrinkage between the top and bottom surfaces of the slab. Since the top surface tends to dry more rapidly, it shrinks further; because the bottom is wetter, the slab curls. With induced joints a certain amount of restraint is offered by aggregate interlock, or special sliding reinforced joints if these have been specified. In less severe cases of curling, the raised sections can be ground down (otherwise shoulder damage may occur under wheeled traffic). However, if more severe, grouting under the slab may be needed to restore strength and prevent the floor cracking or rocking.

Surface defects

Occasionally, particles of organic material or other contaminants may find their way into aggregates (see figure 127). If near to the surface, the soft particles can break down leading to localised pockets of spalling. Such defects can be cored out and filled with a suitable resin. Similarly, small particles of aggregate can rise to the surface during placing; their subsequent disturbance under traffic can lead to small pockmarks in the surface; these are usually not a problem although if bad enough to affect serviceability, the defective areas can be drilled out and filled.

Abrasion resistance

Abrasion resistance of a slab is important and in an area that is in heavy use, severe scratching, scuffing or wear can lead to problems. In aisles, a lack of abrasion resistance can lead to the gradual wearing of shallow grooves into the surface which can affect serviceability and flatness. Severe wear will usually be self-apparent, but if there is any doubt, an accelerated wearing test is set out in BS 8204-2:2002. The test involves a rotating steel wheeled plate which is held in position by a fixed clamp. The machine is operated for a fixed period of time and the difference in level measured and compared with BS 8204, which sets out various categories of wear according to use. Factors affecting abrasion resistance include the concrete strength, water cement ratio, method of finishing and curing.

Steel fibres can be found in the surface of a floor and occasionally they may stick proud. Generally, fibres that are exposed in this way are not a major problem, but any that do protrude can be ground or cut away fairly easily. The incidence of fibre exposure is generally less when dry shake finishing methods have been employed. Steel fibres tend to reduce the amount of bleed water as their high surface area impedes the settlement of aggregate particles and thus the displacement of water.

Figure 127 Organic material (part of an acorn) and the pocket from which it was extracted – typical example of localised damage.

Figure 128 Abrasion test in progress.

Figure 129 Floor following abrasion test. In this case, the results showed a very good level of abrasion resistance – in excess of the specification. This is surprising given the extent of scuffing and striations apparent on the floor surface.

Chemical effects on floors

Aside from the risks associated with sulphate attack (see section 3.5), concrete industrial ground floors can be affected by a variety of chemicals contained within common products. Beer, cider, sour milk and sugar for example all cause gradual disintegration of concrete, whilst damp coal can cause rapid deterioration in damp conditions on account of soluble sulphides producing sulphuric acid. Further details of potential harmful chemicals are given in Technical Report 54 published by the Concrete Society.[155]

21.4 POWER FLOATING

The process of power floating and steel trowelling is common in industrial buildings and retail warehouses where for operational and cost reasons the provision of a sand and cement screed is undesirable. Sometimes termed panning, the process involves the use of a machine fitted with rotary paddle blades. Some large machines are ride-on whilst other types are walk-behind – the latter impart less energy to the floor than the ride-on types. The rotation of the blades embeds aggregate particles just beneath the surface, removes slight imperfections, humps, and voids, and compacts the mortar at the surface in preparation for additional finishing operations. Sometimes swirl marks from the rotating blades can be seen in the finished surface. If the surface is overworked during panning, a burnished appearance can result.[156] Where a smooth, hard, dense surface is desired, floating is usually followed by steel troweling. Careful quality control is needed to achieve a good power floated floor, the process relying upon the skill of the float operatives and appropriate judgment as to the optimum time after laying as well as various other factors.

Problems develop when the floor is power floated too early and consequently overworked.[157]

21.5 DELAMINATION

Whilst failures with applied cementitious screeds are well documented, problems with the delamination of power floated slabs are perhaps less frequently encountered; indeed, with no applied finish as such it is at first sight difficult to understand why delamination should occur in the first place. However, the problem is not unknown, with areas extending from a few tens of millimetres to several square meters in area and of a thickness ranging from microns to 8mm or thereabouts. For a modern, highly specified high-bay warehouse floor, the consequences of delamination can be severe – not only creating a health and safety hazard due to rocking of fork lift trucks but also causing damage to the trucks themselves. In extreme cases, blisters of concrete can be formed during laying, thus putting the stability and durability of the floor into question.

The problem can affect floors of conventional concrete mixes as well as those constructed from lightweight aggregates such as expanded clay.

Delamination (within the context of this chapter) is essentially a separation within the slab, mainly occurring horizontally and within 4–9mm of the surface (although under 1mm is not unknown). Whilst delamination can also occur as a result of the corrosion of reinforcement bars within the slab, the damage occurring from this cause would usually be at greater depth and possibly very localised – clearly following the pattern of the reinforcement. Reinforcement corrosion is dealt with in chapter 3.

Aside from corrosion problems, there are two possible mechanisms of delamination failure – both occur during placing. The first mode is due to overlayering – drawing fresh layers of concrete over concrete that has already achieved an initial set. The second mode is a debonding process caused by bleed water.

Overlayering

To be successful, power floating of a floor slab needs to be carried out at the optimum time – usually when no bleed water or sheen is visible on the newly placed concrete and, as a rule of thumb, when the top surface allows an indentation underfoot of around 4–6mm. If the surface is too wet, the floating machine will do no more than tear up the surface, whilst if too dry, it will not be able to trim high spots and fill lows effectively. Exposure to sun, wind or rain and temperature and humidity changes during placing can affect the optimum time.

By failing to level off the concrete sufficiently before power floating, the float will tend to draw concrete from high spots into hollow areas. Since a degree of stiffening will have already occurred, the 'new' layer in the hollow area may not bond effectively, thus creating a plane of weakness.

Sometimes a contractor may be tempted to add water and cement powder to the surface of a slab that has already began to stiffen. This is bad practice as whilst it works up a slurry coat (which enables a measure of smoothing off, it also creates a possible zone of weakness between the main body of the concrete and the slurry, leaving it vulnerable to delamination under traffic or as a result of shrinkage.

Adjacent areas of concrete can sometimes stiffen at different rates, for example at the perimeter of an area laid at a slightly different time or due to variations in workability. One area may be ready for floating before a neighbouring area in which case there will be a time interval between floating operations. In some cases the boundary between the two areas can be merged successfully, but if the concrete becomes too hard this may not be possible. Thus, there is a temptation to float the softer area before it is ready, and thus to draw a layer of softer cement over harder areas.

Overlayering is therefore the process of drawing one layer of concrete over another, with the risk that the bond between the two may not be adequate.

Identification of overlayering:

- The contact surface of the delaminated area may be open in texture but reflect the profile of the underlayer, i.e. it is moulded to suit it.
- The surface of the under layer will have a grey mortar finish with markings and other characteristics that were present at the time that it was overlayered.
- The surface of the under layer may have a slight lime bloom.
- In the case of overlayering due to added surface water, the delaminated layer will probably be fine grained and weak.

- With other modes, unless the layer is very thick, it will be finer grained than the main body of the concrete.
- In the case of thicker layers, the delaminated section might be stratified, with a less compacted layer beneath the surface.

Bleeding

The particles of cement and aggregate in freshly mixed concrete are (or should be) distributed evenly through the mix. Once it has been placed, the effects of gravity mean that aggregates gradually settle down until they bridge one another (achieve point to point contact), whilst the smaller cement particles continue to settle between the aggregate particles. The gradual settling displaces water – bleed water – which rises to the surface.

In certain circumstances the surface of the concrete may set before the process of bleeding has been completed. If this happens, the presence of trapped bleed water could again give rise to a plane of weakness. Such a problem could even occur prior to power floating. Water trapped below the surface could be disturbed by the action of the paddle blades from the floating machine dragging over the surface, extending the water below the surface and pulling up a blister. The blister may burst, releasing some of the water and letting air in.[158]

Investigations by Klieger[159] attempted to create conditions in a laboratory to reproduce the effect – primarily by trying to achieve a surface set by exposure to higher temperatures and drying wind velocities over the surface. Klieger's experiments were unsuccessful, but it is generally recognised that concrete with a propensity to bleed slowly over a long period of time is more susceptible to this type of defect given changing climatic conditions. Low humidity and breezy conditions will cause rapid evaporation and this will result in stiffening if the bleed rate is low. The nature of the aggregate and size of particle can also have an influence on the process. The table below summarises the factors that could affect the risk of delamination due to bleed water.

Identification of delamination due to bleeding:

- Some aggregate surfaces will be clean owing to the presence of water.
- The surface of the underside of the delaminated layer will be granular in nature and in some cases particles of aggregate will be covered with cement paste.
- There may be lime bloom and silty material on the surface of the exposed face.
- Shrinkage and/or curling may have occurred.

Problems with air-entrained concrete

Air-entrained concrete was developed in the 1930s with a view to improving resistance to freezing water and de-icing chemicals. By using air entraining Portland cement or by introducing air-entraining admixtures, millions of microscopic voids are formed in the concrete. These voids can often account for five per cent of the volume of the concrete and create small expansion chambers thus relieving the effects of the expansion of water when it freezes.[160] Besides the increase in freeze-thaw resistance, air-entrained concrete is more workable than non-entrained concrete whilst bleeding and segregation is less than with conventional mixes.[161]

However, the use of air-entrained concrete for floors that are to be power floated is known to be problematic and is best avoided. The reason for this is that the surface of the concrete can become very dry, whilst a few millimetres below the surface the concrete behaves normally, with gradual stiffening taking place in the way one would expect. As the paddle blades are drawn across the surface, they create a shearing action, which in turn tends to unite the small air bubbles into a much larger, elongated void. These voids can extend to form quite large sheets of debonded concrete surface, which subsequently fail a few days or weeks later as a result of drying shrinkage or mechanical action.[162]

Identification of delamination in air-entrained concrete:

- The flat surfaces of the bubble may be visible.
- The surfaces along the line of failure will be grey and granular in nature.

Factors that could affect the risk of delamination due to bleed water

Factors that could affect bleeding	Factors that could affect the rate of setting at the surface
The temperature of the subgrade	Exposure to high wind or air currents
The existence of a DPM or plastic slip layer membrane beneath the slab	Low relative humidity
A thick slab	Exposure to sunlight
A high water/cement ratio	Rising air temperature
Entrained air or water reducing admixtures	

22 Multi-storey car parks

At a glance

- There are over 4,500 multi-storey car parks in the UK, many dating back to the 1960s, a time when design standards were less onerous than they are today. There exists a legacy of poorly constructed and equally poorly maintained structures which demand expensive and extensive remedial work. Too often, car parks are treated with little respect for the serious potential problems that may exist.
- The engineering professions regard multi-storey car parks with the same level of caution as they do bridge structures; both are exposed elements subjected not only to the weather but to aggressive forms of deterioration from chlorides in road salts and other chemicals.
- Whilst steel framed car parks are not unusual, a majority are of either precast or reinforced concrete. All structures are vulnerable to decay arising from water brought in on vehicles or from the weather; intermediate floors and access ramps are often worse affected than top floors as a degree of regular washing under rainfall can help to reduce the build up of harmful chemicals. On intermediate floors chemicals become more concentrated as there is less dilution by rain.
- Deflection of structures leads to ponding and difficulties with rainwater disposal. In turn, water then works its way into the structure via poorly designed or maintained movement joints (or cracks arising through lack of movement joints) where it may assist in the corrosion of reinforcement. Often, leaks occur in the areas that are structurally most vulnerable such as around balustrade fixings, half joints, etc.
- The Standing Committee on Structural Safety recommends that owners of car parks should:
  - Commission inspections by suitably experienced engineers before commissioning repairs; these inspections should go beyond those areas that are visibly defective.
  - Identify the risk of progressive collapse.
  - Establish the suitability of edge barriers to resist accidental vehicle loads and to protect small children from falling.

22.1 INTRODUCTION

Multi-storey car parks (MSCPs) are unlikely to arouse much passion other than a general belief that most are not very pleasant places to visit. Many early versions were poorly built and continue to be poorly maintained; they are often rated as a low priority for expenditure and command only cursory inspection during sale or purchase. There are in excess of 4,500 MSCPs in the UK, virtually all having been constructed since the 1940s and many free-standing types constructed during the 1960s. Many early car parks were procured as cheaply as possible, creating a legacy of poor construction and a lack of durability. Construction to the lowest cost per bay did not necessarily lead to good construction.

For a surveyor, there are significant risks in underestimating the complexity of forces and agencies of deterioration that are at work. Fortunately, there are relatively few collapses (see section 22.9 on lift slabs), but MSCPs are large structures and repair bills can be very high. The apparent simplicity of a car park structure often means that its inspection is not treated in the same light as for example the shopping centre or office building to which it is attached. However, the conditions that the car park has to endure will be far more onerous than the adjoining enclosed buildings and if the various modes of deterioration are not understood or appreciated there remains the risk that something may be missed.

Following the collapse of a car park floor slab at Pipers Row, Wolverhampton in 1997, the design and condition of MSCPs was placed in the spotlight. The collapse spawned various reports and investigations, leading, in September 2002, to the publication of an OPDM sponsored publication *Enhancing the Whole Life Structural Performance of Multi-Storey Car Parks*[163] together with *Recommendations for the inspection and maintenance of car park structures*[164], which was published by the Institution of Civil Engineers in the same year.

Most MSCPs in the UK are concrete column and slab structures with the main types being:

- in-situ reinforced concrete column and slab;
- in-situ frame, precast prestressed deck;
- precast frame and deck;
- lift-slab construction;
- steel frame with in-situ or composite deck; and
- in-situ concrete with bonded post-tensioning.

All of the above types are vulnerable to deterioration due to chemical agencies, but the design of the car park will affect its durability to a large extent. For example, a prestressed deck will have a much lower tolerance to ingressed chlorides than one cast in-situ simply because the tendons are of smaller diameter and the threshold chloride for prestressed concrete is much lower.

Early car parks are most likely to suffer the greatest levels of deterioration due to their age and exposure to the elements, but also as a result of poor construction or inadequate design. Structural and design codes were less onerous in the 1960s than they are today, meaning that matters such as shear reinforcement, depth of cover, joint design, etc. were often inadequate, whilst higher water to cement ratios and lower strength concrete created more permeable structures that would be more vulnerable to chemical attack and carbonation. Early design codes for flat slab construction are now regarded as being particularly suspect as evidenced by the Pipers Row collapse. The simplified rules set out in BS 8110 do not necessarily ensure sufficient robustness with all forms of flat slab construction.

22.2 MODES OF DETERIORATION

All buildings move; they are not static structures. The Institute of Structural Engineers considers car park construction to be similar in many respects to bridge construction because of, amongst other things, the large unobstructed spans (frequently 16m) but also the transient loadings of vehicles parking and moving about. Furthermore, the structures are exposed to the elements; not only are the top decks usually large areas exposed to solar radiation and resultant thermal movements, but also the ramps, parking zones and main traffic routes are subject to chemical attack arising from salts – usually calcium chloride or urea being brought into the building by vehicles. Being in the main open-sided and unheated structures, car parks are also subject to frequent wetting and drying, with the attendant risks of frost action and concentration of aggressive chemicals.

The Standing Committee on Structural Safety (SCOSS) in its 10th Annual report, warned of a threat to structural safety as a result of neglected corrosion, especially of bonded or unbonded prestressing tendons or of reinforcement to cantilevers.[165] SCOSS went on to draw attention to the potential lack of precautions against progressive collapse as this subject was not initially addressed by Building Regulations in buildings of less than five storeys in height. The Committee concluded:

> '... that the premature deterioration of multi-storey car parks if ignored, could lead to serious structural distress, and even collapse, in some cases. Attention should be paid to zones likely to be affected by drainage water contaminated by road salt. This could be particularly damaging at half-joints or bearings. Corrosion may, in some cases, be advanced without conspicuous cracking and in other cases cosmetic repairs may have masked signs of deterioration'.

Movements in the building due to loading and thermal expansion can cause a number of effects. Bowing of a large tee ribbed deck will create the risk of cracking of surface finishes around the perimeter. Furthermore, water can then collect at the perimeter and penetrate cracks. Water entering at a bearing joint is unsatisfactory since this is where shear reinforcement will be located; the resultant spalling or deterioration in the bearing could cause unwelcome structural effects.

Similarly, the effects of insufficient falls on the decks, or actions such as regular car washing, create pools of water which can gradually percolate the structure to affect reinforcement.

So, deterioration in car parks in not simply the result of one agency over another, it is more likely to be the result of a combination of factors each leading to a further problem. For example:

- expansion causing bowing;
- bowing causing ponding;
- ponding causing water ingress; and
- water ingress permitting deterioration in the concrete.

Figure 130 Water ingress through a defective perimeter joint has caused corrosion of steel reinforcement and subsequent spalling of the bearing to this column head.

It is easy to underestimate the effects of water ingress into a structure; however a brief reflection reveals that the areas that are most likely to leak are those either at mid-span or at or close to the perimeter. At the perimeter will be found bearing conditions for precast structures, prestressing anchorages, reinforcement for edge beam and vehicle protection balustrades, structural joints and so on – precisely the details that need to be protected against corrosion.

As noted earlier, a MSCP is not generally a controlled temperature environment and in particular the top deck can suffer from the effects of thermal expansion. BS 5400 suggests that the top deck of a car park structure can be expected to endure a temperature range of circa 45°C (within the UK).[166] In general, steel and concrete frames have a similar coefficient of expansion which may be taken as 12×10^{-6}/°C. So, for a 75m length deck, movement due to temperature effects could be as much as 40mm; this must be accommodated by movement joints. If the ends of the slab are restrained by elements of structure (for example shear walls, cores or ramps), load reversal can take place whereby elements that are not designed to take high internal stresses become damaged.

Similarly, the restraint of the perimeter of the deck will result in the bowing of the deck in a way which could reverse drainage falls and gradually damage movement joints and bearings. A method of estimating the bow of a deck due to the restraining influence of the structure is given in the IStructE Report (3rd edition), as shown in the box below.[167]

$$\frac{\text{Coefficient of thermal expansion} \times \text{temperature difference} \times \text{length}^2}{8 \times \text{thickness}}$$

For example a 200mm thick deck spans 16.5m and is finished with 50mm asphalt. In summer, the temperature differential between the top surface and the underside of the deck will subjected to a temperature differential of 13°C. Taking the same coefficient of thermal expansion as before gives:

$$\frac{12 \times 10^{-6} \times 13 \times 16500^2}{8 \times 200} = 26.5\text{mm}$$

A similar calculation this time using a double tee beam 600mm deep finished with a thin, dark coloured waterproofing coating produces the following result:

$$\frac{12 \times 10^{-6} \times 23°\text{C} \times 16500^2}{8 \times 600} = 15.6\text{mm}$$

However, bowing effects caused by temperature are not the only problem since the effects of long-term creep must be considered. Creep is generally more significant (and severe) in car parks than in enclosed buildings because the ratio of dead load to total load is higher. Creep modifies the elastic modulus of concrete beams over time such that over a period of say 25 years, the amount of deflection in a beam could increase by 25–50 per cent[168]. Such deflections mean that drainage falls and outlet positions that worked perfectly well when the building was constructed may no longer be effective, giving rise to ponding.

Figure 131 Deflections in top deck giving rise to poor falls and ponding remote from rainwater outlets.

As noted earlier, ponding is undesirable in car parks, as it can create conditions in which contaminated water can percolate into the structure.

A span of 16m (which is the usual design module) leads to a high dead/live load ratio by comparison with other structures together with a greater live load being experienced for longer periods.

As an alternative, some early designs reduced the span between columns to around 9m, with cantilevered portions extending to approximately 3m each side.[169] The performance of the cantilever is therefore critical, particularly as it will in all probability carry the heavy dead loads of a robust balustrade or edge beam system. From a structural point of view, deterioration of the concrete 3m back from the edge of a slab could be serious, as this is where shear between the column heads and the slab will be highest, and where the bending moments for the cantilever will be concentrated.

22.3 UREA AND CHLORIDES

Until 1977, calcium chloride was routinely added to in-situ and precast concrete as an accelerator, either to enable rapid removal of moulds or to permit concreting in cold weather. Thus, older structures could be at risk of corrosion damage arising from cast-in chlorides coupled with increasing carbonation. For prestressed components the risk is greater as the threshold for acceptable levels is much reduced (0.1 per cent chloride by weight of cement compared with 0.4 per cent for reinforced concrete).

However, the structures are also at risk from chloride salts brought into the car park on vehicles during cold weather. Analysis of half cell potential surveys on car park decks reveals high risk areas on ramps and traffic routes as well as the wheel positions in car parking bays. Ramps leading up (or down) to the first two levels of the car park are at risk of salt or other contaminants being brought in. Cars are not the only problem though as salt spreading on pedestrian areas can cause contamination. On the top level of a car park, the effects of rainfall will wash the structure and dilute some of the salt. At lower levels, where there is no washing effect and water dries by evaporation, chlorides can become concentrated, leading to a greater risk of corrosion damage.

Chloride attack is an insidious problem and can be difficult to treat unless electrochemical methods such as cathodic protection are used (see section 2.12 on cathodic protection). Pitting corrosion is a particularly dangerous condition, particularly in prestressed or post-tensioned structures, since the reinforcement can effectively be eaten away without the formation of expansive rust. Pitting corrosion usually occurs in localised pockets of high chlorides and saturated conditions.

Mild steel or galvanised steel fixing bolts for balustrades or vehicle impact barriers can be vulnerable to corrosion due to chlorides. Of particular concern at perimeters is the fixing of precast concrete balustrade panels. These may themselves not be intended to provide impact protection, but water that accumulates at the perimeter could find its way in to what may have been poorly grouted construction joints leading to corrosion of locating dowels or gravity fixings.

Problems of vehicles crashing through perimeter balustrades are rare but not unknown; an incident in Canterbury in 1996 revealed the barrier to be of insufficient strength to prevent a car from plunging over the side of a car park. In this case a car fell some 20m from the fourth floor of the car park, seriously injuring the driver.

The risks to safety of defective edge protection were addressed by SCOSS in their 10th and 11th annual reports, which highlighted in particular the fact that many of the country's 4,000 car parks were built outside the requirements for protection that featured in the Building Regulations current at that time.

Figure 133 Chloride concentrations at the base of this ramp (in this case the second ramp from the entrance) have resulted in localised corrosion and spalling

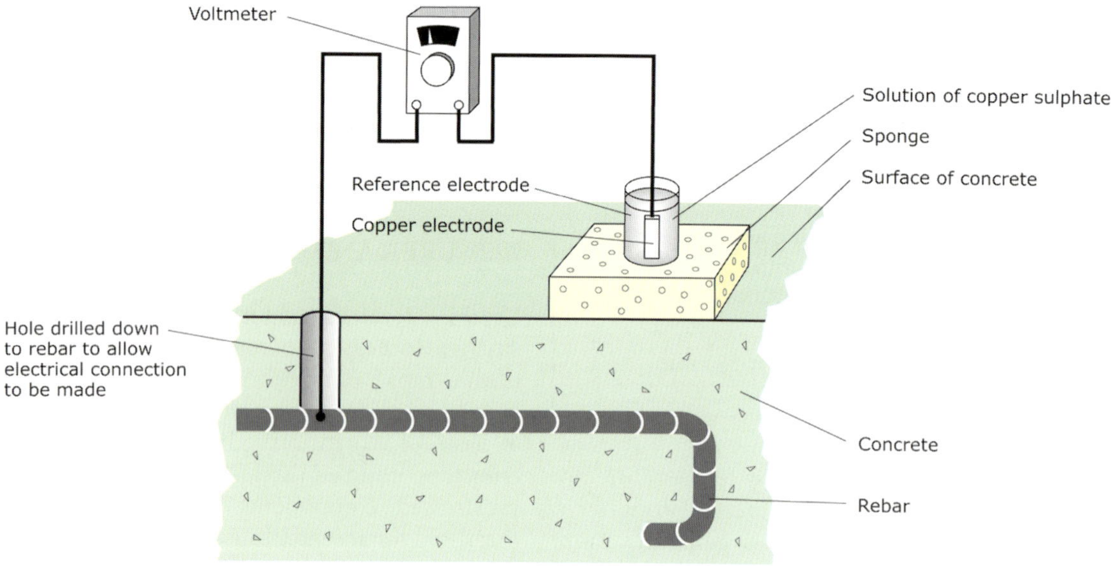

Figure 132 Half cell apparatus.

Figure 134 Rapid corrosion of mild steel balustrade as a result of chloride contaminated water.

SCOSS[170] recommended that the owners of MSCPs should:

- commission inspections by suitably experienced engineers before commissioning repairs;
- inspect beyond the areas that are visibly deteriorated;
- identify the risk of progressive collapse;
- establish the suitability of edge barriers to resist vehicle loads and modify the barriers if necessary; and
- ensure that the height and design of barriers is appropriate to safeguard small children.

Because of the damaging effects of chloride on structures (and also the pollution hazards associated with its use), several alternative de-icing agents have been developed. Some proprietary systems include corrosion inhibitors, and others, such as urea, calcium magnesium acetate-CMA or potassium, all have the capability to be harmful to the environment and are all generally more expensive than conventional rock salt complying with BS 3247, some two million tonnes of which is used annually on highway structures within the UK.[171] Rock salt and its substitutes effectively lower the freezing point of ice to around −10°C, although some systems are more effective than others at different temperatures.

Whilst alternatives such as urea may not be as damaging to structures, they are salt based and when present in concentration can lead to the formation of salt crystals within the pore spaces in a concrete structure. Crystal growth will eventually disrupt the concrete, leading to spalling; in turn, salt solutions can then percolate further into the body of the concrete thus perpetuating the process and permitting side effects such as increased rates of carbonation.

Good quality, well compacted concrete is less susceptible to salt crystalisation but the identification of affected areas serves as a good locator for core tests or chemical analysis.

22.4 CARBONATION

Carbonation is dealt with in more detail in section 3.1. Essentially, carbon dioxide in the atmosphere reacts with cement hydrates to form calcium carbonate. The process lowers the alkalinity of the concrete which then leads to the depassivation of the steel reinforcement. Corrosion does not necessarily follow, but if the steel is depassified it is most certainly at risk of corrosion given exposure to moisture and oxygen.

Carbonation spreads from the outside of the concrete inwards, so it follows that good quality, well compacted concrete will carbonate at a much slower rate than poor quality materials. If poor compaction and strength (perhaps caused by too high a water:cement ratio) is coupled with misplaced reinforcement with little cover, the risk of corrosion is high.

Corrosion caused by carbonation will be accelerated in damp conditions, whilst in dry areas, the rate of carbonation will be higher. Carbonation also liberates a certain number of chloride ions that may have been incorporated in the original mix and which would otherwise be chemically bound in. Thus, there is a risk that electrical conductivity in the concrete could increase, with the consequent increase in corrosion currents and therefore corrosion.

In certain cases carbon dioxide released from vehicle exhausts can lead to greater carbonation rates in concrete than would be experienced in well ventilated conditions. Such problems would usually occur in crowded, poorly ventilated city centre car parks where vehicles are left running for longer periods.

22.5 FROST DAMAGE

The expansion of water when it freezes is well known, and whilst the deterioration of concrete is often due to spalling of reinforcement through corrosion, general damage to the surface can often occur on the top deck of a car park or in the ground or access floor and first two ramps as a result of prolonged saturation. Tyre abrasion and the freezing and thawing of the water in the surface pores gradually roughens the surface, making it friable and more susceptible to further water ingress. In extreme conditions, potholing of the concrete can occur. Frost damage can be associated with ponding due to deflections in the slab or poor drainage arising from badly located or blocked outlets.

The effects of freeze-thaw action can result in the following types of defect:

- scaling in the surface;
- formation of conical 'pop-outs' (this occurs when bleed water has accumulated under a layer of flaky aggregate); and
- cracking or spalling of lumps of concrete.

The phase change from water to ice is accompanied by a volumetric expansion of around 1.09 per cent, which generates a bursting stress in the concrete.

Figure 135 Surface deterioration due to frost damage in an area that is regularly saturated.

Concrete surfaces that are heavily tamped create conditions where water is more likely to be held in the surface; whilst frictional resistance is good, deterioration may eventually become more pronounced. Smooth, power floated floors perform better in this respect, although they do give rise to tyre squeal.

It would be easy to conclude that the UK climate is unlikely to create the same risk of freeze-thaw damage as might be experienced in North America or Siberia, for example. However, the reverse is probably true, since in cold environments the low temperatures are maintained for far longer whereas in the UK or northern European areas, the frequency of freezing and thawing will be far greater – especially where de-icing salts are used on a regular basis.

22.6 USER DAMAGE

Whilst the effects of user damage will generally be localised, they can nonetheless be severe. For example, oil spills or drips from vehicles can disrupt and soften asphalt surfaces and lead to their degradation, whilst gypsum based oil absorbent materials that are used for cleaning or mopping up after oil spills have been known to initiate sulphate attack in concrete.

Similarly, car washing activities could have an unfortunate consequence as not only do these lead to increased risk of saturation of a concrete deck but concentrations of salt and other contaminants washed off vehicles will be much greater in localised areas. Concentrations of chlorides can give rise to serious corrosion problems.

22.7 PROBLEMS WITH MOVEMENT

As noted earlier, car parks are usually large-span structures subjected to frequent load changes as well as the effects of long-term creep, deflection and temperature movements. A failure to make sufficient provision for movement will lead to cracking and disruption of joints, which, if severe enough, can damage waterproof coverings and lead to water ingress.

Joints through decks are particularly troublesome and often badly maintained. It is not unusual to find accumulations of short-life materials and various bodged attempts to seal a joint, usually leading to failure and at worst the entrapment of reservoirs of water which gradually leeches out.

Movement joint designs are many and varied, ranging from simple membrane types (EPDM for example) or more sophisticated bolted joints. These systems can be effective but do rely upon correct installation and sealing. Often the holding down bolts perforate the waterproof membrane and give rise to leakage problems. To prevent water from penetrating the joint it may be preferable to contrive a raised joint to minimise the flow of water over it. Such a detail would of course only apply if adequate drainage can be provided elsewhere.

The detailing of reinforcement at the joint between a column and a flat slab is difficult to engineer but is also critical in terms of performance and the avoidance of problems of progressive collapse. In 1972 and again in 1984 design codes were modified to improve shear reinforcement around column heads (see section 22.9 on lift slab floors). However, these changes in code would obviously not apply to structures constructed during the 1960s where design was often carried out to CP114, CP115, CP110 and BS 449 – all of which were found to be inadequate for the task. For this reason particular regard should be given to any signs of cracking around column head positions.

22.8 PRECAST STRUCTURES

Precast systems of construction can have benefits in terms of speed of erection and quality control. For savings to be made however, the concrete must usually be capable of being struck from the moulds on a 24-hour cycle. The addition of chlorides was one method of achieving this production rate although the practice was ceased in the late 1970s.

There are several different configurations of precast as detailed in the following table.

Frame clear span	A precast portal frame is erected having a clear span of 16m, with flooring spanning onto each frame in the longitudinal direction. Since the frames are cast as single assemblies they possess their own rigidity in one direction only.
Hollow core clear span	400mm depth hollow cored units spanning the width of the car park. The flooring units are relatively heavy. Water entrapment within the voids is possible and so they must be drained to prevent risks of disruption through freezing.
Double tee clear span	Usually 500–600mm depth and able to give a 16m clear span. Usually, the planks are covered with a concrete topping.
Beam clear spans	In which the beams (either in-situ or precast) span 16m and support shorter spans (possibly 6m) of flooring.

Figure 136 Joints at bearing positions are notoriously troublesome. In this case, the bearing of a trimmer beam supporting a double tee deck has failed as a result of water ingress and corrosion of reinforcement.

Figure 137 Water ingress above this beam/column connection has lead to corrosion of reinforcement and spalling of the bearing and reduced support.

Figure 138 The end of a double tee beam shows a halved joint at its bearing end. However, due to movements in the structure the original bearing was compromised and an additional steel beam has been retrofitted in order to provide adequate support.

Figure 139 Worrying amount of slippage in beam bearing, coupled with shear cracking.

Thermal movements, particularly in clear span structures can be severe and must be accommodated in joint design. Sliding joints are often very problematic and quite often lead to weeping problems. For this reason, split columns are desirable, although from an aesthetic point of view they may be less welcome.

22.9 LIFT SLAB FLOORS

Lift slab construction is a method of constructing buildings with repetitive floor layouts by casting them on the ground and jacking them into place. There are several different methods but commonly a series of jacks are fitted to the top of precast columns, connected to threaded lifting rods and lifting collars cast within the floor slabs. After reaching a satisfactory strength the floors are jacked up into position and then fixed with wedges attached to the columns. The void between the floor and the column is then filled with fine aggregate concrete.

The system is fairly popular in the United States, but there have been several cases of collapse during construction due to insufficient lateral bracing during the hoisting operations. One characteristic of these types of floor is that they provide little by way of lateral bracing to a building and this has to be compensated by the construction of reinforced shear walls or braced frames with steel diagonal members.

In the UK, lift slab construction achieved notoriety when, on 20 March 1997, a 120 tonne section of the top deck at the Pipers Row multi-storey car park in Wolverhampton collapsed. Although open, the car park was unoccupied and no-one was injured. (Whilst Pipers Row is often cited, a further collapse of three floors occurred in Coventry in 1988 when another lift slab car park was being demolished.)

Following detailed analysis as to the cause of the failure, HSE published their findings in a detailed report[172] which concluded, amongst other things, that if it had been properly maintained, the car park ought to have performed satisfactorily in service. However, whilst the design provided good falls for drainage, the construction tolerances, deflection and creep of the slab created areas of ponding on the surface. Waterproofing leakage and ponding saturated the slab and frost damage developed to fissure and microcrack the concrete, particularly in local areas of high porosity and low strength.

Deep carbonation developed into the microcrack system and zones of high porosity, possibly aggravated by carbon dixoxide released from the breakdown of urea used for de-icing during the previous decade. When microcracking reached the top reinforcement, the degradation, sufficient to destroy most of the bond from steel to concrete, had developed for most of the upper top layer of reinforcement This degradation, and the depth to which it penetrated, lead to punching shear around the column heads and, of course, the removal of support to the slab which then collapsed.

A brief description of moment connections and shear

Shear can be described as a 'slicing' force; it is an internal force that can act at any point along the length of a structural member. A short beam with a heavy load close to one of the supports would be subjected to a slicing type force; this could be significantly greater than the forces generated in bending.

A simple tent frame constructed of two members with a pivot joint is unstable until the ends of the frame are securely located or tied. If the two ends were to be joined by a third member, the frame would be rigid. Similarly, a square frame could easily be distorted into a parallelogram if a horizontal load were applied to one corner. If a diagonal member were to be inserted, the square would be rigid – in effect the insertion of a single diagonal 'brace' has created two triangular structures.

A framed building behaves in exactly the same way; unless it can be suitably braced it would be free to move under lateral loads such as wind pressure. By putting diagonal structural members in place, the frame can be given rigidity. However, diagonal braces are not always satisfactory; they can intrude upon the internal space and may detract from the architectural appearance of the building.

An alternative to the provision of diagonal bracing is the provision of a shear wall – a stiff vertical cantilever that is able to resist lateral loads. Constructing a rigid reinforced concrete core for the lifts and stairs is a similar solution.

However, both shear walls and diagonal bracing have disadvantages and if a large uninterrupted space is needed internally, the option is to create very rigid connections within the frame and floors that are able to provide sufficient stiffness to the frame to prevent buckling. Such connections are termed moment connections – they permit no relative rotation between the columns and slabs and their connections.

A form of failure related to lift slabs is punching shear. Again, punching shear is a slicing action, acting in the same way as a pen being pushed through a sheet of card. When the force of the pen exceeds the shear strength of the card, it pushes through to the other side.

The lack of maintenance to the waterproofing layer led to deterioration of the concrete but the failure generated further concerns over the design codes used for the structure. The structural integrity of this form of lift slab structure requires cautious engineering judgment. Interpretation of the unusual characteristics of the column/slab connection using current design codes can result in an under-estimation of load acting on the structure, and an overestimation of strength.

Once deterioration, corrosion or cutting out exposes reinforcement, it must be considered as a potential structural problem requiring a check by a structural engineer to establish if a structural repair, as distinct from a corrosion control, infill or cosmetic repair, is required. For structural repairs a detailed method statement with reinforcement drawings marked up with limits of cutting out and support requirements should be prepared by a structural engineer.

Punching shear anchorage zones are particularly difficult to patch repair reliably and if effective anchorage has been damaged it may be better to bolt completely through the slab or recast to full depth.

Identification

Flat slabs without thickening around the column heads, together with evidence of infill around the column head/floor slab, should alert the surveyor to the possibility of lift slab construction. Early forms adopted columns at 4m centres across a bay width of 16m with a typical floor slab depth of 225mm.

By the late 1960s the method had changed to columns at 9.8m centres with cantilevers of 3.1m each side. Columns would support a slab area of $50m^2$ and a slab of 280mm thickness.

In the early 1980s the same structural configuration was used, but with post-tensioned slabs of 250mm thickness. Car parks constructed of this method may not have waterproofing layers on the top level.

22.10 WATERPROOFING

The weatherproofing of early car park structures was somewhat hit and miss, with some having protection simply to the top deck, others to the ground (entry deck) and roof deck and some having no waterproofing at all.

Asphalt was common for roof decks, although roofing grade asphalt is generally insufficient in itself to deal with traffic. For this reason paving grade asphalt was often applied as a wearing course on top of the actual waterproof membrane. Slip and skid resistance is often provided in the wearing course by sand rubbing or surface crimping – sometimes pre-coated chippings were used. Surface dressings are often found; these are usually post-applied epoxy resin coatings incorporating bauxite chippings to provide good frictional resistance.

In cases where heavy point loads were expected (for example the car park and loading deck of a covered shopping centre) steel grillages were set into the paving grade asphalt to resist indentation.

Although asphalt is a durable material, particularly when laid direct to a concrete deck which can act as a heat sink, it is still vulnerable to defects arising in the main from poor detailing of upstands, lack of provision for movement, perforation by the fixings for car park barriers and lighting bollards as well as general neglect.

Aside from keeping the water out of a structure, the waterproofing membrane also, and possibly more importantly, protects the concrete against the ingress of chlorides or other harmful substances. Neglecting the covering may permit water penetration (which given the utility nature of a car park might be acceptable) but it carries with it the risk of more serious deterioration to reinforcement.

Figure 140 Plastic deformation in asphalt skirting has resulted in the perforation of the asphalt, water logging and plant growth.

At normal temperatures asphalt is relatively inflexible but solar radiation results in a build up of heat in the covering and although this is mitigated by the ability of the roof deck to act as a heat sink, high temperatures during summer can cause the asphalt to soften and degrade. When heated, asphalt is able to accommodate imposed movements by plastic deformation but at colder temperatures it is prone to fracture if it is subject to shock or impact loading.

Exposure to UV light has the effect of hardening the asphalt, effectively promoting the stabilisation of the bitumen binder after melting and cooling so that it loses tensile strength. Consequently, the roofing may not be able to accomodate temperature and loading changes that affect the deck, resulting in tensile failure. Temperature related stresses act in three dimensions and whilst this should not affect flat surfaces, upstands and vertical work are prone to damage.

Asphalt surfacing can also be damaged by the following:

- water ponding on the surface, which causes crazing and gradual deterioration; and
- oil spillages from vehicles, particularly in the mid-levels of car parks which are not subject to the degree of regular washing that occurs at roof level.

Defects in traditional asphalt roofs are often dealt with using traditional asphalt repair methods (see section 9.4 on mastic asphalt). In recent years however, the car park refurbishment industry has been influenced by the use of a variety of resin based roofing systems or generically cold applied liquid roofing systems. (See section 9.11 on cold-applied liquid membranes.) These systems can prove to be a very useful method of upgrading waterproofing systems without necessarily going to the expense or disruption of replacing the original asphalt. Being lightweight, they can also be used to upgrade intermediate floors and by introducing contrasting colours, define pedestrian walkways, parking spaces and traffic routes as part of a refurbishment programme.

Cold applied systems are now generally applied by spray to ensure consistency and fall into two broad camps:

- polyurethanes; and
- acrylics.

The systems usually comprise a priming coat which is used to seal the substrate and ensure adequate bond – a one or two coat membrane incorporating a reinforcement fleece, a wearing layer and a finish coat giving a total thickness of around 4.5–5mm.

One of the key attractions of this type of system is that it can be applied over complex details. Imperfections in an asphalt substrate need to be cut out and made good, although given the inherent flexibility of most materials a certain degree of crack bridging can be provided by the membrane.

However, the apparent attractiveness of these systems does not mean that they can be used as a substitute for a lack of attention to detail or indeed a lack of proper preparation. (See case study overleaf.)

Figure 141 Failure in membrane – asphalt has insufficient tensile strength to accommodate the level of movement experienced.

Figure 142 Joint in precast planks immediately below the failed joint in figure 141. Water penetration has led to the severe corrosion of a retained lifting lug.

Figure 143 Profound adhesion failure of a cold applied waterproofing system probably caused by its application over a contaminated surface or application in unsuitable conditions.

Case study

The premises comprised a roof top car park located above a provincial shopping centre. The deck provided access for servicing the shop units as well as visitor car parking. The development dated from the 1960s.

In 1996, due to progressive defects in the asphalt surfacing of the car park, the then managing agents concluded that out of a number of options, the most suitable would be the provision of a liquid applied polyurethane resin based waterproofing wearing surface incorporating a layer of quartz granules. The selected system had an Agrément Certificate, was supplied by a European manufacturer and laid by licensed operators.

Work started in July 1997 on the first of what was intended to be four separate contracts, the intention being that the majority of the works would be completed in the first three phases, with the car park ramps forming the final fourth phase. The works involved the retention of the original asphalt wearing course save for any defective areas, skimming the asphalt to achieve a level surface and then spraying a number of layers of various resins, which when cured would provide an impervious barrier to water, whilst permitting moisture to escape from below the surface.

Defects in the new surfacing soon began to appear in the first phase works and these were repaired whilst the contractor was still on site. However, by the end of the final phase, other similar defects had begun to appear in a number of previously treated areas (mainly in the running aisles), by which time it was apparent that the scale of defects was greater than first thought.

A dispute then arose between the employer and the contractor over payments and retentions. Further expert evidence revealed the problem to be with the asphalt substrate over which the membrane had been laid rather than with the liquid applied membrane itself. The cause of the asphalt failure was due to water held within the concrete slab under the asphalt which, due to the effects of heat (from solar gain), had caused water vapour to create blisters within the asphalt. Over time the asphalt above the blister thinned and collapsed under traffic load.

After a protracted dispute over liability, the contractor returned to site to execute remedial works, but before they were complete went into liquidation (coincidentally along with the managing agent). Given uncertainties in the performance of the new coatings, the employer took the decision to replace the waterproofing in its entirety with new asphalt.

It is interesting to note the final outcome of the case. The cold applied membrane was a highly engineered product and capable of working under extreme conditions. Nevertheless, it was no substitute for a poor substrate. At the time of its repair, only one UK based operator was licensed to use the product and this left the employer vulnerable to the effects of poor trading conditions.

Glossary

AAMA Architectural Aluminium Manufacturer's Association.

ABICS Asbestos Building Inspection Certification Scheme (run by BOHS).

Abrasion Wear or removal of the surface of a solid material as a result of relative movement of other solid bodies in contact with it (BS CP3).

Absolute pressure Gauge pressure plus atmospheric pressure.

Absorption Entry and/or retention (of a fluid) into the bulk of a solid material by virtue of the porosity/capillarity (BS 892).

ACAD Asbestos Control and Abatement Division of TICA (Thermal Insulation Contractors Association).

ACM Asbestos-containing materials.

ACOP Approved Code of Practice.

Acrylic resin One of a group of thermoplastic resins formed by polymerisation of esters or amides of acrylic acid – used in concrete or masonry construction as a bonding agent or surface sealer.

Adhesion The force that resists the separation of two bodies in contact (BS 5168).

Adhesion failure A failure in the adhesive bond between two materials, usually following the interface between them (known as the interfacial adhesive bond).

Admixture A material, other than aggregate, cementitious material or water, added in small quantities to the mix in order to produce some (desired) modifications, either to the properties of the mix or of the hardened product (BS 4049).

Adsorption Attachment of a substance to the surface of a solid by virtue of forces arising from molecular attraction (BS 892).

AIC Asbestos Information Centre Ltd.

Alkaline hydrolysis A process of deterioration in HAC concrete whereby alkali solutions can dissolve the hydration products of HACC to form calcium hydroxide and alkali aluminate. By combining these chemicals with water and carbon dioxide, calcium carbonate and hydrated alumina are formed. In turn, these react once more with the hydrated cement compounds, thus creating a vicious circle that ends with the decomposition of the calcium aluminate hydrates and the crumbling of the concrete. Alkaline hydrolysis is accompanied by a significant loss of strength.

Ångstrom Molecular measure of measurement, 1 ten-thousandth of a micron (1 micron = 10,000Å).

Anion A negatively charged ion.

Anisotropic The display of unequal physical properties in different directions.

Anisotropy in toughened or heat strengthened glass *See* leopard spots.

Anode The part of a corrosion cell where reacting metals lose electrons (become oxidised).

ANSI	American National Standards Institute.
ARCA	Association of Asbestos Removal Contractors Association.
ASLIC	Asbestos (Licensing) Regulations 1983.
Asphalt	Black petroleum residue, which can be anywhere from solid to semisolid at room temperature. When heated to 100°C, it becomes pourable.
ASTM	American Society for Testing and Materials.
ATAC	Asbestos Testing and Consultancy wing of ARCA.
Atomic number	The number of protons in a nucleus. It determines the chemical properties of an element.
Atomic weight	The sum of the number of neutrons and protons in the nucleus of an atom.
AWPD	Asbestos Worker Protection Directive.
Axial load	Load applied along or parallel to and concentric with the primary axis.
BAT	Best available technique.
Bauxite	Impure hydrated aluminum oxide, the commercial source of aluminum and its compounds.
Bearing	The linear or area dimension over which a higher component transmits load to a lower component.
Bedding	A layer, usually of concrete or mortar, for providing continuous support to such items as bricks, slabs, pipes (BS 892, 5385).
Bimetallic corrosion (galvanic corrosion)	Corrosion of two different metals in electrical contact and immersed in the same electrolyte.
Biocide	A chemical that kills life forms, usually in the form of an additive.
BIOH	The British Institute of Occupational Hygienists (now known as BOHS).
Birefringent	The characteristic of double refraction – the splitting of light into two rays but unequally and in different directions. Often observed in toughened glass viewed in polarised light.
Bleeding	The separation of water from an unhardened concrete mix (BS 4049).
Blister	A local delamination occurring between two layers of a membrane or between the membrane and its substrate. The defect forms a raised area that may be filled with fluid or gas. Blisters are usually caused by moisture penetration and poor adhesion.
BOHS	The British Occupational Hygiene Society (formerly BIOH, the British Institute of Occupational Hygienists).
BRAC Report or BRAC Rules	Advice published by subcommittee P of the Building Regulations advisory committee following the collapse of a swimming pool roof in Stepney, London, on 8th February 1974. The guidance is the most appropriate for use in the assessment of prestressed X or I beams – it is not applicable to other HACC uses.
Brewster's fringes	Optical effect producing an 'oily' appearance on high quality optical glass caused by wavelengths of light meeting up with each other exactly 180 degrees out of phase.
Calcareous scale	A chalk-like scale precipitated from salts dissolved in hard waters.
Calcium Aluminate Cement (CAC)	Modern name for High Alumina Cement (HAC).
Capillarity	Absorption of a liquid due to surface tension.
Carbonation	The gradual reduction in alkalinity of a concrete due to exposure to atmospheric carbon dioxide. Carbon dioxide is diluted in rainwater to form carbonic acid which reduces alkalinity.

Cathode The part of a corrosion cell where positively charged ions (cations) gain electrons – a process known as reduction.

Cathodic protection A process whereby the potential of a metal structure is made more negative in order to decrease corrosion rates.

Cation A positively charged ion.

Cavitation In a broad sense, cavitation means the formation of vapour or gas-filled cavities in liquids. In more specific terms it is used to describe the formation of vapour bubbles as a result of mechanical action (such as a mixing screw) and their sudden or explosive collapse.

CAW Control of Asbestos at Work Regulations (2002 unless stated otherwise).

Cem-Fil Trade name for alkali resistant glass fibre used in the manufacture of glass reinforced cement products or as a crack reducing fibre in reinforced concrete floors.

Cementitious foam insulation A magnesium oxide-based material blown with air to create an inert, effective insulation. It may be especially helpful for people with chemical sensitivities.

CFC Chlorofluorocarbon. Containing chlorine and possessing a large ODP. CFCs are generally found in refrigeration and air-conditioning systems, e.g. centrifugal chillers.

Ciment Fondu Trade name for High Alumina Cement, registered by the Lafarge Cement Company in 1914.

Cohesion failure Splitting or loss of integrity of a formerly intact material such as a mastic sealant, characterised by deterioration remote from the bonding faces and extending through the body of the material. Cohesion failure can be a symptom of an overstressed joint or inadequate tensile strength of the sealant, perhaps through ageing and embrittlement.

Composite woodwool slab WWCB bonded to a layer of insulation such as rigid cellular glass, EPS or mineral wool. The panel can be either double or triple layer.

Conversion Mineralogical change in concrete made from HAC. The process involves the conversion of the metastable hydration products into stable products. This process is accompanied by a change in volume and increased porosity. As a result, strength is reduced.

Corrosion allowance An additional thickness of metal over and above that needed for the actual design to compensate for loss of metal due to corrosion.

Corrosion current The electric current flowing in the metal between the electrodes of a corrosion cell due to the corrosion reaction.

Corrosion fatigue Cracking occurring as a result of fluctuating stresses within a metal that is subject to corrosive conditions.

Corrosion monitoring Any method used to observe or measure the progress of corrosion. Often used to describe a process of remote examination of the performance of a structure that is undergoing cathodic protection to determine the effectiveness of the system and the need for adjustments.

Corrosion potential The overall potential of a metal in a corrosion cell.

Corrosion-erosion (erosion-corrosion) The effects of a combination of corrosion and erosion that serves to damage or remove a metal's passive layer causing localised material loss through corrosion.

Crazing Network of surface cracks (BS 3446).

Creep Slow deformation of a stressed material at temperatures which may be within or above the normal working range of the material (BS 1755).

Crevice corrosion Localised corrosion from corrosive conditions within a fine crevice formed between two surfaces.

De-aeration (de-oxygenation) The removal of oxygen either from gases or aqueous liquids.

De-alloying (selective dissolution); preferential corrosive	A form of corrosion whereby a component of an alloy corrodes preferentially, thus creating weakness.
Deansgate disorder	Local term (Manchester) for corrosion of steel frames in masonry clad buildings – *see also* Regent St disease.
Delamination	The breakdown of a material by separation of the layers of which it is composed.
Deposit attack	A form of crevice corrosion resulting from solids deposited on a surface.
Dezincification	Loss of zinc from brass.
DGU	Double glazed unit (or insulating glass unit – IGU).
Differential aeration	A variation in the amount of dissolved oxygen in a solution that is in contact with the metal surface creating a more vigorous corrosion cell.
Ecor	The electrode potential acquired by a corroding element in a particular environment.
Efflorescence	White salts brought to the surface of porous building materials by water movement (BS 4049). Efflorescence is caused by the dissolution of chemicals (usually salts) inside a product when wet, followed by the deposition of these chemicals on the surfaces of the products when dried.
Elastic strain	Reversible change in length of an object caused by an external force.
Electrode	The anode or cathode in a corrosion cell.
Electrode potential	The difference in electrical potential between the electrode and the electrolyte that it is in contact with.
Electrolyte	An electrically conductive solution in which the passage of electrical charge occurs.
Electron	An electrically charged particle carrying current through a metal conductor. In a corrosion cell, electrons pass along the metal.
EMF (electromotive force) series of metals	A standard list of the electrode potentials of various metals
EPS	Expanded polystyrene, a thermoplastic resin material composed of white beads. The material softens on heating and is flammable, although fire resistant grades are available.
Ettringite	A mineral, high-sulphate calcium sulphoaluminate occurring in nature or formed by sulphate attack on mortar and concrete.
Exfiltration	Leakage out of a material or structure.
Fatigue	The weakening of a material caused by repeated or alternating loads – may be used in conjunction with either static or dynamic loading (BS 2787).
Fg	femtogram (10^{-15}g).
Filiform corrosion	A small corrosion cell that migrates across the surface of a metal, often beneath a coating, leaving a 'snail-trail'.
Filler joist	Early type of fireproof floor comprising a series of steel or iron joists at about 600mm centres, with an infilling of coke breeze concrete.
Foamglass	Trade name for expanded glass insulation product.
Formaldehyde	Pungent, water soluble chemical, turns to gas at room temperature. Naturally occurring and an ingredient in paints, resins and glues as well as fabrics.
Fretting corrosion	A form of corrosion occurring at the site of two metals rubbing together under load.
FRP	Fibre reinforced polymer. A fabric (often carbon fibre) usually bonded to the exterior of a structural member to improve its loadbearing capacity.

Galvanic series	A standard list demonstrating the corrosion potential of various metals and alloys in sea water.
Galvanising	A method of applying a sacrificial protective coating to iron and steel by immersion in a bath of molten zinc.
Glass transition temperature (Tg)	The temperature below which a polymer becomes hard and brittle and above which it becomes soft.
Graphitic corrosion	A form of corrosion of cast iron, in which the iron is selectively dissolved or leached away leaving a porous mass apparently intact but in reality consisting largely of graphite. *See also* dezincification.
GRC	Glass reinforced concrete.
GRG	Glass reinforced gypsum.
GRP	Glass reinforced plastic.
Half-cell (electrode)	One of the electrodes together with its immediate environment in a corrosion cell.
Hatschek process	Process involved in the manufacture of asbestos cement fibre slates using a paper-making sieve cylinder machine on which a very dilute slurry of asbestos fibres (up to about 10 per cent by weight of solids) and ordinary Portland cement (about 90 per cent or more) was dewatered in films of about 0.3mm, which were then wound up to a thickness of about 6mm on a roll.
HCFC	Hydrochlorofluorocarbon. Less saturated with chlorine than CFCs, the hydrogen within these compounds mean that HCFCs have a much shorter atmospheric life span and thus possess a lower ozone depleting potential (ODP).
HEPA filter	High Efficiency Particulate Air Filter – a filter capable of removing very small particles from air.
HFC	Hydroflurocarbon.
Honeycombing	Large interconnecting voids in concrete due to poor compaction.
HSE	Health and Safety Executive.
Hydration	The chemical process that takes place when cement is mixed with water.
Hydrogen embrittlement	Embrittlement of a metal caused by the absorption of atomic hydrogen.
Hysterisis	A property of physical objects that do not instantly follow the forces applied to them, but react slowly, or do not return completely to their original state.
Icor	Corrosion current.
IGU	Insulating glass unit (or double glazed unit – DGU).
Impingement attack	A form of corrosion and erosion caused by fluids or gases impinging on a surface.
Inhibitor	A chemical compound added to an environment to reduce the amount or risk of corrosion to an acceptable level.
Inorganic cementing agent	Binder used for producing WWCB products, usually Portland cement, magnesite or cement and lime
Interfacial adhesion failure	Adhesion failure occurring at the interface of two materials. Usually the result of contamination of one of the bonded surfaces with moisture or other contaminants, lack of primer, incompatibility with the substrate or overstressing of the joint. *See also* interphasal adhesive bond.
Intergranular attack	Corrosion occurring at the boundaries of individual grains of a metal.

Interphasal adhesive bond	Type of adhesion failure between two materials where the line of separation occurs immediately adjacent to the bond line within one of the materials. Usually associated with the saturation of one of the materials and its delamination under load.
Interstitial condensation	Condensation occuring within the thickness of a building element.
Isotropic	Materials in which all characteristics are the same in all directions.
Kinetic energy	The energy of motion. Can be due to vibration, rotation or translational – the movement of an object from one place to another. The amount of energy is a function of its mass and speed. Kinctic energy is directly proportional to the square of its speed.
Leopard spots	Apparent coloured patches visible in toughened or heat treated glass when viewed in polarised light. Different stresses in the glass result in a tendency to produce a birefringent effect. Sometimes referred to as anisotropy.
Lignin	Chemical compound and an integral part of the walls of plant cells. Lignin is the most abundant organic material on earth after cellulose. It is believed that lignin gives wood its stiffness and improves water transport.
Limit state	State at which performance criteria are just met.
MDHS	Method of Determining Hazardous Substances.
Millscale	Thick layer of oxide scale that forms on the surface of steel during hot rolling.
Newton's rings	Optical effect caused by a spherical section of glass in contact with a flat plane of glass. The coloured bands appear because the different wavelengths of light interfere at different thicknesses of the air between the surfaces.
NIACS	National Individual Asbestos Certification Scheme (jointly run by RICS and ARCA).
ODP	Ozone Depleting Potential. Represented as a comparative measurement from the ODP of CFC11, which is taken to have a value of 1.0.
Ozone	1. Stratospheric ozone: in the stratosphere (the atmosphere layer beginning 7–10 miles above the earth), ozone is a form of oxygen found naturally which provides a protective layer shielding the earth from ultraviolet radiation. 2. Ground level ozone: ozone produced near the earth's surface through complex chemical reactions of sunlight, volatile organic compounds and nitrogen oxides.
Passivity	The condition whereby a molecular layer of corrosion products protects the metal against further corrosion or substantially reduces the corrosion rate.
PCB	Abbreviation for polychlorinated biphenyls, a group of synthetic chlorinated organic compounds that are very stable and resist change from extremes of temperature or from alkali or acid environments. PCBs are classed as probably causing cancer in humans by the International Agency for Research on Cancer. Often used as caulking in expansion joints prior to mid-1970s.
PCDD	polychlorinated dibenzo-p-dioxins (also known as 'dioxins').
PCDD/F	Mixture of congeners of PCDD and PCDF.
PCDF	polychlorinated dibenzofurans (also known as 'furans').
Petrography	Branch of geology that deals with the classification and description of different types of stone
pg	picogram (10^{-12}g)
pH value	A value representing the acidity or alkalinity of a solution. pH of 7 = neutral; pH > 7 = alkaline conditions; pH < 7 = acidic conditions.
PIR	Polyisocyanurate foam – a thermosetting polyurethane foam with additional fire retardants.
Pitting (pitting corrosion)	Very localised form of penetrative corrosion.

PRF Phenolresoreinol formaldehyde adhesive, used in the manufacture of laminated timber beams.

PUR Polyurethane foam – a thermosetting insulation foam.

Quartz silica Quartz silica is also known as crystalline quartz silica, crystallized silicon dioxide, free crystalline silica, alpha-quartz, silica, silica flour, silicon dioxide, crystalline silica (non-specific name), and silica (non-specific name). An abundant naturally occurring mineral and one form of a range of naturally occurring minerals composed principally of silicon dioxide (SiO_2).

Rapid Chemical Test A test designed by the BRE to act as an initial indicator to determine the likelihood of HAC in a concrete sample. The production of yellow precipitate is a good, but not conclusive, indicator that HAC is present. Little or no precipitate is a good indicator that HAC is not present. 'False positive' readings are possible if the samples are taken from carbonated Portland cement concrete or if contamination has occurred.

Refractories Heat-resistant materials that constitute the linings for high-temperature furnaces and reactors and other processing units.

Regent St disease London term for corrosion of steel framed, masonry clad buildings constructed in the early 20th century. *See also* Deansgate disorder.

RICS Royal Institution of Chartered Surveyors.

Scale Thick visible oxide film formed during oxidation of a metal at high temperature.

Settlement Downward movement of the soil or of a structure which it supports (BS 892).

Spalling Detachment of fragments, usually of flaky shape, from a larger mass by a blow, or by the action of weather or pressure; chipping of stone, masonry or concrete (BS 2787).

SRP Solar reflective paint; or

Steel reinforced polymer a flexible fabric usually bonded to the exterior of a structural member to improve its loadbearing capacity.

Stitching Insertion of new bricks to replace existing damaged bricks.

Strain A change in length caused by an external force.

Stress-corrosion cracking Cracking occurring as a result of the combination of tensile stresses and corrosion.

Tarnish (patina) Thin film of corrosion product.

TDI Tolerated daily intake.

Thermal hysterisis Thermal expansion or the release of geological strain causing irreversible movement.

Thermal shock Force, arising out of thermal expansion or contraction, which causes disruption of a material on sudden heating or cooling.

Tonderzement High Alumina Cement.

Type A woodwool slab Woodwool slab categorised under BS 1105:1981 as being suitable for non loadbearing applications such as permanent shuttering, partitions, ceilings and wall linings.

Type B woodwool slab Woodwool slab categorised under BS 1105:1981 as being thicker than 50mm and of greater density and strength than Type A.

Type SB woodwool slab Woodwool slab categorised under BS 1105:1981 as being stronger than types A and B and being suitable for flat roof deck and low pitched roofs.

UFFI Urea Formaldehyde Foam Insulation – cavity wall insulation product.

UKAS United Kingdom Accreditation Service.

Vcor Corrosion rate – the amount of substance transferred per unit of time at a specified surface.

VOCs Volatile organic compounds. The term volatile means that the compounds become gases at normal room temperature.

Warm-edge technology An insulated edge spacer within an insulating glass unit. Warm edge spacers are designed to reduce heat losses through the perimeter of the IGU.

WEL Workplace exposure limit.

WW Long shavings of wood.

WW-C Composite woodwool slab.

WWCB Woodwool cement board – a board formed from softwood shavings coated in Portland cement and pressed into an insulating board material.

Yield Permanent deformation which a material undergoes when it is stressed beyond its elastic limit.

Young's modulus Ratio of stress to strain for a particular material in a particular direction.

Zinc whiskers A metallurgical phenomonen whereby tiny filaments of zinc crystals are pushed out of the surface of metals coated with zinc as part of an electroplating process. The filaments can be dislodged and transported in air currents to cause short circuits in electrical equipment.

Further information

2 *Corrosion*

Cathodic protection: Code of practice for land and marine applications, BS 7361-1:1991, BSI, London, 1991 (ISBN 0 580 18964 3)
Code of practice for protective coating of iron and steel structures against corrosion (AMD 4443) (AMD 7898), BS 5493:1977, BSI, London, 1977 (ISBN 0 580 09565 7). Amended by AMD 4443 and AMD 7898. This standard has been declared obsolescent but remains current and is cited in Building Regulations.
Corrosion Control in Engineering Design, National Corrosion Service, Teddington, 2000
Corrosion mechanisms: an introduction to aqueous corrosion, Technical Note 5, Corrosion Prevention Association, Aldershot, 2005
McKenzie, M., *Corrosion of weathering steel under real and simulated bridge decks*, Research Report 233, TRL, Wokingham, 299 (ISSN 0266-5247)
Cracking in buildings, Report 292, BRE, Watford, 1995 (ISBN 1 860 81039 X)
Bull, J.W. and Yu, C.W., *Durability of materials and structures in building and civil engineering*, Whittles Publishing, Caithness, 2006 (ISBN 978 1 87032 558 5)
Fixings and corrosion, Guidance Note, Construction Fixings Association, Oakham, 2002
Guidance on external cathodic protection of underground steel storage tanks and steel pipework at petrol filling stations, Energy Institute, London, 2002 (ISBN 978 0 85293 372 5)
Guide G: Public health engineering, Part 5 - Corrosion and corrosion protection, CIBSE, London, 2004 (ISBN 1 90328 742 1)
The Highways Agency, *Inspection manual for highway structures, Volume 1: Reference Manual*, TSO, London, 2007 (ISBN 978 0 11552 797 5)
Management of accelerated low water corrosion in steel maritime structures, Publication C634, CIRIA, London, 2005 (ISBN 0 86017 634 7)
Maritime structures – Part 1: Code of practice for general criteria (AMD 13445), BS 6349-1:2000, BSI, London, 2000 (ISBN 0 580 33169 5)
Barry, D.L., *Material durability in aggressive ground*, Report R098M, CIRIA, London, 1983 (ISBN 978 0 86017 208 6)
Performance of building materials in contaminated land, Report 255, BRE, Watford, 1994 (ISBN 0 85125 624 4)
Prevention of corrosion on structural steelwork, BSSPC, London, 1996
Protection of metallic materials against corrosion. Corrosion likelihood in atmospheric environment. Classification, determination and estimation of corrosivity of atmospheric environments, BS EN 12500:2000, BSI, London, 2000 (ISBN 0 58036 002 4)
Protection of steel structures against corrosion by coatings, Publication 50, ECCS, Brussels, 1987
Risks of contaminated land to buildings, building materials and services: a literature review, Technical Report P331, Environment Agency, 2000 (ISBN 1 85705 247 1)
Standard Guide for Examination and Evaluation of Pitting Corrosion, ASTM G46–94, ASTM International, West Conshohocken, USA, 2005
The Assessment of Concrete Structures affected by Steel Corrosion, in *Design Manual for Roads and Bridges*, Volume 3, Section 4, Part 13, BA 51/95, Highways Agency, London, 1995
Water mains: guidance on assessment and inspection techniques, Report R162, CIRIA, London, 1996 (ISBN 978 0 86017 449 3)
Water treatment for building services systems, Application Guide 2/93, BSRIA, Bracknell, 1993 (ISBN 0 86022 319 1)
Zinc coated steels in buildings, Report No. 20, BBA, Watford, 1983

3 *Deterioration of concrete*

Accelerated carbonation testing of concrete, Information Paper 20/00, BRE, Watford, 2000 (ISBN 1 86081 454 9)
Alkali silica reaction (ASR), Fact Sheet 4, BCA, Camberley, 2006
Alkali-silica reaction in concrete - Background to the guidance notes (2004 edition), Digest 330 Part 1, BRE, Watford, 2004 (ISBN 1 86081 709 2)

Alkali-silica reaction in concrete - Detailed guidance for new construction (2004 edition), Digest 330 Part 2, BRE, Watford, 2004 (ISBN 1 86081 710 6)
Alkali-silica reaction: minimising the risk of damage to concrete. Guidance notes and model specification clauses (3rd edition), Technical Report 30, The Concrete Society, Camberley, 1999
Analytical and electrochemical investigations of reinforcement corrosion, Contractor Report 30, TRL, Wokingham, 1986
A review of carbonation in reinforced concrete, BCA/BRE, Camberley/Watford, 1987 (ISBN 0 72101 365 1)
Assessing carbonation depth in ageing high alumina cement concrete, Information Paper 11/98, BRE, Watford, 1998 (ISBN 1 86081 217 1)
Assessment of existing high alumina cement concrete construction in the UK, Digest 392, BRE, Watford, 1994 (ISBN 0 85125 625 2)
A survey of site tests for the assessment of corrosion in reinforced concrete, Laboratory Report 953, TRL, Wokingham, 1980
Avoiding deterioration of cement-based building materials and components. Lessons from case studies: 4, Report 441, BRE, Watford, 2001
Avoiding deterioration of cement-based building materials. Lessons from case studies: 1, Report 324, BRE, Watford, 1997 (ISBN 1 86081 579 0)
Avoiding the thaumasite form of sulphate attack: two-year report, Report 385, BRE, Watford, 2000 (ISBN 1 86081 358 5)
Brickwork durability (2nd edition), Design Note 7, BDA, London, 2006
Briefing note for specifiers on the advantages of low alkali sulphate-resisting portland cement, BCA, Camberley, 1996
Carbonation depths in structural-quality concrete: an assessment of evidence from investigations of structures and from other sources, Report 75, BRE, Watford, 1986 (ISBN 8 5125185 4)
Carbonation of concrete and its effects on durability, Digest 405, BRE, Watford, 1995 (ISBN 1 86081 015 2)
Carbonation of concrete made with dense natural aggregates, BRE, Watford, 1995 (ISBN 1 86081 015 2)
Corrosion of steel in concrete: a review of the effect of humidity, Digest 491, BRE, Watford, 2004 (ISBN 1 86081 730 0)
Corrosion of steel in concrete: investigation and assessment, Digest 444 Part 2, BRE, Watford, 2000 (ISBN 1 86081 361 5)
Corrosion of steel in concrete: protection and remediation, Digest 444 Part 3, BRE, Watford, 2000 (ISBN 1 86081 362 3)
Critical review of the structural implications of the alkali silica reaction in concrete, Contractor Report 169, TRL, Wokingham, 1989
Designing quality buildings: a BRE guide, Report 487, BRE, Watford, 2007 (ISBN 1 86081 899 4)
Diagnosis of alkali-silica reaction, BCA, Camberley, 1992
Durability of ageing high alumina cement (HAC) concrete: a literature review and summary of BRE research findings, Report 386, BRE, Watford, 2000 (ISBN 1 86081 370 4)
Durability of pfa concrete, Report 216, BRE, Watford, 1994 (ISBN 85125 520 5)
Durability of pre-cast HAC concrete in buildings, Information Paper 8/00, BRE, Watford, 2000 (ISBN 1 86081 368 2)
Durability of reinforced concrete: effects of concrete composition and curing on carbonation under different exposure conditions, Report 360, BRE, Watford, 1998 (ISBN 1 86081 262 7)
Effects of alkali-silica reaction on concrete foundations, Information Paper 16/93, BRE, Watford, 1993
Electrochemical repair using realkalisation and chloride extraction techniques, CRA, Aldershot, 1999
Enhancing reinforced concrete durability: Guidance on selecting measures for minimising the risk of corrosion of reinforcement in concrete: Part 2 Supplementary technical reports, Technical Report 61: Part 2, The Concrete Society, Camberley, 2004
Influence of cement content on the performance of concrete, Publication CS125, The Concrete Society, Camberley, 1999
Minimising the risk of alkali silica reaction: alternative methods, Information Paper 1/02, BRE, Watford, 2002 (ISBN 1 86081 529 4)
Modelling degradation processes affecting concrete, Report 434, BRE, Watford, 2001 (ISBN 1 86081 153 6)
Mortars for masonry. Guidance on specification, types, production and use, Publication CS159, The Concrete Society, Camberley, 2005
On-site curing of concrete - microstructure and durability, Publication C530, CIRIA, London, 2001 (ISBN 978 0 86017 530 8)
Overview of the thaumasite form of sulphate attack and its significance for highway structures, Technical Paper 1, Concrete Bridge Development Group, Camberley, 2001
Performance of concrete in practice: a field study of highway bridges, Contractor Report 43, TRL, Wokingham, 1987
Performance of pfa concrete in aggressive conditions. 4: Carbonation, Report 297, BRE, Watford, 1995
Regeneration of brownfield sites containing ferrous slags, Report 481, BRE, Watford, 2005 (ISBN 1 86081 759 9)
Reinforced concrete service life design. Part 2. Design for durability, Information Paper 3/06 Part 2, BRE, Watford, 2006, (ISBN 1 86081 907 9)
Repair and maintenance of reinforced concrete, Report 254, BRE, Watford, 1994 (ISBN 0 85125 623 6)
Sea-dredged aggregates in concrete, Information Paper 7/87, BRE, Watford, 1987
Structural effects of alkali-silica reaction: technical guidance on the appraisal of existing structures, IstructE, London, 1992
Structural implications of alkali-silica reaction - effect of natural exposure and freeze-thaw, Report 365, BRE, Watford, 1999 (ISBN 1 86081 276 7)
Structural implications of the alkali silica reaction in concrete, Contractor Report 177, TRL, Wokingham, 1990

Structural performance of reinforced concrete: phase 1, Contractor Report 267, TRL, Wokingham, 1991
Structural performance of reinforced concrete: phase 2, Contractor Report 311, TRL, Wokingham, 1992
Sulphate damage to concrete floors on sulphate-bearing hardcore: identification and remediation, CLG, London, 2008 (ISBN 978 1 85946 301 7)
Sulphate specification for structural backfills (updated), Report 447, TRL, Wokingham, 2001 (ISBN 1 84608 446 6)
Thaumasite expert group report one-year review, Office of the Deputy Prime Minister, London, 2000
Thaumasite expert group report: review after three years experience, Office of the Deputy Prime Minister, London, 2002
Thaumasite form of sulphate attack (TSA), Fact Sheet 2, BCA, Camberley, 2006
The effect of restrained expansion on the assessment of structures with ASR, Report 161, TRL, Wokingham, 1996 (ISBN 1 84608 160 2)
The influence of cement content on the performance of concrete, Publication CS125, The Concrete Society, Camberley, 1999
The non-destructive detection and mapping of ASR cracking in concrete, Contractor Report 156, TRL, Wokingham, 1989
The thaumasite form of sulphate attack: risks, diagnosis, remedial works and guidance on new construction, DETR, London, 1999 (ISBN 1 85112 150 1)

4 High Alumina Cement (HAC) concrete

Assessing carbonation depth in ageing high alumina cement concrete, Information Paper 11/98, BRE, Watford, 1998 (ISBN 1 86081 217 1)
Assessment of existing High Alumina Cement construction in the UK, Digest 392, BRE, Watford, 1994 (ISBN 0 85125 625 2)
BRAC: report by Sub-committee P – High alumina cement concrete (and Addenda), Report 429, BRE, Watford, 2002
Calcium aluminate cements in construction – a re-assessment, Technical Report 46, The Concrete Society, Camberley, 1997
Durability of pre-cast HAC concrete in buildings, IP 8/00, BRE, Watford, 2000 (ISBN 1 86081 368 2)
HAC concrete in the UK: assessment, durability management, maintenance and refurbishment, Special Digest 3, BRE, Watford, 2002 (ISBN 1 86081 591 X)
High Alumina Cement Concrete: BRAC rules – revised 2002, Report 451, BRE, Watford, 2002 (ISBN 1 86081 600 2)
Report on the collapse of the roof of the assembly hall at Camden High School for Girls, HMSO, London, 1973
Structural appraisal of buildings with long-span roofs, Digest 282, BRE, Watford, 1984 (ISBN 11726 861 5)
Structural use of concrete. Code of practice for design and construction, BS 8110-1:1997, BSI, London, 1997 (ISBN 978 0 580 59893 7)
The structural condition of Orlit houses, Report 36, BRE, Watford, 1983 (ISBN 0 85125 056 4)

5 Glulam and box beams

Adhesively-bonded timber connections – adhesives and timber construction (June 2003 revision), Wood Information Sheet 2.3/31, TRADA, High Wycombe, 2003
Developments in timber technology, Construction Papers 22, CIOB, Ascot, 1993
Glued Laminated Timber Association website provides free access to the *Specifiers Guide*, information on *Engineered timber* and Technical Data Sheets on *Design for durability* and *Glulam in sports halls*. The website can be accessed at www.glulam.co.uk
Glued laminated timber (January 2003 edition), TRADA, High Wycombe, 2003
Glued laminated timber. Performance requirements and minimum production requirements, BS EN 386:2001, BSI, London, 2002 (ISBN 978 0 580 39270 8)
How to calculate design values of material properties using Eurocode 5, Eurocode 5 Guidance Document 3, TRADA, High Wycombe, 1994
Multiple fastener timber joints: guidance on BS 5268-2 and Eurocode 5 (January 2003 revision), Guidance Document GD 7, TRADA, High Wycombe, 2003
Nonresidential roof systems, The Engineered Wood Association, Tacoma, USA, 1996
Panelized wood roofs, The Engineered Wood Association, Tacoma, USA, 2006
Substitution of glulam beams for steel or solid-sawn lumber, The Engineered Wood Association, Tacoma, USA, 2007
Timber materials – properties and associated design procedures with Eurocodes (2nd edition), Guidance Document GD 3, TRADA, High Wycombe, 2008 (ISBN 978 1 90051 053 0)
Timber structures. Glued laminated timber. Strength classes and determination of characteristic values, BS EN 1194:1999, BSI, London, 1999 (ISBN 978 0 580 30762 X)
Wide-span wood sports structures, TRADA, High Wycombe, 2007

6 Asbestos cement and cement fibre roofing

Asbestos and man-made mineral fibres in buildings: practical guidance, DETR, Rotherham, 1999

Asbestos cement products, Technical Bulletin 24, NFRC, London, 2000

Asbestos: the licensed contractors' guide, Guidance Booklets 247, HSE Books, Sudbury, 2006 (ISBN 0 71762 874 4)

Bird, bee and plant damage to buildings, Digest 418, BRE, Watford, 1996 (ISBN 1 86081 099 3)

BRE building elements: building services – performance, diagnosis, maintenance, repair and the avoidance of defects, Report 404, BRE, Watford, 2000

BRE building elements: roofs and roofing – performance, diagnosis, maintenance, repair and the avoidance of defects (revised 2000), Report 302, BRE, Watford, 2000

BRE building elements: walls, windows and doors – performance, diagnosis, maintenance, repair and the avoidance of defects, Report 352, BRE, Watford, 1998 (ISBN 978 1 86081 235 X)

Cleaning debris from guttering on an asbestos cement (AC) roof, Asbestos Essentials A10, HSE, Sudbury, 2007

Cleaning weathered asbestos cement (AC) roofing and cladding, Asbestos Essentials A12, HSE, Sudbury, 2007

Code of practice for demolition, BS 6187:2000, BSI, London, 2000 (ISBN 978 0 58033 206 3)

Code of practice for the specification and use of liquid waterproofing systems, ELRA, Haywards Heath, 1999

Code of practice. Work with domestic soffits and rainwater goods containing asbestos cement, Glass and Glazing Federation, London, 2006

Control of asbestos regulations 2006, SI 2006/2739. This legislation can be viewed at www.opsi.gov.uk

Control of asbestos regulations 2006: A guide for safety representatives, HSE Books, Sudbury, 2006

Drilling holes in asbestos cement (AC) and other highly bonded materials, Asbestos Essentials A9, HSE, Sudbury, 2007

Enclosing undamaged asbestos materials to prevent impact damage, Asbestos Essentials A8, HSE, Sudbury, 2007

Explanatory memorandum to the Control of asbestos regulations 2006, DWP. This legislation can be viewed at www.opsi.gov.uk

Fibre-cement profiled sheets and fittings – Product specification and test methods (AMD 16051) (AMD 16726) (AMD 17300), BS EN 494:2004, BSI, London, 2004 (ISBN 978 0 58059 708 4)

GRC, Digest 331, BRE, Watford, 1988 (ISBN 85125 288 5)

Guidance note for safe working on fragile roofs. Covering the designing, planning and carrying out of inspection, maintenance, repair and refurbishment work, ACR [CP] 002:2005, Advisory Committee for Roofwork, London, 2005

Health and safety in roof work (3rd edition), Guidance Booklets 33, HSE Books, Sudbury, 2008

Identification, removal and disposal of asbestos containing materials (ACM) from roofs and walls of buildings, Technical Bulletin 31, NFRC, London, 2004

Inclined preformed GRP valley troughs, Technical Bulletin 28, NFRC, London, 2002

Installation of sheet roof and wall coverings. Profiled fibre cement. Code of practice, BS 8219:2001, BSI, London, 2001 (ISBN 978 0 58038 538 8)

Introduction to asbestos essentials: comprehensive guidance on working with asbestos in the building maintenance and allied trades, Guidance Booklet 213, HSE Books, Sudbury, 2001 (ISBN 0 71761 901 X)

Management of asbestos in non-domestic premises. Regulation 4 of the Control of asbestos regulations 2006. Approved code of practice (ACOP) and guidance (2nd edition), Legal Series L127, HSE Books, Sudbury, 2006

Painting asbestos cement (AC) sheets, Asbestos Essentials A16, HSE, Sudbury, 2007

Performance of fibre cement slates: effects of condensation, Information Paper 19/01, BRE, Watford, 2001 (ISBN 1 86081 519 7)

Richardson, B., *Remedial treatments of buildings* (2nd edition), Butterworth-Heinemann, Oxford, 1995 (ISBN 978 0 75062 158 8)

Removing asbestos cement (AC) debris, Asbestos Essentials A11, HSE, Sudbury, 2007

Removing asbestos cement (AC) products (flues, tanks, etc), Asbestos Essentials A15, HSE, Sudbury, 2007

Removing asbestos cement (AC) sheets, gutters, etc., Asbestos Essentials A14, HSE, Sudbury, 2007

Repairing damaged asbestos cement (AC), Asbestos Essentials A13, HSE, Sudbury, 2007

Roofs and roofing, Chapter 1 – History (2 of 20), Whittles Publishing, Caithness, 1993

Substrates for liquid waterproofing systems, Technical Bulletin 23, NFRC, London, 1999

Substrates for liquid waterproofing systems, Guidance Note 3, ELRA, Haywards Heath, 1998

Technical guide for the assessment of the durability of thin fibre reinforced cement products (without asbestos) for external use, MOAT 48, BBA, Watford, 1991

Working with asbestos cement (2nd edition), Health and Safety: Guidance Booklets 189/2, HSE Books, Sudbury, 1999 (ISBN 0 71761 667 3)

Working with asbestos cement products (revised 2001 and 2007), AIC, Widnes, 2000

Working with asbestos in buildings, INDG 289, HSE, Sudbury, 2004

Work with materials containing asbestos. Control of asbestos at work regulations 2006. Approved code of practice (ACOP) and guidance, Legal Series L143, HSE, Sudbury, 2006

7 *Profiled metal sheet roofing*

Best Practice for the Specification and Installation of Metal Cladding and Secondary Steelwork, Publication P346, Steel Construction Institute, Ascot, 2006 (ISBN 978 1 85942 171 0)

Code of practice for sheet roof and wall coverings. Aluminium, corrugated and troughed (PD 4346) (Obsolescent but remains current), CP 143-1:1958, BSI, London, (ISBN 0 58035 613 2)

Code of practice for sheet roof and wall coverings. Galvanized corrugated steel. Metric units, CP 143-10:1973, BSI, London, 1973 (ISBN 0 580 07621 0)

Code of practice for the use of profiled sheet for roof and wall cladding on buildings. Design, BS 5427-1:1996, BSI, London, 1996 (ISBN 0 58025 738 X)

Composite beams and slabs with profiled steel sheeting, Report R99, CIRIA, London, 1983 (ISBN 978 0 86017 210 9)

Design of twin-skin metal cladding: how to meet the new requirements of Part L of the Building Regulations, Publication 311, Steel Construction Institute, Ascot, 2002

Eurocode 3: Design of steel structures. General rules - Supplementary rules for cold-formed members and sheeting, BS EN 1993-1-3:2006, BSI, London, 2006 (ISBN 0 58049 665 1)

Eurocode 9: Design of aluminium structures. Cold-formed structural sheeting, BS EN 1999-1.4:2007, BSI, London, 2007 (ISBN 978 0 58050 416 7)

Fasteners for metal roof and wall cladding: design, detailing and installation guide, Technical Paper 12, MCRMA, Wirral, 2000

Fire design of steel sheet clad external walls for building - construction performance standards and design, Technical Paper 7, MCRMA, Wirral, 1994

Guidance for the effective sealing of end lap details in metal roofing constructions, Technical Paper 16, MCRMA, Wirral, 2004

Installation of sheet roof and wall coverings. Profiled fibre cement. Code of practice, BS 8219:2001, BSI, London, 2001 (ISBN 0 58038 538 8)

Insulated profiled metal roofs, Good Building Guide 43, BRE, Watford, 2000 (ISBN 1 86081 440 9)

Metal cladding: U-value calculation: assessing thermal performance of built-up metal roof and wall cladding systems using rail and bracket spacers, Publication 312, Steel Construction Institute, Ascot, 2002 (ISBN 1 85942 137 7)

Metal roof systems for slating and tiling, Technical Bulletin 30, NFRC, London, 2002

Profiled metal roofing design guide, Technical Paper 6, MCRMA, Wirral, 2004

Roofing products from metal sheet. Specification for self-supporting products of steel, aluminium or stainless steel sheet. Aluminium, BS EN 508-2:2000, BSI, London, 2000 (ISBN 0 58035 729 5)

Safe installation of profiled metal sheeting to cold rolled purlins, rails and spacers, Technical Bulletin 19, NFRC, London, 1999

Secret fix roofing design guide (Revised edition), Technical Paper 3, MCRMA, Wirral, 1999

Structural use of steelwork in building. Code of practice for design of composite slabs with profiled steel sheeting (AMD 8315) (AMD 9326) (Partially superseded but remains current), BS 5950-4:1994, BSI, London, 1994 (ISBN 0 58021 808 2)

Structural use of steelwork in building. Code of practice for stressed skin design (AMD 8315) (AMD 9326) (Partially superseded but remains current), BS 5950-9:1994, BSI, London, 1994 (ISBN 0 58021 895 3)

Survey of performance of organic-coated roof sheeting, Report 259, BRE, Watford, 199 (ISBN 85125 617 1)

8 *Roof drainage*

Approved Document H – Drainage and waste disposal (The Building Regulations 2000)(2002 edition), ODPM, London, 2006 (ISBN 978 1 85946 208 9)

Guide to siphonic roof drainage systems, BS 8490:2007, BSI, London, 2007 (ISBN 978 0 58050 206 4)

Gullies for buildings. Requirements, BS EN 1253-1:2003, BSI, London, 2003 (ISBN 0 58042 134 1)

Metal gutters: the guide, Metal Gutters Manufacturers Association, Wirral, 2005

The Siphonic Guide: a guide to siphonic roof drainage, Siphonic Roof Drainage Association, Bolton, 2007

9 *Flat roofing*

7. Flat roof construction. AJ 22.4.87, Architects' Journal Element Design Guide: Roofs, Architect's Journal, London, 1987

BLP Construction Durability Database – http://componentlife.com

Code of practice for mastic asphalt roofing, BS 8218:1998, BSI, London, 1998 (ISBN 0 58028 384 4)

Design guide for single ply roofing, SPRA, London, 2003

European Liquid Waterproofing Association – www.elwassociation.org.uk

Flat roofing: a guide to good practice (Blue book) (4th edition), Ruberoid plc, Wigan, 2002

Flat Roofing Alliance – www.fra.org.uk – publish a code of practice for remedial works

Flat roofing: design and good practice, Chapter 2 – History (3 of 24), CIRIA/BFRC Flat Roofing Design and Good Practice, CIRIA, London, 1993
Flat roofing: design and good practice, Chapter 11 – Built-up roofing (12 of 24), CIRIA/BFRC Flat Roofing Design and Good Practice, CIRIA, London, 1993
Flat roofing: design and good practice, Chapter 12 – Mastic asphalt (13 of 24), CIRIA/BFRC Flat Roofing Design and Good Practice, CIRIA, London, 1993
Flat roofing: design and good practice, Chapter 13 – Polymeric single-ply (14 of 24), CIRIA/BFRC Flat Roofing Design and Good Practice, CIRIA, London, 1993
Flat roofs: assessing a roof repair and making repairs – in two parts, Good Repair Guide 16, BRE, Watford, 1998 (ISBN 1 86081 220 1)
Flat roofs with continuously supported coverings. Code of practice, BS 6229:2003, BSI, London, 2003 (ISBN 0 580 41305 5)
Handbook for clerks of works (3rd edition), Butterworth-Heinemann, Oxford, 1983
Mastic asphalt for flat roofs: testing for quality assurance, Information Paper 8/91, BRE, Watford, 1991
Paving, Technical Guide, Mastic Asphalt Council. This guide is available in PDF format at www.masticasphaltcouncil.co.uk/technical-guides
Permanite technical handbook (Red book), Ruberoid plc, Wigan, 2003
Reinforced bitumen membranes for roofing. Code of practice, BS 8217:2005, BSI, London, 2005 (ISBN 0 580 45116 X)
Roofing. AJ Focus 01.2000, AJ Focus, Architect's Journal, London, 2000
Roofing. AJ Focus 01.99, AJ Focus, Architect's Journal, London, 1999
Roofing, Technical Guide, Mastic Asphalt Council. This guide is available in PDF format at www.masticasphaltcouncil.co.uk/technical-guides
Roof overlays using built-up roofing, Roofing Handbook Information Sheet 24, FRA, London, 1999
Screeds, bases and in situ floorings. Mastic asphalt underlays and wearing surfaces. Code of practice, BS 8204-5:2004, BSI, London, 2004 (ISBN 0 58044 350 7)
Single-ply roofing membranes, Canadian Building Digest 235, NRC-IRC, Ottawa, Canada, 1985
Specification for mastic asphalt for building and civil engineering (limestone aggregate) (AMD 7150) (AMD 9582), BS 6925:1988, BSI, London, 1988 (ISBN 0 58016 362 8)
Technical guide introduction, Mastic Asphalt Council. This guide is available in PDF format at www.masticasphaltcouncil.co.uk/technical-guides
Workmanship on building sites. Code of practice for waterproofing, BS 8000-4:1989, BSI, London, 1989 (ISBN 0 58017 540 5)

10 Masonry facades

AAC 'aircrete' blocks and masonry, Digest 468 Part 2, BRE, Watford, 2002 (ISBN 1 86081 563 4)
Attractive masonry: water - friend or foe?, MIA, London, 1983
Accredited construction details. Indicative detail drawings: masonry cavity wall insulation details, DCLG, London, 2007
Accredited construction details. Indicative detail drawings: masonry external wall insulation details, DCLG, London, 2007
Accredited construction details. Indicative detail drawings: masonry internal wall insulation details, DCLG, London, 2007
Analysis of the stability of masonry-faced earth retaining walls, Report 550, TRL, Wokingham, 2002 (ISBN 1 84608 549 7)
Appraisal of existing structures (2nd edition), IstructE, London, 1996
Assessment of repair and strengthening techniques for brick and stone masonry arch bridges, Contractor Report 284, TRL, Wokingham, 1992
BRE building elements: walls, windows and doors - performance, diagnosis, maintenance, repair and the avoidance of defects, Report 352, BRE, Watford, 1998 (ISBN 1 86081 235 X)
Building masonry with lime-based bedding mortars, Good Building Guide 66, BRE, Watford, 2005 (ISBN 1 86081 889 7)
Civil engineer's reference book (4th edition), Part 15 – Load-bearing masonry (16 of 47), EBH, Oxford, 1989
Code of practice for the installation of remedial damp-proof courses in masonry walls, Code of Practice 2, British Wood Preserving and Damp Proofing Association, Derby, 1997
Code of practice for use of masonry. Materials and components, design and workmanship, BS 5628-3:2005, BSI, London, 2005 (ISBN 0 580 46718 X)
Code of practice for use of masonry. Materials and components, design and workmanship (AMD 4974) (No longer current but cited in Building Regulations), BS 5628-3:1985, BSI, London, 1985
Code of practice for the use of masonry. Structural use of reinforced and prestressed masonry, BS 5628-2:2005, BSI, London, 2005 (ISBN 0 58046 717 1)
Code of practice for the use of masonry. Structural use of reinforced and prestressed masonry (No longer current but cited in Building Regulations), BS 5628-2:2000, BSI, London, 2000
Code of practice for use of masonry. Structural use of unreinforced masonry, BS 5628-1:2005, BSI, London, 2005 (ISBN 0 58046 682 5)

Design guide handbook for EN 1996. Design of masonry structures, DCLG, London, 2006
Design of masonry walls subjected to concentrated vertical loads, Information Paper 10/92, BRE, Watford, 1992
Earth masonry. Design and construction guidelines, EP 80, BRE, Watford, 2008 (ISBN 978 1 86081 978 0)
Efflorescence and bloom on masonry, Issue 3, Data Sheets 8, MIA, London, 2005
Eurocode 6: Design of masonry structures. Design considerations, selection of materials and execution of masonry, BS EN 1996-2:2006, BSI, London, 2006 (ISBN 0 58047 822 X)
Eurocode 6: Design of masonry structures. General rules - Structural fire design, BS EN 1996-1-2:2005, BSI, London, 2005 (ISBN 0 58046 334 6)
Eurocode 6: Design of masonry structures. Simplified calculation methods for unreinforced masonry structures, BS EN 1996-3:2006, BSI, London, 2006 (ISBN 0 58047 821 1)
Eurocode 6. Design of masonry structures. General rules for reinforced and unreinforced masonry structures (AMD Corrigendum 16209), BS EN 1996-1-1:2005, BSI, London, 2005 (ISBN 0 58046 719 8)
External masonry walls: assessing whether cracks indicate progressive movement, Defect Action Sheet 102, BRE, Watford, 1987
External masonry walls: eroding mortars - repoint or rebuild?, Defect Action Sheet 70, BRE, Watford, 1986
External masonry walls: repointing, Defect Action Sheet 72, BRE, Watford, 1986
External masonry walls: repointing - specification, Defect Action Sheet 71, BRE, Watford, 1986
Handbook to BS 5628: structural use of masonry: Part 1: Unreinforced masonry, Design Guide 10, BDA, Windsor, 1981
Insulating solid masonry walls, Good Building Guide 50, BRE, Watford, 2002 (ISBN 1 86081 553 7)
Manual for the design of plain masonry in building structures to Eurocode 6, IStructE, London, 2008 (ISBN 978 1 90633 502 1)
Masonry arch bridges: condition appraisal and remedial treatment, Publication C656, CIRIA, London, 2006 (ISBN 978 0 86017 656 5)
Thomas, K., *Masonry walls - specification and design*, Butterworth-Heinemann, Oxford, 1996 (ISBN 978 0 75062 465 7)
Observations on the use of reclaimed clay bricks, Properties of Bricks and Mortar Generally 1.4, BDA, Windsor, 2001
Performance specifications for wall ties, Report 45, BRE, Watford, 1984
Practical building conservation, Volume 1: Stone masonry, English Heritage, Swindon, 1988
Principles of masonry conservation management, Digest 502, BRE, Watford, 2007 (ISBN 978 1 86081 966 7)
Repair and maintenance of brickwork, BDA, Windsor, 1988
Repair and preservation of historic masonry, Technical Conservation Note 1, CADW, Cardiff, 2003
Repairing damage to brick and block walls, Good Repair Guide 3, BRE, Watford, 1996 (ISBN 1 86081 110 8)
Repairing brick and block freestanding walls, Good Repair Guide 28, BRE, Watford, 2000 (ISBN 1 86081 393 3)
Repairing brick and block masonry, Digest 359, BRE, Watford, 1991 (ISBN 85125 485 3)
Repairing flood damage, Part 3: foundations and walls, Good Repair Guide 11, BRE, Watford, 1997
Repairing frost damage, Part 2: walls, Good Repair Guide 20, BRE, Watford, 1998
Replacement ties in cavity walls: a guide to tie spacing and selection, Report R117, CIRIA, London, 1989 (ISBN 978 0 86017 298 7)
Repointing external brickwork walls, Good Repair Guide 24, BRE, Watford, 1999 (ISBN 1 86081 037 3)
Repointing of brickwork, BDA, Windsor, 1976
Repointing stone and brick, Technical Note 37, EHS, Belfast, 2006
Retention of masonry facades - best practice guide, Publication C579, CIRIA, London, 2003 (ISBN 978 0 86017 579 7)
Retro installation of bed joint reinforcement in masonry, Good Building Guide 62, BRE, Watford, 2004 (ISBN 1 86081 724 6)
Rising damp in walls. Diagnosis and treatment (new edition), Digest 245, BRE, Watford, 2007 (ISBN 978 1 84806 012 8)
Specification for ancillary components for masonry. Bed joint reinforcement of steel meshwork, BS EN 845-3:2003, BSI, London, 2003 (ISBN 0 58042 608 4)
Specification for ancillary components for masonry. Lintels, BS EN 845-2:2003, BSI, London, 2003 (ISBN 0 58042 607 6)
Specification for ancillary components for masonry. Ties, tension straps, hangers and brackets (AMD Corrigendum 14736) (AMD 15539), BS EN 845-1:2003, BSI, London, 2003 (ISBN 0 58042 606 8)
Specification for masonry units. Aggregate concrete masonry units (dense and light-weight aggregates) (AMD 16001), BS EN 771-3:2003, BSI, London, 2004
Specification for masonry units. Autoclaved aerated concrete masonry units (AMD 16000), BS EN 771-4:2003, BSI, London, 2003 (ISBN 0 580 42782 X)
Specification for masonry units. Calcium silicate masonry units (AMD 15974), BS EN 771-2:2003, BSI, London, 2003 (ISBN 0 58042 721 8)
Specification for masonry units. Clay masonry units (AMD 15998), BS EN 771-1:2003, BSI, London, 2003 (ISBN 0 580 42722 6)
Specification for masonry units. Manufactured stone masonry units (AMD 15999), BS EN 771-5:2003, BSI, London, 2003 (ISBN 0 58043 558 X)
Specification for masonry units. Natural stone masonry units, BS EN 771-6:2005, BSI, London, 2006 (ISBN 0 58047 276 0)
Specification for mortar for masonry. Masonry mortar, BS EN 998-2:2003, BSI, London, 2003 (ISBN 0 580 42781 1)

Beckmann, P. and Bowles, R., *Structural aspects of building conservation* (2nd edition), Butterworth-Heinemann, Oxford, 2004 (ISBN 978 0 75065 733 4)
Testing in construction, Volume 2: Testing of ceramics in construction, Whittles Publishing, Caithness, 2005 (ISBN 978 1 87032 543 1)
UK National Annex to Eurocode 6 – Design of masonry structures. General rules for reinforced and unreinforced masonry structures, NA to BS EN 1996-1-1:2005, BSI, London, 2007 (ISBN 978 0 58050 753 3)
UK National Annex to Eurocode 6: Design of masonry structures. Simplified calculation methods for unreinforced masonry structures, NA to BS EN 1996-3:2006, BSI, London, 2007 (ISBN 978 0 58059 810 4)
Use of traditional lime mortars in modern brickwork, Properties of Bricks and Mortar Generally 1.3, BDA, Windsor, 2001
Walls: repair and repointing brickwork, Historic Buildings Factsheet T 4.03, Defence Estates, London, 1997
Walls: repointing brickwork, Historic Buildings Factsheet T 4.0, Defence Estates, London, 1998
Wall technology, Volume B: Loadbearing small units, Special Publication SP 87/B, CIRIA, London, 1992 (ISBN 0 86017 337 2)
Wall technology, Volume C: Small units on framed buildings, Special Publication SP 87/C, CIRIA, London, 1992 (ISBN 0 86017 337 2)

11 Applied exterior finishes

Accredited construction details. Indicative detail drawings: masonry external wall insulation details, DCLG, London, 2007
AJ handbook of building enclosures, Part 4 - External envelope - external walls (5 of 12), Butterworth-Heinemann, Oxford, 1975
BRE building elements: walls, windows and doors - performance, diagnosis, maintenance, repair and the avoidance of defects, Report 352, BRE, Watford, 1998 (ISBN 1 86081 235 X)
Slate and stone products for discontinuous roofing and cladding. Product specification, BS EN 12326-1:2004, BSI, London, 2004 (ISBN 0 58044 150 4)
Wall and floor tiling. Design and installation of external ceramic and mosaic wall tiling in normal conditions. Code of practice, BS 5385-2:2006, BSI, London, 2006 (ISBN 0 58049 232 X)
Wall technology, Volume G: Applied finishes, Special Publication SP 87/G, CIRIA, London, 1992 (ISBN 0 86017 337 2)
Workmanship on building sites. Code of practice for wall and floor tiling. Ceramic tiles, terrazzo tiles and mosaics (AMD 8623), BS 8000-11.1:1989, BSI, London, 1989 (ISBN 0 58017 306 2)
Workmanship on building sites. Code of practice for wall and floor tiling. Natural stone tiles, BS 8000-11.2:1990, BSI, London, 1990 (ISBN 0 58018386 6)

12 Thin stone cladding

Aspects of cladding, IstructE, London, 1995
Cast in concrete: reconstructed stone and precast concrete – a guide for Architects, ACLA, Leicester, 1995
Cladding fixings: good practice guidance, Publication C524, CIRIA, London, 2000 (ISBN 978 0 86017 524 7)
Code of practice for design and installation of natural stone cladding and lining, BS 8298:1994, BSI, London, 1995 (ISBN 0 58023 152 6)
Code of practice for design and installation of non-loadbearing precast concrete cladding (AMD Corrigendum 11064) (AMD Corrigendum 13018), BS 8297:2000, BSI, London, 2000 (ISBN 0 58033 146 6)
Conservation of building and decorative stone, Part 1, Chapter 3 – Igneous rocks (4 of 9), Butterworth-Heinemann, Oxford, 1998
Design and appearance. Part 1, Digest 45, BRE, Watford, 1964
External cladding: how to determine the thickness of natural stone panels, Information Paper 7/98, BRE, Watford, 1998
External cladding using thin stone, Information Paper 6/97, BRE, Watford, 1997 (ISBN 1 86081 138 8)
Lightweight veneer stone cladding panels, Information Paper 10/01, BRE, Watford, 2001
Stone cladding panels: in-situ weathering, Information Paper 18/98, BRE, Watford, 1998
Stone in building - its use and potential, Stone Federations Great Britain, Kent, 1984
Wall technology, Volume E: Large heavy units on framed buildings and in-situ concrete, Special Publication SP 87/E, CIRIA, London, 1992 (ISBN 0 86017 337 2)

13 Concrete cladding

Aspects of cladding, IstructE, London, 1995
Cast in concrete: reconstructed stone and precast concrete – a guide for architects, ACLA, Leicester, 1995

Cladding, AJ Focus 02.2002, AJ Focus, Architect's Journal, London, 2002
Cladding bibliography of selected publications 1970–1990, and current national and international standards concerning cladding, Report 194, BRE, Watford, 1991 (ISBN 85125 480 2)
Cladding fixings: good practice guidance, Publication C524, CIRIA, London, 2000 (ISBN 978 0 86017 524 7)
Code of practice for design and installation of natural stone cladding and lining (No longer current but cited in the Building Regulations), BS 8298:1989, BSI, London, 1989
Code of practice for design and installation of non-loadbearing precast concrete cladding (AMD Corrigendum 11064) (AMD Corrigendum 13018), BS 8297:2000, BSI, London, 2000 (ISBN 0 58033 146 6)
Code of practice for the safe erection of precast concrete cladding (2nd edition), ACLA, Leicester, 1999
Design of composite beams using precast concrete slabs, Publication 287, SCI, Ascot
Fixings for precast concrete panels, ACLA, Leicester, 1994
Precast concrete floors in steel framed buildings, Publication 351, SCI, Ascot, 2007
Precast concrete frame buildings: design guide, BCA, Camberley, 1992
Precast concrete products. Wall elements, BS EN 14992:2007, BSI, London, 2007 (ISBN 978 0 58052 952 8)
Precast concrete structures, Part 0 – Contents and preface (1 of 12), Butterworth-Heinemann, Oxford, 2002
Precast concrete structures, Part 2 – Materials used in precast structures, Butterworth-Heinemann, Oxford
Wall technology, Volume E: Large heavy units on framed buildings and in-situ concrete, Special Publication SP 87/E, CIRIA, London, 1992 (ISBN 0 86017 337 2)

14 GRC cladding

GRC, Digest 331, BRE, Watford, 1988 (ISBN 85125 288 5)
GRC in action (2006 revision), GRCA, Wigan, 2006
Guide to fixings for glassfibre reinforced concrete cladding (1999 revision), GRCA, Wigan, 1998
Non-Portland cement GRC, Information Paper 7/82, BRE, Watford, 1982
Properties of GRC: ten year results, Information Paper 36/79, BRE, Watford, 1979
Specification for alkali resistant glassfibre rovings and chopped strands for reinforcement of cements and concretes, GRCA, Wigan, 1986
Specification for the manufacture, curing and testing of GRC products (2nd edition), Concrete Society, Camberley, 2000
The use of glass reinforced cement in cladding panels, Information Paper 5/84, BRE, Watford, 1984
The use of glass reinforced cement in cladding panels, Report 49, BRE, Watford, 1984 (ISBN 85125 068 8)
Wall technology, Volume D: Large lightweight units on framed buildings, Special Publication SP 87/D, CIRIA, London, 1992 (ISBN 0 86017 337 2)

15 Curtain walling

Application of fire safety engineering principles to the design of buildings. Structural response and fire spread beyond the enclosure of origin (Sub-system 3), PD 7974-3:2003, BSI, London, 2003 (ISBN 0 580 40304 1)
Aspects of cladding, IStructE, London, 1995
BRE building elements: walls, windows and doors - performance, diagnosis, maintenance, repair and the avoidance of defects, Report 352, BRE, Watford, 1998 (ISBN 1 86081 235 X)
Cladding Component Failures, CWCT, Bath, 1997
Commercial offices handbook, Part 11 – Facades (12 of 19), RIBA Publishing, London, 2003
Condensation on glazing frames, Technical Update 3, CWCT
Curtain walling. Air permeability. Performance requirements and classification, BS EN 12152:2002, BSI, London, 2002 (ISBN 0 58039 416 6)
Curtain walling. Air permeability. Test method, BS EN 12153:2000, BSI, London, 2000 (ISBN 0 58035 789 9)
Curtain walling and large composite windows. Notes for guidance, AWA, 1980
Curtain walling. Impact resistance. Performance requirements, BS EN 14019:2004, BSI, London, 2004 (ISBN 0 58044 017 6)
Curtain walling. Product standard, BS EN 13830:2003, BSI, London, 2003 (ISBN 0 58042 601 7)
Curtain walling. Resistance to wind load. Performance requirements, BS EN 13116:2001, BSI, London, 2001 (ISBN 0 580 37503 X)
Curtain walling. Watertightness. Performance requirements and classification, BS EN 12154:2000, BSI, London, 2000 (ISBN 0 58034 232 8)
Curtain walling. Watertightness. Site test, BS EN 13051:2001, BSI, London, 2001 (ISBN 0 58037 155 7)
Defects in buildings, HMSO, London, 1989
EIFS Restoration Guide, Dow Corning Corporation, Midland, USA, 2002

Rousseau, M.Z., *Facts and fictions of rainscreen walls*, article in *Construction* Canada, Vol. 32, no. 2, National Research Council, Ottawa, Canada, 1990
Layzall, J., *Failure mode and effects analysis in the cladding industry*, University of Bath, Bath, 1997
Fire resistance tests for non-loadbearing elements. Curtain walling - Full configuration (complete assembly), BS EN 1364-3:2006, BSI, London, 2006 (ISBN 0 58049 390 3)
Fire resistance tests for non-loadbearing elements. Curtain walling - Part configuration, BS EN 1364-4:2007, BSI, London, 2007 (ISBN 978 0 58052 967 2)
Nicastro, D., *Glazing system leaks*, article in *The Construction Specifier*, January 1996 issue, The Construction Specifications Institute, Alexandria, USA, 1996
Rain penetration and its control, Canadian Building Digest 40, NRC-IRC, Ottawa, Canada, 1963
Site testing: method statement for the simplified cabinet test, Technical Update 2, CWCT, 2000
Spandrel Panel Glazing, ATS-124, Pilkington Glass, 2005
Specification for aluminium alloy curtain walling, AWA, 1980
Structural Silicone Glazing, Dow Corning Corporation, Midland, USA, 1999
The principle of pressure equalisation, Technical Note 6, CWCT, Bath, 2000
The rain screen wall system, Ontario Association of Architects
Thermal performance of curtain walling. Calculation of thermal transmittance, BS EN 13947:2006, BSI, London, 2007 (ISBN 978 0 580 56531 1)
Choi, E., *Velocity and Impact Direction of Wind Driven Rain on Building Faces*, Nanyang Technological University, Malaysia, 1997
Wall cladding defects and their diagnosis, Digest 217, BRE, Watford, 1978 (ISBN 85125 160 9)
Wall technology, Volume F: Glazing, curtain walls and overcladding, Special Publication SP 87/F, CIRIA, London, 1992 (ISBN 0 86017 337 2)
Day, R., *What effect will the publication of the new Standard and Guide to Good Practice have on Curtain Walling failures in the UK?*, 1993
Nicastro, D., *Zipper gaskets*, article in *The Construction Specifier*, September 1996 issue, The Construction Specifications Institute, Alexandria, USA, 1996

16 Insulating glass units

Catering equipment specification, Specification 42, Defence Estates, London, 2004
Code of practice for fire resistant glazing, FRGGSA, Princes Risborough, 1994
Code of practice for glazing for buildings (AMD 4063) (AMD 4582) (AMD 8279), BS 6262:1982, BSI, London, 1982
Code of practice for glazing for buildings. Safety related to human impact, BS 6262-4:1994, BSI, London, 1994
Code of practice for installation and application of security glazing, BS 5357:2007, BSI, London, 2007 (ISBN 978 0 58056 791 9)
Directives for the approval of insulating glazing, MOAT 35, BBA, Watford, 1985
Double glazing for heat and sound insulation, Digest 379, BRE, Watford, 1993 (ISBN 85125 552 3)
Double-glazing units: a BRE guide to improved durability, Report 280, BRE, Watford, 1995 (ISBN 1 86081 005 5)
Environmental design manual: summer conditions in naturally-ventilated offices, Report 86, BRE, Watford, 1988 (ISBN 85125 209 5)
Environmental performance toolkit for glazed facades, Technical Memoranda TM 35, CIBSE, 2004
Factory glazed windows: introduction and the glazing process, Digest 497 Part 1, BRE, Watford, 2005 (ISBN 1 86081 885 4)
Fire resistant glazing, Glazing Manual Data Sheet 2.8, GGF, London, 2003
Glass in building - Determination of luminous and solar characteristics of glazing, BS EN 410:1998, BSI, London, 1998 (ISBN 0 58030 154 0)
Glazing for buildings - Code of practice for energy, light and sound, BS 6262-2:2005, BSI, London, 2005 (ISBN 0 58046 775 9)
Glazing for buildings - Code of practice for fire, security and wind loading, BS 6262-3:2005, BSI, London, 2005 (ISBN 0 58046 773 2)
Glazing for buildings - General methodology for the selection of glazing, BS 6262-1:2005, BSI, London, 2005 (ISBN 0 58046 774 0)
Glazing – safety in relation to impact, opening and cleaning (incorporating 2000 amendments), Building Regulations 2000: Approved Documents N, ODPM, London, 2006 (ISBN 978 1 85946 212 6)
Guidance on glazing at height, Publication C632, CIRIA, London, 2005 (ISBN 978 0 86017 632 9)
Guide to best practice in the specification and use of fire-resistant glazed systems (awaiting copyright approval for Issue 2, April 2008), GGF, London
Guide to best practice in the specification and use of fire-resistant glazed systems, GGF, London, 2005
Guide to the selection of glass and plastics glazing sheet materials for overhead glazing in conservatories, Glazing Manual Data Sheet 5.7.1/1, GGF, London, 2001

Handbook 2005: Fundamentals, Chapter 31 – Fenestration (32 of 41), ASHRAE, Atlanta, USA, 2005
Highly glazed buildings: assessing and managing the risks, Report 482, BRE, Watford, 2005 (ISBN 1 86081 760 2)
Insulating glazing units, Digest 453, BRE, Watford, 2000 (ISBN 1 86081 435 2)
Patent glazing and sloping glazing for buildings. Code of practice for design and installation of sloping and vertical patent glazing, BS 5516-1:2004, BSI, London, 2004 (ISBN 0 58044 711 1)
Patent glazing and sloping glazing for buildings. Code of practice for sloping glazing, BS 5516-2:2004, BSI, London, 2004 (ISBN 0 580 44710 3)
Recommendations for safety in glazing, GGF, London, 1985
Requirements and tests for fire resistant glazing systems, Issue 2.1 dated 10.96 , Loss Prevention Standard 1158, LPCB, London, 1996
Steel supported glazing systems: interfaces, Publication 193, SCI, 1997
Systems design and glazing considerations for insulating glass units, Glazing Manual Data Sheet 4.2, GGF, London, 2004
Systems design and glazing considerations for insulating glass units, Glazing Manual Data Sheet 4.2, GGF, London, 1995
Thermal performance of windows, doors and shutters - Calculation of thermal transmittance. General, BS EN ISO 10077-1:2006, BSI, London, 2006 (ISBN 0 580 49527 2)
Wall technology, Volume F: Glazing, curtain walls and overcladding, Special Publication SP 87/F, CIRIA, London, 1992 (ISBN 0 86017 337 2)
Workmanship on building sites. Code of practice for glazing, BS 8000-7:1990, BSI, London, 1990 (ISBN 0 58019 000 5)

17 Upper floor and roof decks

Silber, R., *Conversions – general guidelines*, article in *The structural engineer*, Vol. 56, Issue 2, ISTructE, London, 1978
Structural renovation of traditional buildings (1994 reprint with amendments), Report R 111, CIRIA, London, 1994
Swailes, T., *Nineteenth Century fireproof buildings, their strength and robustness*, article in *The structural engineer*, IStructE, London, 2003

18 Floor screeds and finishes

AJ handbook of building enclosures, Part 7 – Internal division – suspended floors (8 of 12), Butterworth-Heinemann, Oxford, 1975
BRE building elements: floors and flooring - performance, diagnosis, maintenance, repair and the avoidance of defects (2nd edition), Report 460, BRE, Watford, 2003 (ISBN 1 86081 631 2)
BRE screed tester: classification of screeds, sampling and acceptance limits, Information Paper 11/84, BRE, Watford, 1984
Code of practice for installation of laminate floor coverings, BS 8425:2003, BSI, London, 2003 (ISBN 0 58042 114 7)
FeRFA guide to the specification and application of synthetic resin flooring (revised November 2007), FeRFA, Farnham, 1999
Flooring (3rd edition), Health Technical Memorandum 61, DOH, Leeds, 2006
Floor levelling screeds, BCA, Camberley, 1997
Floors: cement-based screeds – mixing and laying, Defect Action Sheet 52, BRE, Watford, 1984
Floors: cement-based screeds – specification, Defect Action Sheet 51, BRE, Watford, 1984
High Alumina Cement Concrete: BRAC rules – revised 2002, Report 451, BRE, Watford, 2002 (ISBN 1 86081 600 2)
Screed material and floor screeds. Definitions, BS EN 13318:2000, BSI, London, 2000 (ISBN 0 58036 768 1)
Screed material and floor screeds. Screed material. Properties and requirements, BS EN 13813:2002, BSI, London, 2002 (ISBN 0 58040 710 1)
Screeds, bases and in situ floorings. Cementitious terrazzo wearing surfaces. Code of practice, BS 8204-4:2004, BSI, London, 2004 (ISBN 0 58044 480 5)
Screeds, bases and in-situ floorings. Concrete bases and cement sand levelling screeds to receive floorings. Code of practice, BS 8204-1:2003, BSI, London, 2003 (ISBN 0 58043 051 0)
Screeds, bases and in situ floorings. Concrete wearing surfaces. Code of practice, BS 8204-2:2003, BSI, London, 2003 (ISBN 0 58043 052 9)
Screeds, bases and in-situ floorings. Polymer modified cementitious levelling screeds and wearing screeds. Code of practice, BS 8204-3:2004, BSI, London, 2004 (ISBN 0 58044 351 5)
Screeds, bases and in situ floorings. Pumpable self-smoothing screeds. Code of practice, BS 8204-7:2003, BSI, London, 2003 (ISBN 0 58041 250 4)
Screeds, bases and in situ floorings. Synthetic resin floorings. Code of practice, BS 8204-6:2008, BSI, London, 2008 (ISBN 978 0 58059 735 0)
Screeds, flooring and finishes: selection, construction and maintenance, Part 1, Report R 184, CIRIA, London, 1998

Screeds, flooring and finishes: selection, construction and maintenance. Part 3 – Flooring systems, Report R 184, CIRIA, London, 1998
Screeds, flooring and finishes: selection, construction and maintenance, Part 4 – Structural bases, Report R 184, CIRIA, London, 1998
Screeds, flooring and finishes: selection, construction and maintenance, Part 5 – Screeds Report R 184, CIRIA, London, 1998
Screeds, flooring and finishes: selection, construction and maintenance, Part 6 – Toppings and in-situ applied floorings, Report R 184, CIRIA, London, 1998
Screeds, flooring and finishes: selection, construction and maintenance, Part 7 – Applied floorings, Report R 184, CIRIA, London, 1998
Screeds, flooring and finishes: selection, construction and maintenance, Part 8 – Testing, Report R 184, CIRIA, London, 1998
Screeds, flooring and finishes: selection, construction and maintenance, Part 9 – Failure, repair and refurbishment, Report R 184, CIRIA, London, 1998
Screeds, flooring and finishes: selection, construction and maintenance, Part 12 –Appendices, Report R 184, CIRIA, London, 1998
Screeds with underfloor heating: guidance for a defect-free interface, Publication CS150, CS, Camberley, 2003
Specification and guidelines for polymer-modified cementitious flooring as wearing surfaces for industrial and commercial use, EFNARC, 2001
Specification and guidelines for synthetic resin flooring, EFNARC, 2001
Testing cement-sand screeds using the BRE screed tester, BCA, Camberley, 1986
Thin floor coverings on cement: sand screeds, Concrete Advice 15, CS, Camberley, 2003
Thin flooring on solid floors – visual imperfections, Special Publication SP 37, CIRIA, London, 1985
Wall and floor tiling. Design and installation of internal and external ceramic and mosaic floor tiling in normal conditions. Code of practice, BS 5385-3:2007, BSI, London, 2007 (ISBN 978 0 580 53243 6)
Workmanship on building sites. Cementitious levelling screeds and wearing screeds. Code of practice, BS 8000-9:2003, BSI, London, 2003 (ISBN 0 58042 811 7)

19 Raised access floors

8. Proprietary elements. AJ 12.8.87, Architects' Journal Element Design Guide: Interior Elements, AJ, London, 1987
AJ handbook of building enclosures, Part 3 – External envelope – lowest floor and basement (4 of 12), Butterworth-Heinemann, Oxford, 1975
Court standards and design guide 2004, Department for Constitutional Affairs (This document is no longer current but is cited in Building Regulations.)
Fire resistance tests for service installations. Raised access and hollow core floors, BS EN 1366-6:2004, BSI, London, 2004 (ISBN 0 580 44799 5)
Focus on: ceilings, partitions and raised floors, AJ Specification 06.2006, AJ, London, 2006
NASA Electronic Parts and Packaging Programme (NEPP) – homepage of material on metal whiskers – http://nepp.nasa.gov/whisker
Platform floors (raised access floors) – performance specification, Method of Building Performance Specifications PF2 PS/SPU, PSA, London, 1992
Raised access floors, BS EN 12825:2001, BSI, London, 2001 (ISBN 0 58038 699 6)
Space requirements for plant access, operation and maintenance, Design and Maintenance Guide 8, MOD, London, 1996

20 Vibration in composite floors

Beside the track side: Vibration isolation AJ 23.10.91, AJ, London, 1991
BRE building elements: building services – performance, diagnosis, maintenance, repair and the avoidance of defects, Report 404, BRE, Watford, 2000
Commercial offices handbook, Part 10 – Structure (11 of 19), RIBA Publishing, London, 2003
Composite slabs and beams using steel decking: best practice for design and construction, Publication 300, SCI, Ascot, 2000
Defects in buildings, HMSO, London, 1989
Design guide on the vibration of floors, Publication 076, SCI, Ascot, 1989
Design guide on the vibration of floors in hospitals, Publication 331, SCI, Ascot, 2004
Design of floors for vibration: a new approach, Publication 354, SCI, Ascot, 2007
Effects of traffic induced vibrations on heritage buildings – further case studies, Research Report 207, TRL, Wokingham 1989
Evaluation and measurement for vibrations in buildings. Guide for measurement of vibrations and evaluation of their effects on buildings, BS 7385-1:1990, BSI, London, 1990

Evaluation and measurement for vibration in buildings. Guide to damage levels from groundborne vibration, BS 7385-2:1993, BSI, London, 1993
Evaluation of human exposure to vibration in buildings (1Hz to 80Hz), BS 6472:1992, BSI, London, 1992
Ground-borne vibrations arising from piling, Technical Note TN 142, CIRIA, London, 1992
Ground vibrations from impact pile driving during road construction, Supplementary Report 544, TRL, Wokingham, 1980
Loads generated by jumping crowds: experimental assessment, Information Paper 4/02, BRE, Watford, 2002 (ISBN 1 86081 541 3)
Steel buildings, Publication 35/03, BCSA, London, 2003
Structural vibration and damage, Report 69, BRE, Watford, 1974
Traffic-induced ground-borne vibrations in dwellings, Research Report 102, TRL, Wokingham, 1987
Traffic vibration and building damage - TRRL papers presented at Acoustics 87, Research Report 146, TRL, Wokingham, 1988
Vibration in timber floors, Timber Engineering Guidance Document GD 6, TRADA, High Wycombe, 1998

21 Ground bearing floor slabs

BRE building elements: floors and flooring - performance, diagnosis, maintenance, repair and the avoidance of defects (2nd edition), Report 460, BRE, Watford, 2003 (ISBN 1 86081 631 2)
Concrete floors, finishes and external paving, Butterworth-Heinemann, Oxford, 1993
Concrete industrial ground floors – a guide to their design and construction (3rd edition), Technical Report 34, CS, Camberley, 2003
Crazing: Power trowelled concrete floor slabs, Concrete Advice 08, CS, Camberley, 2003
Gaber, G., *Design and construction of concrete floors* (2nd edition), Butterworth-Heinemann, Oxford, 2006

22 Multi-storey car parks

Code of practice for ground floor, multi storey and underground car parks (2nd edition), APEA, 1995
Design of buildings and their approaches to meet the needs of disabled people. Code of practice (AMD 15617) (AMD Corrigendum 15982), BS 8300:2001, BSI, London, 2001
Design recommendations for multi-storey and underground car parks (3rd edition), IStructE, London, 2002
Enhancing the whole life structural performance of multi-storey car parks, ODPM, London, 2002
Multi storey buildings in steel – car parks, Publication 84, ECCS, London, 1999
Parking: dynamic capacities for car parks,

Endnotes

1 Pullar-Strecker, P., *Building Defects : Are they avoidable?*, article in *Building Maintenance and Modernization Worldwide*, Vol. 1, Quah L.K (ed.), Longman, Singapore, 1990
2 Addleson, L., *Building failures: A guide to diagnosis, remedy and prevention*, Butterworth Architecture, 1992
3 Douglas, J. and Ransom, W., *Understanding Building Failures*, Taylor and Francis, 2007
4 Addleson, L., *Building failures: A guide to diagnosis, remedy and prevention*, Butterworth Architecture, 1992
5 Kahles, H., *Compatibility – the Rocky Path to Making Sure Things Do Not Go Wrong!*, Glass Performance Days 2005 – www.gpd.fi
6 Derived from data produced by Craig, B. and Pohlman, S.L. in *ASM Handbook*, Vol.13 *Corrosion*, ASM International, 1987
7 Nimmo, B., and Hinds, G., *Beginners Guide to* Corrosion, NPL, February 2003
8 Rothwell, G. P., *Corrosion Phenomena – An Introduction*, NPL
9 *Commentary on corrosion at bimetallic contacts and its* alleviation, BS PD 6484:1979, BSI, London, 1979
10 *Filiform corrosion – Preventing filiform corrosion on finished aluminium parts*, Guidance Sheet Al.01, Verband der Fenster- und Fassadenhersteller e.V., October 1997
11 *Corrosion*, Technical Note 24, CWCT, 2000
12 Lloyd, G. O., *Atmospheric Corrosion*, NPL
13 *Corrosion of steel in concrete: a review of the effect of humidity*, Digest 491, BRE, Watford, 2004
14 *Determination of the chloride and cement contents of hardened concrete,* IP 21/86, BRE, Watford, 1986
15 www.germann.org – instrument makers
16 www.mg-assoc.co.uk – half cell testing of concrete to detect reinforcement corrosion
17 *Corrosion of reinforcement in concrete: electrochemical monitoring*, Digest 434, BRE, Watford, 1988
18 *Corrosion of steel in concrete: investigation and assessment*, Digest 444 Part 2, BRE, Watford, 2000
19 *Corrosion of steel in concrete: protection and remediation*, Digest 444 Part 3, BRE, Watford, 2000
20 *Sulphate damage to concrete floors on sulphate-bearing hardcore: identification and remediation*, CLG, London, 2008
21 *Thaumasite form of sulphate attack (TSA)*, Fact Sheet 2, BCA, Camberley, 2006
22 *Structural effects of alkali-silica reaction: technical guidance on the appraisal of existing structures*, IstructE, London, 1992
23 Neville, A. with Wainwright, P.J., *High Alumina Cement Concrete*, The Construction Press, 1975
24 www.cement.ca/cement.nsf – Cement Association of Canada
25 *Durability of pre-cast HAC concrete in buildings*, IP 8/00, BRE, Watford, 2000
26 Williamson, N., *Failure of prestressed concrete beam and hollow block roof at Sheephouses Reservoir*, June 1962
27 Morris, A.E.J., *Precast Concrete in Architecture*, George Godwin, London, 1978
28 Neville, A., *Concrete: Neville's Insights and Issues*, Thomas Telford, 2006
29 *Materials and workmanship: Approved Document to Support Regulation 7* (incorporating 2000 amendments), ODPM, London, 2000
30 *Calcium Aluminate Cement. Composition, specifications and conformity criteria*, BS EN 14647:2005, BSI, London, 2006
31 Neville, A., *Concrete: Neville's Insights and Issues*, Thomas Telford, 2006
32 *HAC concrete in the UK: assessment, durability management, maintenance and refurbishment*, Special Digest 3, BRE, Watford, 2002
33 *Conversion Preventing Additive for High Alumina Cement Products*, Research Council Canada
34 *Assessing carbonation depth in ageing high alumina cement concrete*, Information Paper 11/98, BRE, Watford, 1998; *Durability of pre-cast HAC concrete in buildings*, IP 8/00, BRE, Watford, 2000
35 Bloomfield, J.P. and Rodriguez, J., *Corrosion rate measurement and life prediction for reinforced concrete structures*, Structural Faults and Repairs Symposium, June 1993, pp.155–163

36 A definition of ettringite as found in the American Concrete Institute's *Cement and Concrete Terminology* (ACI 116R)

37 *HAC concrete in the UK: assessment, durability management, maintenance and refurbishment*, Special Digest 3, BRE, Watford, 2002

38 The Glued Laminated Timber Association *Glulam Specifiers Guide* can be freely accessed at www.glulam.co.uk

39 Hanhijärvi, A., and Kevarinmäki, K., *Timber failure mechanisms in high-capacity dowelled connections of timber to steel. Experimental results and design*, VTT Technical Research Centre of Finland, 2008

40 *Glued laminated timber* (January 2003 edition), TRADA, High Wycombe, 2003

41 *Design for* durability, Technical Data Sheet 2, Glued Laminated Timber Association. This can be freely accessed at www.glulam.co.uk

42 Frese, H.J., *Failure analysis on timber structures in Germany: A contribution to COST Action E55 M*, Blaß Lehrstuhl für Ingenieurholzbau und Baukonstruktionen, Universität Karlsruhe, Germany

43 Gary, W. and Gilham, P., *Glulam Beam Repair and Reinforcement, Timora High School*, article in *Structure Magazine*, September 2006

44 Sitar, M., *Glulam beam repair and strengthening*, article published on www.gostructural.com, May 2006

45 Asbestos Cement Products – TB24 September 2000

46 *Cleaning weathered asbestos cement (AC) roofing and cladding*, Asbestos Essentials A12, HSE, Sudbury, 2007

47 *Installation of sheet roof and wall coverings. Profiled fibre cement. Code of practice*, BS 8219:2001, BSI, London, 2001

48 *Durability of non-asbestos fibre reinforced* cement, Information Paper 1/91, BRE, Watford, 1991

49 Rushton, T., *Investigating hazardous and deleterious materials*, RICS Books, Coventry, 2006

50 Ryan, P.A, Wolstenholme, R.P. and Howell, D.M., *The Durability of Cladding*, Thomas Telford, 1994

51 *Design for Movement in* Buildings, CIRIA Technical Note 107, CIRIA, London, 1981

52 Ryan, P.A, Wolstenholme, R.P. and Howell, D.M., *The Durability of Cladding*, Thomas Telford, 1994

53 Akzo Nobel Industrial Coatings Ltd STEELSEAL Agrément Certificate No 04/4117

54 *Colourcoat in building – A guide to architectural practice*, British Steel Strip Products, 1993

55 *The Profile Metal Roofing Design Guide*, MCRMA Technical Paper No. 6, MCRMA Ltd, 2004

56 *BRE building elements: roofs and roofing – performance, diagnosis, maintenance, repair and the avoidance of defects* (revised 2000), Report 302, BRE, Watford, 2000

57 *Flat roofs with continuously supported coverings. Code of practice*, BS 6229:2003, BSI, London, 2003

58 *The Profile Metal Roofing Design Guide*, MCRMA Technical Paper No. 6, MCRMA Ltd, 2004

59 *Code of practice for the control of condensation in buildings*, BS 5250:2002, BSI, London, 2002

60 *Guidance note on the use of rooflights with profiled cladding systems*, Technical Bulletin 17, National Federation of Roofing Contractors, London, 1997

61 Davies, J., *Lightweight sandwich construction*, Blackwell Science, 2001

62 *Roof drainage tables*, Technical Bulletin 29, National Federation of Roofing Contractors, London, 2002

63 *The Siphonic Guide: a guide to siphonic roof drainage*, Siphonic Roof Drainage Association, Bolton, 2007

64 Arthur, S. and Swaffield, J., *Siphonic roof drainage*, article in *Urban Water* Vol. 3, Issue 1–2, 2001, pp. 43–52

65 *Mastic asphalt for flat roofs: testing for quality assurance*, Information Paper 8/91, BRE, Watford, 1991

66 *Defects in buildings*, HMSO, London, 1989

67 *Solar reflective paints*, Information Paper 26/81, BRE, Watford, 1981

68 *Flat roofing: design and good practice*, Chapter 7 – Design selection (13 of 24), CIRIA/BFRC Flat Roofing Design and Good Practice, CIRIA, London, 1993

69 Liu, K.K.Y., Paroli, R.M. and Smith, T.L., *Blistering in SBS Polymer Modified Bituminous Roofs*, Construction Technology Update No. 38, June 2000, National Research Council Canada – http://irc.nrc-cnrc.gc.ca/pubs/ctus/index_e.html

70 Paroli, R.M. and Booth, R.J., *Ways to Reduce Blistering Built-Up Roofs*, Construction Technology Update No. 4, May 1997, National Research Council Canada – http://irc.nrc-cnrc.gc.ca/pubs/ctus/index_e.html

71 *Maintenance and renewal in Educational Buildings. Flat Roofs – Criteria and methods of assessment, repair and replacement*, Design Note 46, Department of Education & Science Archtects & Building Group, 1985

72 *BRE building elements: roofs and roofing – performance, diagnosis, maintenance, repair and the avoidance of defects* (revised 2000), Report 302, BRE, Watford, 2000

73 *Asphalt and built-up felt roofing*, Digest 144, BRE, Watford

74 *Stability under wind load of loose-laid external roof insulation boards*, BRE Digest 295, 1985

75 *Wind Scour of Gravel Ballast on Roofs*, Digest 311, BRE, 1986

76 Liu, K.K.Y., Paroli, R.M. and Smith, T.L., *Blistering in SBS Polymer Modified Bituminous Roofs*, Construction Technology Update No. 38, June 2000, National Research Council Canada – http://irc.nrc-cnrc.gc.ca/pubs/ctus/index_e.html

77 Russo, M., *TPO Problems require a closer look*, Roof Consultants Institute, Raleigh, USA, 1 July 2001

78 Wallace, T.J. and Rossiter, W.J., *ASTM Committee D-8 on Roofing, Waterproofing and Bituminous Materials*, ASTM International, 1990

79 *Code of practice for design and installation of DPCs in masonry construction*, BS 8215:1991, BSI, London, 1991

80 *Estimation of thermal and moisture movements and stresses: part 2*, Digest 228 Part 2, BRE, Watford, 1979

81 *Wall cladding: designing to minimise defects due to inaccuracies and movements*, Digest 223, BRE, Watford, 1979

82 *Cracking in buildings*, Report 292, BRE, Watford, 1995

83 *Why do buildings crack?*, Digest 361, BRE, Watford, 1991

84 *Brickwork: Prevention of sulphate attack*, DAS Defect Action Sheet (Design) 128, BRE, Watford, 1989

85 Rushton, T., *Investigating hazardous and deleterious materials*, RICS Books, Coventry, 2006

86 *Corrosion of metal components in walls*, Digest 461, BRE, Watford, 2001

87 British Urethane Foam Contractors Association – www.bufca.co.uk/applications/cavity_wall.asp

88 *Better Exploitation Of Materials Research: Interfaces & Connections*, BRE – http://newprojects.bre.co.uk/condiv/cpm/portal/guides/Interfaces%20and%20Connections%20flyer.html

89 *Further observations on the design of brickwork cladding to multi-storey rc frame structures*, Technical Note 9, BDA, 1979

90 *Design for movement in buildings*, Technical Note 107, CIRIA, London, 1981

91 *Further observations on the design of brickwork cladding to multi-storey rc frame structures*, Technical Note 9, BDA, 1979

92 *Designing for movement in brickwork*, Design Note 10, BDA, 1986

93 *Movement and cracking in long masonry walls*, Special Publication 44, CIRIA, London, 1986

94 *Code of practice for use of masonry. Materials and components, design and workmanship*, BS 5628–3:2005, BSI, London, 2005

95 *Why do buildings crack?*, Digest 361, BRE, Watford, 1991

96 *Movement and cracking in long masonry walls*, Special Publication 44, CIRIA, London, 1986

97 *Bracing for non-domestic timber trussed rafter roofs*, Timber Engineering Guidance Document 8, TRADA, 1999

98 *Repairing brick and block masonry*, Digest 359, BRE, Watford, 1991

99 *Wall technology, Volume G: Applied finishes*, Special Publication SP 87/G, CIRIA, London, 1992

100 *Wall and floor tiling. Design and installation of external ceramic and mosaic wall tiling in normal conditions. Code of practice*, BS 5385–2:2006, BSI, London, 2006

101 *Cladding fixings: good practice guidance*, Publication C524, CIRIA, London, 2000

102 *External cladding using thin stone*, Information Paper 6/97, BRE, Watford, 1997

103 Nicastro, D., *Stone stains*, article in *The Construction Specifier*, November 1995 issue, The Construction Specifications Institute, Alexandria, USA, 1995

104 *Wall technology, Volume E: Large heavy units on framed buildings and in-situ concrete*, Special Publication SP 87/E, CIRIA, London, 1992

105 *Code of practice for design and installation of non-loadbearing precast concrete cladding* (AMD Corrigendum 11064) (AMD Corrigendum 13018), BS 8297:2000, BSI, London, 2000

106 *Guidance note on the security of the outer leaf of large concrete panels of sandwich construction*, IStructE, 1989

107 *Guide to fixings for glass fibre reinforced concrete cladding* (1999 revision), GRCA, Wigan, 1998

108 Endean, K., *Weatherproofing problems in modern glazing sytems*, Construction Paper 1, CIOB, 1992

109 *Watts Pocket Handbook 2008*, RICS Books, Coventry, 2008

110 Choi, E., *Velocity and Impact Direction of Wind Driven Rain on Building Faces*, Nanyang Technological University, Malaysia, 1997

111 *Rain penetration and its control*, Canadian Building Digest 40, NRC-IRC, Ottawa, Canada, 1963

112 Nicastro, D., *Zipper gaskets*, article in *The Construction Specifier*, September 1996 issue, The Construction Specifications Institute, Alexandria, USA, 1996

113 *Structural Silicone Glazing*, Dow Corning Corporation, Midland, USA, 1999

114 *The principle of pressure equalisation*, Technical Note 6, CWCT, Bath, 2000

115 Rousseau, M.Z., *Facts and fictions of rainscreen walls*, article in *Construction* Canada, Vol. 32, no. 2, National Research Council, Ottawa, Canada, 1990

116 Layzall, J., Failure mode and effects analysis in the cladding industry, University of Bath, Bath, 1997

117 Day, R., *What effect will the publication of the new Standard and Guide to Good Practice have on Curtain Walling failures in the UK?*, 1993

118 *Cladding Componenet Failures*, CWCT, Bath, 1997

119 *Site testing: method statement for the simplified cabinet test*, Technical Update 2, CWCT, 2000

120 *The rain screen wall system*, Ontario Association of Architects

121 *Condensation on glazing frames*, Technical Update 3, CWCT

122 Nicastro, D., *Glazing system leaks*, article in *The Construction Specifier*, January 1996 issue, The Construction Specifications Institute, Alexandria, USA, 1996

123 Nicastro, D., *Glazing system leaks*, article in *The Construction Specifier*, January 1996 issue, The Construction Specifications Institute, Alexandria, USA, 1996

124 *Defects in buildings*, HMSO, London, 1989

125 *EIFS Restoration Guide*, Dow Corning Corporation, Midland, USA, 2002

126 *Wall cladding defects and their diagnosis*, Digest 217, BRE, Watford, 1978

127 *Spandrel Panel Glazing*, ATS-124, Pilkington Glass, 2005

128 Lichtenberger, W., *Calculating the Theoretical Life Expectancy of an Insulating Glass Unit*, Glass Performance Days 2005 – www.gpd.fi

129 *Double-glazing units: a BRE guide to improved durability*, Report 280, BRE, Watford, 1995

130 Wolf, A.T., *Silicone Sealed Insulated Glass Units*, originally presented at International Symposium on the Application of Architectural Glass, 2004

131 Bally, M., *Warm Edge Insulating Glass Technology for Structural* Glazing, article in *Architects' Guide to Glass & Metal*, Vol. 19, Issue 2, Winter 2005, Key Communications Inc

132 *Unger, G., The Thermoplastic Edge Seal System – A New Insulating Glass Generation*, Glass Performance Days 1999 – www.gdp.fi

133 *Assessing the appearance of glass*, Technical Note 35, CWCT, 2003

134 *Glass breakage*, Technical Note 13, CWCT, 1999

135 Kasper, A., *Fundamentals of Spontaneous Breakage Mechanism caused by Nickel Sulphide*, Glass Performance Days, 2000 – www.gdp.fi

136 *SCOSS Bulletin 3*, February 1999

137 Matsushita, F., Aonon, Y. and Shibata, S., *Microstructure Changes of Autoclaved Aerated Concrete during Carbonation under Working and Accelerated Conditions*, Journal of Advanced Concrete Technology, Vol. 2, No. 1, Japan, Concrete Institute, 2004, pp. 121–129

138 *Screeds, flooring and finishes: selection, construction and maintenance*, Part 5 – Screeds, Report R 184, CIRIA, London, 1998

139 *Screeds, flooring and finishes: selection, construction and maintenance*, Part 5 – Screeds, Report R 184, CIRIA, London, 1998

140 *BRE building elements: floors and flooring – performance, diagnosis, maintenance, repair and the avoidance of defects* (2nd edition), Report 460, BRE, Watford, 2003

141 *Mortar. Methods of test for mortar. Chemical analysis and physical testing*, BS 4551:2005, BSI, London, 2005

142 *Screeds, bases and in-situ floorings. Concrete bases and cement sand levelling screeds to receive floorings. Code of practice*, BS 8204-1:2003, BSI, London, 2003

143 *BRE screed tester: classification of screeds, sampling and acceptance limits*, Information Paper 11/84, BRE, Watford, 1984

144 *Screeds, bases and in-situ floorings. Concrete bases and cement sand levelling screeds to receive floorings. Code of practice*, BS 8204-1:2003, BSI, London, 2003

145 *Domestic floors: repairing or replacing floors and flooring – magnesite, tiles, slabs and screeds*, Good Building Guide 28 part 4, BRE, Watford, 1997

146 Beul, W. and Menzel, K., *Corrosion Damages Caused By Cast Magnesite Floor Screed*, Otto-Graf-Journal, Vol. 14, 2003

147 www.theafa.com/specs.html; *Platform floors (raised access floors) – performance specification*, Method of Building Performance Specifications PF2 PS/SPU, PSA, London, 1992

148 *Raised access floors*, BS EN 12825:2001, BSI, London, 2001

149 Lindborg, U., *Observations on the Growth of Whisker Crystals from Zinc Electroplate*, article in *Metallurgical and Materials Transactions A – Physical Metallurgy and Material Science*, Vol. 6, No. 8, TMS, 1975, pp. 1581–1586

150 Brusse, J. and Sampson, M., *Zinc Whiskers: Hidden Cause of Equipment Failure*, article in *IT Professional Magazine*, Vol. 6, No. 6, IEEE, 2004, pp. 43–47

151 Lahtinen, R. and Gustafsson, T., *The driving force behind whisker growth* – www.metalfinishing.com

152 *Design guide on the vibration of floors*, Publication 076, SCI, Ascot, 1989

153 *Evaluation of human exposure to vibration in buildings (1Hz to 80Hz),* BS 6472:1992, BSI, London, 1992

154 Allen, D.E. and Pernica, G., *Control of floor* vibration, Construction Technology Update No. 22, December 1998, National Research Council Canada – http://irc.nrc-cnrc.gc.ca/pubs/ctus/index_e.html

155 *Diagnosis of deterioration in concrete structures*, Technical Report 54, Concrete Society, 2000

156 *Floated and trowelled* finishes, The Concrete Society – www.concrete.org.uk/fingertips_nuggets.asp?cmd=display&id=880

157 *Finishing lightweight concrete floors*, Publication 4640, Expanded Shale, Clay and Slate Institute, December 2003

158 Plimmer, J., *Delamination of Power Floated Floors*, Concrete, July 1988

159 Klieger, P., *Effect of Atmospheric Conditions During the Bleeding Period and Time of Finishing on the Scale Resistance of Concrete*, Journal of the American Concrete Institute, Vol. 27, No. 3, November 1955

160 *Air-entrained concrete*, Portland Cement Association – www.cement.org/basics/concretebasics_airentrained.asp

161 *Air-entrainment*, Federal Highway Administration, US Department of Transportation – www.fhwa.dot.gov/infrastructure/materialsgrp/airentr.htm

162 Hulett,T. and Plimmer, J., *Delamination of concrete floor surfaces*, Concrete Society Advice No.18, Concrete Society, 2003

163 *Enhancing the whole life structural performance of multi-storey car parks*, ODPM, London, 2002

164 *Recommendations for the inspection, maintenance and management of car park structures*, Institution of Civil Engineers, 2002

165 *Standing Committee on Structural Safety Tenth report July 1992 to June 1994*, SCOSS, London, October 1994

166 *Steel, concrete and composite bridges: Specification for loads*, BS 5400–2:1978, BSI, London, 1978

167 *Design recommendations for multi-storey and underground car parks* (3rd edition), IStructE, London, 2002

168 *Design recommendations for multi-storey and underground car parks* (3rd edition), IStructE, London, 2002

169 *Concrete Car Parks – Dersign and maintenance issues*, Conference Papers, BCA, September 1997

170 *Standing Committee on Structural Safety 11th report 1994–1996*, SCOSS, London, January 1997

171 *Pollution Prevention Guidelines Highway Depots PPG10*, Environment Agency and SEPA

172 Wood, J., *Pipers Row Car Park, Wolverhampton: Quantitative Study of the Causes of the Partial Collapse on 20th March 1997*, available at www.hse.gov.uk

Index